McGraw Hill CONSTRUCTION Sweets ⑤

Green Building
Square Foot Costbook 2011

BNi® Building News

GREEN Building Square Foot Costbook 2011

ISBN 9781557017048

Cover photography provided by shutterstock.com & istock.com

Table of Contents

Introduction 5

Part One: Green Articles 7

Part Two: Green Case Studies...............................19

Part Three: Unit-in-Place Costs.............................133

Part Four: Metro Area Multipliers.......................191

Index...197

Introduction

Design & Construction Resources proudly presents the 2nd Annual *Green Building Square Foot Costbook*. Following are actual LEED® rated case studies that have been published by Design Cost & Data (DCD) and which are included in DCD's National Historical Building Cost Database (NHBCdb) online at DCD.COM.

DCD is based on the philosophy that actual buildings, when coupled with up-to-date cost indices, are the most reliable basis for future building costs. The buildings appearing in this guide reflect square foot costs that have been escalated to January 2011.

The square foot cost guides appearing in this manual are provided as a service of *Design Cost Data* magazine, which has served the industry since 1958.

The 2011 *Green Building Square Foot Costbook* looks at the growing field of sustainable design in Commercial, Public and Residential construction. As noted in the 2010 edition, adaptive-reuse of public buildings continued to be a cost-effective practice for 2010. Furthermore, expanding its popularity and adaptability into the Multi-Residential sector. Some examples of Multi-Residential adaptive-reuse projects are loft and studio conversions.

LEED continues to be the measure for Greener development in high density metropolitan areas. The enhancement and urban revitalizations accompanied by the retrofit of existing buildings made 2010 a perfect opportunity for LEED Certified Adaptive-Reuse and Mixed-Use development.

The first quarter of 2010 also showed strength in the Public Building sector by a 2.3 percent increase. The stimulus funding has aided tremendously by increasing construction activity by almost $2 billion between February and March. The stimulus funds have mainly affected public infrastructure and power. However, the Public Building sector has seen significant results in building maintenance, expansions and infrastructure upgrades.

Part One

Green Articles

ARTICLE 1   ESTIMATING PROJECTS SEEKING LEED® CERTIFICATION

ARTICLE 2   STATES REQUIRING LEED® SILVER CERTIFICATION

ARTICLE 3   CONVINCING THE OWNER TO GO LEED®

ARTICLE 4   RED GOES GREEN

ARTICLE 5   THE SEARCH FOR SUSTAINABLE ROOFING

ARTICLE 6   KNOWING HOW TO MEASURE A GREEN BUILDING

Estimating Projects Seeking LEED® Certification - An Update

"Four years and what have estimators learned?"
By Joseph J. Perryman MRICS MAPM LEED-AP

There is no doubt that, encouragingly, the LEED® certification system is here to stay. Construction projects of all different types are now LEED® certified and so many projects currently about to start design, and those already in design, are registered with the USGBC for certification. So what have I learned from budgeting for LEED® certification during the design stages over the years on numerous projects?

It has proved to be crucial that a LEED® consultant be engaged as early as the concept design stage. The earlier the LEED® consultant can engage the Owner, the Design Team and the estimator to meet to discuss the LEED® goals and opportunities the better especially from a cost perspective. Certain points can be ruled out immediately based purely upon the project's location and nature. For example, point opportunities exist for renovation projects such as reusing the facade etc. which, of course, can be ruled out completely if your project is a brand new building on an open site. Similarly, a site location in a rural area cannot seek the points that reward a dense downtown location with established public transportation nearby. An experienced team with LEED® experience can very quickly ascertain the points that can be ruled out immediately based upon the project's inherent characteristics.

It is after this first session that the Owner's expectation of a certain level of LEED® certification can somewhat be confirmed as being at least achievable. The next steps are to establish whether the costs to reach that particular level of LEED® certification are attainable within the Owner's budget.

The result, or deliverable, from that first LEED® session should be a preliminary LEED® checklist that records the Owner's and Design Team's first pass of potential and realistic available points for the Project. At this time, the experienced estimator can take the checklist and start to analyse and establish realistic cost implications of pursuing each point.

It has become clear that the cost of certain specific building elements are influenced more than others by the LEED® system. Mechanical, plumbing, landscape, and interior finishes are four examples of building elements that are likely to be considerably affected by LEED® certification. Therefore, we have learned that instead of just applying a percentage to your overall projected construction cost to cover the cost of seeking LEED® certification, we can now start to evaluate exactly what elements, and by how much, are affected by the selected LEED® points. The experienced LEED® estimator can also now advise on what points will likely cost more than other points. One should think of the cost of seeking LEED® like a menu at a restaurant where prices vary for each meal the appetizers are the landscape items with all different costs, the entrees are the mechanical all with

differing prices etc. and at the end of the meal, depending on what you decided to eat, it can turn out to be either been an expensive meal or an economical meal. The experienced LEED® estimator will ensure their Owners always eat economically! Once the estimator takes the LEED® checklist and sorts the points into the relevant construction elements it starts to become easier to see how each building element may be affected in terms of cost.

When this early analysis is complete, the estimator can start to allocate real values to relevant sections of their estimate and at the earliest stages of design. This makes for a much more manageable and realistic estimate during the ensuing design stages instead of just always revisiting an arbitrary percentage on your front page summary sheet of your estimate.

Many LEED® points relate to the materials used on a project and we have found that experienced design consultants have started to select their materials much earlier in the design process than perhaps they would if LEED® was not part of the project. Materials such as stone (for interior and exterior application) are being finalized earlier in design due to the need to utilize local products if those points are being sought. Other interior finishes such as carpet, paint, millwork etc. are also being specified earlier all of which helps the estimator not only price the cost of LEED® certification but also allows for the estimate to be established with more cost certainty much earlier in the design. We have seen stone and brick materials selected as early as Schematic Design on certain LEED® projects compared to much later in the design stage on Projects in years gone by. Who would have thought that a potential benefit of LEED® would be that estimates can be more certain and realistic at an earlier stage of the design!

Constant communication and collaboration with the Design Team during the design stages will ensure that the Client can be constantly made aware of how the budget can be affected by dropping or adding points. As such, the estimator should ideally participate at all meetings that review the LEED® checklist as the design progresses.

As the LEED® system constantly evolves with new versions being released by the USGBC so evolves the estimator's experience and approach as to how LEED® can be realistically allowed for in estimates at early design stages. Estimators can now also compile their estimates in such a way that the LEED® costs are managed throughout the design process from the earliest of stages.

Joe Perryman MRICS MAPM LEED-AP is Principal at Donnell Consultants Incorporated, a firm based in Tampa, Florida providing professional construction cost management services on performing arts and museum projects all over North America. Joe can be reached at joseph@dcicost.com or 813.875.8074

States Starting to Require Architects and Contractors to Design and Construct Public Buildings to Achieve LEED® Silver Certification

By Angela Stephens, Stites & Harbison, PLLC

While many local jurisdictions and cities across the country have started passing regulations which implement and require sustainable design and construction practices, relatively few states have taken steps to mandate that certain public buildings achieve certain levels of LEED® Certification.

Eighteen states (Arizona, California, Connecticut, Hawaii, Illinois, Indiana, Kentucky, Massachusetts, Maryland, New Jersey, New Mexico, Nevada, Rhode Island, South Carolina, South Dakota, Utah, Virginia, and Washington) have adopted laws and regulations mandating that the construction of public buildings achieve LEED® Silver Certification. Although the majority of States do not yet require that public buildings be designed and constructed to achieve a LEED® Silver Certification, many of these States encourage their agencies to use green building practices or use LEED® as a guideline.

The Kentucky law is illustrative of those states that have enacted a LEED® requirement on public buildings. Kentucky requires that, after July 1, 2009, all public buildings (for which fifty percent (50%) or more of the total capital cost is paid by the Commonwealth of Kentucky) shall be designed and constructed in accordance with Kentucky's new High Performance Building Regulations.

Under Kentucky's new regulations, all public buildings (as defined above) worth $25 million or more in budget "shall be designed, built, and submitted for certification to achieve a rating of Silver Level or higher" using LEED® 2009. Public buildings between $5 million and $25 million shall be designed, built, and submitted for certification to achieve a rating of LEED® Certified or higher. Additionally, public buildings greater than $5 million shall achieve a minimum of 7 points under the LEED® Energy and Atmosphere Credit 1, Optimize Energy Performance. Public buildings between $600,000 and $5 million in budget shall be designed and built using the LEED® rating system as guidance.

There are two exceptions to the new regulations. The first exception applies when a public building fails to achieve the LEED® rating due to the sole failure to receive a point for Material and Resource Credit 7 regarding certified wood. Under this first exception, the building will be deemed to meet the LEED® rating required, if the project used wood products certified under the American Tree Farm System or the Sustainable Forestry Initiative.

Under the second exception, a building which is required to meet the high performance building standards may be granted an exemption if there is an "extraordinary undue burden on the agency if project compliance is required." Factors that will be considered in determining if such an extraordinary undue burden exists include whether (1) the cost of compliance exceeds a building's life cycle cost savings, (2) compliance increases costs beyond the funding capacity for the project, (3) compliance compromises the historic nature of a building, (4) compliance will violate any laws, (5) the unique nature of a project makes it impractical, or (6) the building will use another high performance building program such as Energy Star or Green Globes.

In addition to the requirements mentioned above, all public buildings (as defined above) shall be designed and constructed so that they are capable of being rated as Energy Star buildings. However, unlike the requirements discussed above, an exemption cannot be granted from this requirement.

For more information about these regulations or other green initiatives which may impact your business, please contact Angela Stephens at astephens@stites.com or at 502-681-0388.

About the Author: Stites & Harbison, PLLC has a Green Law Practice Group which is devoted to the unique legal issues associated with green initiatives and new regulations addressing environmental stewardship and sustainability. The Green Law Group is comprised of attorneys including construction attorneys, business and finance attorneys, tax attorneys, and environmental attorneys. Within this group, 10 attorneys are LEED® Accredited Professionals (APs), and one attorney is a LEED® Green Associate (GA). LEED® AP and GA attorneys have demonstrated a thorough understanding of green building practices and principles and have the tested ability to apply the LEED® Rating System standards to designing and building projects. For more information visit www.stites.com.

Notes:

1. See http://www.usgbc.org/DisplayPage. aspx?CMSPageID=1852 (last visited October 14, 2009).
2. KRS 56.775.
3. 200 KAR 6:070 Section 2(1) High Performance Building Standards.
4. Id. at Section 2(2).
5. Id. at Section 2(3).
6. Id. at Section 2(4).
7. Id. at Section 3(2).
8. Id. at Section
4.9. Id.

STITES & HARBISON PLLC

ATTORNEYS

Convincing the Owner to go LEED®
By Robert Miller

hat does LEED® cost? You know it is coming: dollars always seem to be the foremost concern. It inevitably becomes the focus within moments of entering a discussion about Green design. There is an ironic hope in the problem itself; rising fuel costs and the increased awareness of global warming has made sustainability a much easier sell.

Financial payback comes in many forms. Educational institutions and commercial ventures are seeing a rising number of promising candidates who will base their choice on the institutions value systems. Recycling programs, Green building, healthy lifestyles, and even responsible portfolio investments are important criteria for the new generation of applicants.

Frenzied marketing campaigns attempt to spin everything from vinyl to Hummers as Green solutions, but as people are becoming educated about Green design they no longer accept hollow promises. Owners increasingly demand sustainable strategies are integrated thoughtfully into strong design solutions that will be appreciated and cared for over generations. The soft costs associated with LEED® are lowering due to simplified documentation, increased knowledge, and availability of information. Commissioning, once seen as a cost deterrent, now demonstrates value to the Owner. These changes have allowed increased access to LEED® as a valuable tool to help Owners validate their efforts within the public arena.

LEED® promotes a whole-building approach to sustainability by recognizing performance in five key areas of human and environmental health: sustainable site development, water savings, energy efficiency, materials selection, and indoor environmental quality.

Familiarity with these sustainable strategies provides architects with necessary tools to create thoughtfully integrated design. We believe the most important of all LEED® credits are those for innovation. We ask: How will this project affect the future? Buildings that educate, generate, and are valued by the community are the current golden ring.

The cost of building sustainably is lowering and Green buildings now have a proven track record of increased property values. Every year more products are available at prices competitive with traditional products. Contractors are becoming familiar and comfortable with sustainable strategies. Through greater affordability and faster paybacks through decreased operating costs we hope that sometime soon we will need to ask "what would it cost NOT to build Green?"

Is LEED® necessary? In a perfect world we would all be making educated decisions based on a strong code of ethics and moral values that would assure Green design as the norm for any project. For any set of guidelines it is difficult to keep up with the rapidly expanding product availability and knowledge base. It is important to remind ourselves that LEED® was designed as a tool to compare buildings based on specific criteria. Unfortunately many people have tried to utilize the system as a design checklist to obtain as many points as possible for the least effort and lowest cost. While this has caused some to be skeptical of the system, I believe all agree that LEED® has served us well in raising the public's awareness. It has simplified the comparison of individual buildings in a way that the competitive nature of the industry has taken over to propel it forward. Each institution wishes to meet or exceed the rating of the last. This energy needs to be harnessed by industry professionals and utilized to bring the next generation of solutions to Green design. Owners are increasingly aware that Green design is rapidly becoming a datum point upon which they are being judged. Even owners skeptical of the LEED® system and associated costs are asking that Green design be incorporated when possible. They are recognizing that integrity equals profitability in the new Green economy.

LEED® criteria will become increasingly stringent as our awareness, options, and experience increase. It is important to grow our knowledge as the system grows to maintain our place as leaders in our fields. Like any other practice, efficiency comes with familiarity. Familiarity with sustainable strategies provides architects with additional tools toward thoughtful design. Green solutions should rise to the top as second nature.

For us, we chose to place our soap box in Seattle. Regional influences undoubtedly open opportunities for Green buildings like the Ballard Library to take root. Public support for Green design has a majority voice and resonates throughout the City. Successes in regions like this provide opportunities for projects in more conservative locations to review these successes and compare dollars spent against real results to build confidence to move forward and ahead of the pack.

What if my client cannot support a cutting edge design? No impact is too small. Simple choices can make incremental impacts on the environment. There are simple "no cost" alternatives that provide greener buildings. Thoughtful material choices can provide healthier indoor environments. Selecting materials that are manufactured locally keeps money local, builds community pride in a project, reduces carbon production, and can save money, including decreased shipping costs. Strategies such as prefabricating building components in a factory where off-cuts and waste can be effectively recycled in reasonable quantities can also benefit the owner with increased quality, faster construction, and decreased manpower requirements on site. Impacts that are more difficult to quantify are productivity and health benefits, however research continues to prove surprising results.

So how do I convince an Owner to build LEED®? First, consider the alternative viewpoint to build thoughtful responses to the inevitable questions that will come up during your discussions. A prepared architect not only will be able to maintain his or her values but also open additional possibilities that are beyond the immediate perceptions of the project. Making value judgments with the Owner that have the most beneficial impact on the environment and inhabitants will inevitably lead to certification. The strength of your commitment and your ingenuity will determine the level of impact you have on future generations. We need the power of a grass roots effort. One individual educating a dozen others, who each educate a dozen more, resulting in awareness of the necessity for Green design to be spread exponentially.

Lastly, we all watch with keen interest the rising phenomenon of globalization. The United States is currently trailing Europe and Asia in the sustainable design movement. The majority of what we discuss as Green design is already commonplace in Europe. China is consulting with the leading experts from all international sources as well as continually pushing themselves to do more. We have boldly labeled our system of measurement "Leadership in Environmental and Energy Design." For us to tout ourselves as leaders we must truly lead. The world can only benefit by all nations innovating, consulting and sharing information with each other. We have the resources and knowledge available to innovate and set a higher bar for both environmental design and values. Change begins with us; it is our professional and social responsibility to lead our clients to make informed and responsible decisions in how we choose to build.

About the author: Robert Miller, AIA, LEED® AP is a senior associate of Bohlin Cywinski Jackson, of Seattle, Washington. The company website is www.bcj.com.

Red Goes Green
Sustainable Design Becomes Part of Triage for Fire and Rescue Stations
By Lynn Murray

When it comes to green building initiatives, most fire and rescue stations have not been first responders. But as larger municipalities across the country begin to see their investments pay off in long-term cost savings, more cities are putting green on the radar.

It is no surprise that green building for fire stations (and municipal facilities in general) has been a slow process. While public structures traditionally strive to set an example for their communities, the government building process is often slow, mired in political red tape, and restricted in budget. And while few would dispute that sustainable design and green building practice are good ideas, changing established protocol to accommodate these ideas simply wasn't an emergency.

In the past four years, however, cities large and small have taken deliberate steps to make "green" a more integral part of new construction and renovation of fire stations.

"Government entities are becoming more aware of sustainable design and are showing much more interest in it," said Ken Newell, AIA, LEED®AP, senior principal, Stewart-Cooper-Newell Architects in Gastonia, N.C. "The Federal Government and many state governments now require some level of sustainable certification on all building projects. Many municipalities are following suit."

The industry standard is LEED® (Leadership in Energy & Environmental Design), a voluntary rating system established a decade ago by the U.S. Green Building Council, which identifies sustainable design elements in six categories: sustainable sites, water efficiency, energy and atmosphere, materials and resources, indoor environmental quality and innovation and design. Certified, Silver, Gold, and Platinum levels of LEED® green building certification are awarded based on the total number of points earned within each LEED® category.

Of Stewart-Cooper-Newell's 50 or more active municipal projects at a given time, Newell said approximately 25 percent or more are now requesting some level of sustainable design up to and including LEED®, with this number increasing annually. Cities ranging from Dallas and Phoenix to Carrboro, N.C., (population 18,000) are part of the movement.

In Maryland and Virginia, J. Lynn Reda, AIA, LEED® AP, senior associate, LeMay Erickson Willcox Architects in Reston, Va., has seen a significant increase in municipalities now mandating green. Both Montgomery County and Howard County, Md., require LEED® Silver. Fairfax County, the City of Alexandria and Arlington County, Va., also require LEED® Silver. In Prince William County, Va., community volunteers have taken the lead in constructing new facilities and LEED® is encouraged, but not mandated by the county.

Municipal green building initiatives are under consideration in many, if not most, communities in Northern California, said Dennis Dong, AIA, CSI, ARA, LEED®AP, principal partner, Calpo Hom & Dong, in Sacramento, Cal. "Most communities are not mandating total compliance to LEED® but are strongly urging the incorporation of green methods in new design and construction," said Dong.

More prominent are specific areas of green initiatives being driven by environmental and energy regulations, such as mandates requiring the treatment and collection of storm water and construction material recycling to minimize waste sent to landfills. California also mandates that all buildings comply with the Title-24 California Energy Efficiency Standards for Residential and Nonresidential Buildings.

"Green building initiatives, along with these standards for building and a forthcoming state energy code, are becoming inherent in all building projects, much like the Americans with Disabilities Act (ADA) made an inherent consciousness of accessibility as an essential design feature in projects," said Dong.

A dialogue with clients

A municipal project, such as a fire station, usually begins with a series of dialogues between the client and the design team. And experts say what happens in these conversations is a foundation for integrating sustainable design into the project.

"If they don't bring it up first, we discuss what is happening in the industry regarding sustainability and find out what their desired approach is to it," said Newell.

"The question, with most clients, is the amount of green, and the benefits derived. Title-24 California Energy Standards already mandate certain levels that must be met, and this is readily accepted by all projects. The push for going beyond the standards, however, is often discussed between client and the design team, and requires an open mind and education on both sides," said Dong.

"I explain the benefits of green building to clients as not only potentially environmental and financial, but from the point of view of employee satisfaction," said Reda.

"Fire and rescue personnel essentially live in those buildings for 24-hour shifts (sometimes longer). The facilities historically have been dark and claustrophobic. Given the incredibly stressful nature of the work, the facility they return to after an emergency should be an inviting, healthy, and comfortable environment," she added.

The green of green

Newell's firm conducted a cost analysis of five typical fire stations in Texas, North Carolina, South Carolina and Florida, ranging from 9,000 to 15,000 square feet with sustainable design elements ranging from minimal (rainwater harvesting) to full LEED® certification. On average, green initiatives including design, construction, third-party commissioning and LEED® registration added approximately 4.5 months and an 18-21 percent cost increase.

"While these are significant impacts to the project, most or all of the sustainable efforts will result in significant 'pay-back' over the life of the facility, especially when you consider that the life span of a fire station is expected to be 50 years plus," said Newell.

If the designer already charges very high fees and designs very expensive buildings, then incorporating LEED® may not cost the client any more, said Newell. But in most cases, incremental costs will be incurred. In the future, these programs may be so standard that the cost increases will diminish. But they are not likely to disappear.

Dong and Reda suggest that fairly standard green initiatives raise costs by between 2 and 5 percent, with the most tangible cost savings in utilities and energy use.

"I have one fire station project currently under construction that should see a yearly energy savings of approximately $21,500 per year," said Reda.

"Green building costs may initially be higher, but can you afford to NOT incorporate them? There are numerous studies by institutions, showing the long-term cost savings in utilities, and the time required for the payback. What cannot be quantifiably measured is the improved mental attitude of the building's users, in a healthier environment," said Dong.

Elements of design

"As a rule, fire stations and municipal projects target the straightforward initiatives that have relatively low first-cost impacts. This is certainly in part due to the fiduciary responsibility of the municipality," said Reda.

Common design elements incorporated in fire and rescue stations include reflective roofs and concrete pavement, to minimize the "heat island" effect; construction waste management practices; low emitting materials to improve air quality; and low flow plumbing fixtures and other systems to recycle and conserve water. Additionally, the use of materials generated from local sources saves money, time and gas and reduces pollution.

Durability and low maintenance are other considerations when selecting green materials.

"We tend to stay away from finish materials that have not been thoroughly tested (wheat board, cork, bamboo). Energy-efficient HVAC system, as well as automatic lighting controls and natural daylight are all standard targets," added Reda, who is currently incorporating these elements in pursuing LEED® Silver on the Germantown-Milestone fire station in Montgomery County, Va., as well as on The Station at Potomac Yard in The City of Alexandria, Va.

Another important factor to keep in mind is that current building codes, along with good design practices, already satisfy significant elements of sustainable design.

"Most building codes require that issues like insulation values, lighting controls, plumbing fixture controls, HVAC efficiencies, etc., be addressed in some level of sustainability. We have evaluated some of the stations designed in our firm without LEED® and found that more than half of the necessary points for 'Certification' were achieved simply due to code requirements and good design practices," said Newell.

Sometimes, the satisfaction lies wholly in the pursuit of LEED®. The Town of Carrboro, N.C. completed all necessary requirements to comply with LEED® Silver for its Fire and Rescue Station No. 2, but elected not to submit the project for official LEED® certification. Instead, the town channeled the costs of filing for LEED® back into the project an estimated $60,000, said Newell.

A shared decision

It's easy to talk green, but making it happen is a shared responsibility.

"One of the most important things to do when considering green initiatives is to get a real commitment from all parties involved from planning through operations and maintenance. This will ensure decisions are made with common priorities, and ideally more consideration will be given to "new" technologies that haven't necessarily been tested," said Reda.

"In addition, include funding at the very initial budgeting exercises for both additional professional fees as well as construction costs. For example, many municipalities are interested in vegetated roofs, but don't include the up-charge in the budget and the idea never makes it past schematic design. Consider holding a LEED® charrette while developing the budget for any given project to identify potential high-cost strategies," she added.

"Think in terms of the long-term investment, not only in the building efficiencies, savings in utility costs, and well-being for your users, but also your contribution to the global environment," said Dong.

"Being green is not limited to exotic and costly energy-generating systems but is often a matter of common sense, and good, sound planning and design."

Sustaining The Momentum: The Search for Sustainable Roofing

By Brian Lambert

The ability to endure, to keep on going, is at the heart of sustainable design. When applied to buildings, sustainable design refers to product solutions that can conserve, recycle, and even help renew natural resources over time. Many products that accomplish that goal are here already, and new ones are being introduced every day.

For a design community faced with the rapid proliferation of new products and evolving performance standards, today's challenge is maintaining the momentum for design solutions enriched by sustainable design concepts. This article will help you understand and explore some of the ways in which sustainable design can help you reinvent the modern building, with roofing solutions that keep on going.

The Green Revolution

Sustainability is not a style... it is nothing less than a revolution in how we design, construct, and operate buildings. In roofing, sustainability can be accomplished in any of five ways:

- Through the use of recycled materials (i.e., materials that are being reused)
- Through the use of materials that are, in themselves, recyclable (i.e., materials that can be reused in the future)
- Through extended service life
- By promoting the more efficient use of energy and other natural resources
- By actually renewing our natural resources

Some of today's sustainable roofing solutions perform only one of these objectives; others perform several. Today's most popular sustainable roofing solutions include:

- Cool, highly reflective roofing
- Metal roofing
- Modified bitumen membranes that incorporate post-consumer recycled materials
- Adhesives and other roofing materials and products that eliminate or reduce hazardous fumes
- Photovoltaic panel systems
- Green roofing

In their efforts to promote sustainable design, government and industry groups are rapidly evolving methods of applying uniform standards of measurement to certify performance. The two standards most often used to evaluate sustainable roofing solutions are:

- The Leadership in Energy and Environmental Design (LEED®) Green Building Rating SystemTM, a voluntary, consensus- building national standard that was initiated by the U.S. Green Building Council (USGBC)
- Energy Star, a collaboration between the U.S. Department of Energy, the U.S. Environmental Protection Agency, and private industry.

Reflective Solutions

According to the Environmental Protection Agency (EPA) web site, buildings with white roofs that reflect ultraviolet rays typically require 40% less energy for cooling. Add to that the fact that a highly reflective roof resists ultraviolet degradation, and the sustainability factor is even higher.

Energy Star certification is an accurate indication of a roof's ability to reduce fossil fuel usage. When products that comply with this performance standard are used in urban areas, peak cooling demand can be reduced by up to 15%. It's no wonder that states like California are creating incentive programs to promote the use of Energy Star roofing solutions.

The rapid development of new Energy Star approved roofing products is making it possible for architects and designers to choose, from a variety of reflective solutions, a roof that will retain the integrity of their original design concept. Knowing that some of these products have been demonstrating their ability to resist UV-related failures for 20 years or longer, is an added incentive for specifying a product that can dramatically reduce a client's energy costs.

Roof reflectivity can be achieved in a variety of ways. One of the most lasting methods is the application of a highly reflective topcoat or mineral surfacing. Although Energy Star solutions are available for all types of roofing, their energy payback benefits are particularly significant when they are installed on single-story, air-conditioned buildings with large roof surfaces and older buildings with insufficient insulation.

Metal Solutions

Metal roofing solutions are perhaps the fastest growing segment of the sustainable roofing market. Their sustainability derives from the fact that typically as much as 100% of their material components are recyclable. In other words when the day comes that your client has to tear off the old roof and put up a new one virtually the entire roof can be reused to create new metal products.

In addition, many of today's metal systems have long track records for lasting performance. A copper roof that has protected a cathedral for over a century is sustainable not only by virtue of its recyclability. It is sustainable because it keeps on going year after year after year.

Metal roofs also eliminate the fume and kettle concerns associated with some types of roofing, for easy and eco-friendly installation. And, with so many manufacturers introducing new products, finishes, colors, profiles, and textures today's recyclable metal roofing is offering architects and designers a diversity of aesthetic features to support a wide spectrum of design concepts.

Recycled-Material Solutions

Yet another category of sustainable roofing is roofing that reuses materials that might otherwise be overflowing our landfills. For example, some built-up, multi-ply modified bitumen roofing systems replace conventional filler with post-consumer crumb rubber from recycled tires. With over 250 million tires discarded yearly in the U.S., our landfills are rapidly running out of space. In addition, discarded tires create several health and environmental hazards, including the potential for mosquito infestation, water contamination, and fires emitting hazardous fumes.

In the roofing industry, innovative manufacturers are also helping to reduce landfill problems by using recycled plastics or rubbers to create roofing that simulates the look and feel of natural slate. Manufacturing new products out of recycled materials is becoming increasingly common with other building components, such as carpeting, as well.

Low-Fume Solutions

In the maintenance and restoration arena, more products are being introduced each year that promote a healthy ecology by eliminating or reducing hazardous fumes. Some BUR roofing can be applied "cold," for eco-friendly installation. Cold adhesives allow multiple layers of built-up roofing to be applied without hot kettles or torches. These adhesives are VOC compliant and significantly reduce odor.

There are also new adhesives available for hot-applied systems that can reduce volatile emissions by as much as 50%, while maintaining all the self-healing, elongation, and performance properties of hot asphalts. Another innovation in this area is the fume-recapturing kettle, which significantly reduces the environmental and health impact of hot-application processes.

Photovoltaic Solutions

Building-integrated photovoltaic (BIPV) materials integrate photovoltaic panels into a building to create power from the sun. The power is generated in the form of DC current that can be used directly or converted into AC current for future use. Many photovoltaic building solutions are available, offering architects and designers a wide variety of distinctive design solutions. Some solutions enable architects to make BIPV part of a building's original design. But retrofit BIPV materials are also available.

Roof-mounted systems have the distinctive advantage of using the expansive and frequently under-utilized roof surface for placement. Whether integrated into the original building design, or added later as an accessory, roof-mounted photovoltaic systems should be looked at as an integral part of a roofing solution.

BIPV solutions increase building sustainability in at least two ways. First, by creating new power from a renewable energy source, thereby reducing peak energy loads and reducing energy costs. Secondly, they are likely to last 25 years or even longer. Recognizing the community value of such solutions, many green pricing programs are available to help offset the costs of investing in BIPV solutions.

Green Roofing

Arguably the most exciting development in sustainable building design is green roofing. Although new green roofing products are being introduced more frequently than ever before, the best designs are integrated solutions that combine a:

- High-performance waterproofing layer
- Root-resistant compound
- Drainage system that draws away excess moisture

- Filter that prevents drain-system clogging
- Specially formulated lightweight soil
- A surface layer of plant life

Aesthetically, green roofing opens up an entire new world of design options to architects and other design professionals. Depending on load capabilities and other application-driven requirements, green roofs can be planted with herbs, grasses, flowers, even trees, in an exciting variety of colors, textures, scents, and heights. Patios and walkways can become a usable part of the roof environment. Aside from their obvious aesthetic and psychosocial advantages in applications such as nursing homes, day care centers, healthcare facilities, and office parks green roofs offer tremendous sustainability benefits:

- Reduced energy costs in hot urban environments
- The ability to reduce storm water run-off, reducing stress on urban sewer systems and decreasing run-off related pollution of natural waterways
- Dust reduction
- Air quality improvement
- Noise pollution reduction

In addition to the inherent recyclability of natural plant materials, some green roof systems use recycled materials in their underlying membranes, for added sustainability. Green roofing is also compatible with many of the other eco-friendly solutions discussed in this article.

Lasting Solutions

Perhaps the single most critical contribution to sustainability is extended life. Today, many roofing solutions are designed to last 15 or 20 years; some, 30 years and longer. The longer a roof can be kept in use, the more it contributes to the overall sustainability of the building by prolonging the impact of eventual tear-off.

Measuring Sustainability

The USGBC LEED®certification is making it easy for architects and building owners to objectively assess the sustainability of an entire building over its life cycle. LEED®uses a point rating system to evaluate various factors that contribute to a building's overall environmental performance. These include:

- The sustainability of the building site
- Water efficiency
- Energy use and atmospheric quality
- The eco-friendliness of various materials and resources
- Indoor environmental quality
- Innovations in the design process%

Conclusion

The choice of roofing system can significantly impact the sustainability factor of a building's overall design. Innovative roofing systems and ancillary products that positively impact the natural environment are rapidly becoming more varied, easier to finance, and more aesthetically pleasing. Architects and designers can sustain the momentum for building solutions that keep on going by investigating exciting new roofing technologies that promote sustainable design.

About the author: Brian Lambert, marketing manager for The Garland Company, Inc., has been active in industry initiatives promoting green roofing and other sustainable design solutions since 1996. He serves on the board of directors for the Toronto-based Green Roofs for Healthy Cities Coalition and frequently promotes sustainable design as a guest lecturer to professional organizations in the U.S. and Canada. You can reach The Garland Company at 800-741-3157.

Knowing How to Measure a Green Building can Help Sell Renewable Energy
By Paul Nutcher, CSI CDT

The renewable energy industry could benefit greatly by targeting green building projects. This is true for project teams setting up systems for distributed renewable energy generation on building sites, the manufacturers of the system components, and many other ancillary businesses with a stake in a building's energy performance. Furthermore, if the diffusion of renewable energy technology has been slow to gain market traction, the LEED® program could be a significant driving force for expediting the growth of BIPV and utility scale installations, as well as other renewable generation systems and energy storage systems.

Commercial buildings in the United States consume 18% of the nation's energy and are responsible for 18% of the nation's carbon dioxide emissions, according to the Department of Energy. The residential building sector consumes 22% of electricity and contributes 20% of the carbon emissions, as reported during the Energy Programs Consortium presentation February 2008: "Income, Energy Efficiency and Emissions: The Critical Relationship." With renewable energy providing less than 10% of the electricity in the US, the building sector of the economy has as much growth potential for renewable energy product manufacturers as the transportation and industrial sectors of the economy.

A realization that buildings needed to become more energy efficient and healthier for their occupants, while reducing their impact on the environment, sparked the earliest stakeholders at the non-profit US Green Building Council (USGBC) to take steps toward addressing more sustainable buildings. The result of those first steps was a program to measure a building's energy and environmental performance. Today, the most widely recognized national rating tool for third-party verification of a sustainably designed, built, operated, and maintained building is the LEED®Green Building Rating System from the USGBC, which first introduced the first of its family of rating programs in 1999. Marketing to the green building industry will take effort from stakeholders in the renewable energy business because they will first need to understand how such energy systems can enhance a high-performance building project which under the USGBC program results in "LEED® certification."

LEED® Score

LEED® stands for Leadership in Energy and Environmental Design, and measures the environmental impacts and energy performance of buildings as two of its fundamental goals. The LEED® family of green building rating systems contains sets of voluntary performance standards, many of which have become mandated by city and state governments, as well as most federal agencies. The Government Services Administration (GSA), possibly the world's largest landlord overseeing courthouses, offices, agencies, among other federal buildings, requires a LEED® Silver certification level. The LEED® for New Construction v. 3.0 (LEED®NC) rating system is divided into five main categories including sustainable sites, water management, energy and atmosphere, and indoor environmental quality. There are also two other categories for exemplary performance including the Innovation in Design and Regional Environmental Priorities categories for capturing ways in which the building significantly outperforms the benchmarks in the first five categories or

On-site generation of electricity from the sun, wind, or other natural sources have the greatest opportunity for assisting building project teams within the Energy and Atmosphere Credit Category of LEED®. Most points in the rating system assigned to any of the categories are in the Energy and Atmosphere category, with a possible 1-19 points for Optimizing

Photo courtesy IB Roof Systems

addresses a regionally specific environmental issue (i.e. water conservation or diverting materials from landfills).

On-site generation of electricity from the sun, wind, or other natural sources have the greatest opportunity for LEED® points in the Energy and Atmosphere Credit Category. IB SolarwiseTM, a photovoltaic (PV) system converts sunlight directly into electricity and works with commercial, industrial, and residential applications. Energy Performance, another seven points are possible for incrementally larger percentages of energy usage from renewables, and the project team can gain even more points toward LEED® certification for purchasing green energy credits. Because many renewable energy systems measure the output of the energy system, other credits dealing with the commissioning and verification of a renewable energy sources can offer further LEED® points. The amount of possible points in the Energy and Atmosphere category is significant because it only takes 40 points for certification. So, the LEED® point totals typically shoot high up the scorecard when renewable energy equipment is installed on LEED® buildings. Many gain LEED® Silver, Gold, or Platinum certification levels. For reference, the highest LEED® Platinum certification is between 80-100 points.

But there is still a catch. Before even considering going for LEED®, there are prerequisites in all but the Water Management, Innovation in Design and Regional Environmental Priorities categories. The important one here is the energy efficiency benchmark, ASHRAE 90.1-2007. Plus, the energy systems must be commissioned by a third-party to verify the performance of the HVAC system and no CFCs are permitted in the cooling system. Once the prerequisites for the energy systems have been met, the higher the percentage of energy optimization the building is designed to achieve the more points it will gather. Should the building go beyond 40% energy efficiency above the benchmark, there is another point available in that sixth category: Innovation and Design, for example.

Retrofitting Renewables

LEED® Green Building Rating System has separate programs for covering the two phases of a building's lifecycle, one for new construction (Schools, Core and Shell, Retail, Homes, Neighborhoods, and there are more in development), and one for ongoing maintenance and operations of a building after the construction phase. The LEED® for Existing Buildings Operation and Maintenance (LEED®-EBOM) rating system follows a sustainably built structure to the end of its useful lifespan and the rating tool can be applied once the building has been occupied for at least a year. The point structure and categories are a bit different as the LEED® EBOM rating system is more focused on post-occupancy issues, such as efforts to reduce peak energy demand, save on utility bills, and reduce the carbon footprint of a building or real estate portfolio.

Though teams constructing new buildings are the focus of the LEED® NC program, property managers, real estate portfolio owners, and other building industry practitioners looking for a larger return on investment after a renewable energy retrofit installation will decide whether to go for LEED®-EBOM. The rationale for LEED®-EBOM to this market segment is that properties certified by the USGBC stand out in annual performance reviews, annual budget planning sessions, and add a premium to leasable space. The LEED®-EBOM certification process can also identify deficiencies in standard practices that need correction, and then provide the best steps toward operational improvements. An annual re-certification of LEED®-EB status from the USGBC may extend the lifespan of a structure because the rating tool can help in the development of a system to avoid a decline in building maintenance.

Renewables & LEED® For Existing Buildings

Between 1-18 points are available in the Optimized Energy Efficiency Performance category in the LEED®-EBOM rating program. During commissioning and energy audits, inefficient areas of the structures' operations are pinpointed and monitoring of energy consumption on major mechanical systems is initiated and a program for ongoing monitoring begins. The number of points earned is contingent upon a facility's EPA ENERGY STAR rating or the percentile level above the national median.

LEED® points are earned when increased levels of operating energy efficiency can be documented and, as a result, reduce the environmental impact of excessive energy use. A facilities manager can also purchase Green-e certificated energy (a similar option for project teams on new construction), to gain points for helping reduce greenhouse gases and by helping to develop a market for renewable energy.

During this process, the building's capabilities, advantages, and weaknesses are assessed by the commissioning agent so financial costs and benefits are calculated. Training programs are developed and building operating plans are updated. Using an ongoing commissioning program that reflects changes in an existing facility's occupancy, usage, maintenance, and repair needs ensures adherence to stated revisions and upgrades. Periodic adjustments and reviews ensure optimal energy efficiency. The on-site and off-site generation of renewable energy can also gain from 1-6 points toward LEED®-EBOM certification. It is up to the energy consultant to determine whether solar, wind, biomass, geothermal, and biogas technologies are the most advantageous.

Conserving Through Greener Site Development

The amount of radiant heat from the sun penetrating the building envelope and the thermal transfer properties of the hardscape can all impact the energy efficiency of a building. LEED® programs recognize this strategy for overall reduction of energy usage on the property. The impact a rooftop solar array has on the performance of the building

New industry products can be applied to new construction and re-roofing low-sloped roofs maintaining the natural look of the property. The IB SolarwiseTM PV system installs over existing roofs with no demolition and disposal costs and includes a 20-year warranty covering roof material and energy performance.

Photo courtesy IB Roof Systems

envelope is good to know when trying to sell a building owner on the systems. In fact, a tax credit from the Energy Policy Act of 2005 provides up to $1.80 per square foot for the developer, based on improvements to the building envelope and another 60 cents each for installing more energy efficiency for lighting and HVAC systems. Renewables can have an impact on gaining the tax credit by powering a portion of the energy needed for lights and conditioning the interior space.

Other site credits with implications for the renewable energy product manufacturer include the designation of parking areas for plu-in electric hybrid cars, among other strategies for reducing carbon emissions from fossil fuels. Indirectly, less energy is needed when building project teams design and build for a reduced impact on the heat island effect. The rooftop configuration of a solar array can have a significant effect on how the thermal gradient differences in urban areas versus undeveloped areas will be addressed. Again, any assistance here from a renewable energy installation should be communicated to the building team prior to the pre-design phase of a construction project to improve the chances that the associated products and services will be specified in the construction documents.

The Future of LEED®

The LEED® rating system (or an equivalent rating tool, i.e. Green-Globes, ICC 700) will one day become the standard that most governments set for project teams wanting a permit to build. Looking at the current building codes in most of the country, with the exception of governments that have already adopted LEED® as the standard for getting a building permit and there are many already, some people may ask why we didn't build like that all along? Much of the intent of the LEED® credits are a throwback to the way the building industry was headed decades ago, but a mindset of unlimited energy and materials took over. LEED® appears to be here to stay and not a moment too soon as the world gears up to address climate change. All of these efforts could help the long term prospects of renewable energy product manufacturers, especially those who take the time to understand how their products enhance LEED® green building projects.

Paul Nutcher, CSI CDT, is the President of Green Apple Group, LLC, a marketing and public relations firm for the building industry. He serves as a technical and sustainability consultant to building product manufacturers. Nutcher writes and speaks frequently on LEED® and BIPV, among other sustainability and construction topics. He holds leadership positions in the U.S. Green Building Council and the Construction Specifications Institute in Central Florida, and he is a member of the Construction Writers Association. Contact Paul Nutcher at 407-517-4748 or pnutcher@greenappleconsult.com.

"Knowing How to Measure a Green Building Can Help Sell Renewable Energy" reprinted from the Nov/Dec 2009 issue of North American Clean Energy www.nacleanenergy.com.

Part Two
Green Case Studies

Every case study is described in a short summary. These summaries are given to provide insight into the circumstances and requirements behind the design. A building's function or location often influence the choice of building materials and thus the cost. Site limitations and local building and zoning codes are factors that have to be taken into consideration. Budget constraints, material availability and personal expertise of an individual builder all affect a project's outcome. Wherever appropriate, these types of issues are explained in the descriptions that accompany each case study. Furthermore, when costs within one or more of the CSI divisions for the project are abnormally high or low, an explanation is usually provided.

Building Case Studies Included In This Edition

Description *Civic/Government Projects* Page No.

Courthouse 20
Public Library 22
Administration Building 24
Environmental Education Center 26
Nature Center 28
Public Library 30

Commercial Projects

Retail Landmark Building *(Adaptive Re-Use)* 32
Bank 34
Children's Center 36
Warehouse Building 38
Retail Building 40

Educational Projects

Science Complex 42
Elementary School 44
Elementary School 46
Elementary School 48
Elementary School 50
Elementary School 52
University Building *(Expansion & Renovation)* 54
Community Center 56
Student Center 58
Elementary School 60
Academic Center 62
Science Center 64
University Facility 66
University Building *(Expansion & Renovation)* 68
High School 70
Middle School 72

Industrial Projects

Fuel Cell Manufacturing Facility 74
Manufacturing Facility 76

Description *Medical Projects* Page No.

Medical Complex 78
Clinical & Research Bldg 80
Hospice Care Complex 82
Health Center 84
Health Center 86
Pediatrics Clinic 88

Office Projects

Headquarters Building 90
Office Building 92
Office Building *(Adaptive Re-Use)* 94
Office Building 96
Office Building 98
Office Building 100
Office Building *(Shell Bldg.)* 102
Office Building 104
Office Building *(Tilt-Up)* 106
Office Building 108

Recreational Projects

YMCA Athletic Facility 110
University Athletic Facility 112
Recreation & Visitor Center 114
University Athletic Facility 116
YMCA Athletic Facility 118

Religious Projects

Church Building 120
Christian Center 122

Residential Projects

Green Housing 124
Green Housing 126
Sustainable Condos 128
Solar House 130

ARCHITECT
DLR GROUP, INC. (A FLORIDA CORP.)
100 East Pine Street, #404
Orlando, FL 32801
www.dlrgroup.com

GENERAL CONTRACTOR
PPI CONSTRUCTION MANAGEMENT, INC.
8200 N.W. 15ᵗʰ Place, #B,
Gainesville, FL 32606

FILE UNDER
CIVIC
Gainsville, Florida

CONSTRUCTION TEAM

STRUCTURAL ENGINEER:
Blum, Schumacher & Associates, Inc.
14260 W. Newberry Road, #347, Newberry, FL 32669
ELECTRICAL & MECHANICAL ENGINEER:
TLC Engineering For Architecture
1717 South Orange Avenue, #300, Orlando, FL 32806
LANDSCAPE ARCHITECT: JCR Consulting
331 White Oak Circle, Maitland, FL 32751
COST ESTIMATOR: Cost Management, Inc.
5507 Alhambra Drive, Orlando, FL 32808

GENERAL DESCRIPTION

SITE: 6.2 acres. **NUMBER OF BUILDINGS:** One.
BUILDING SIZE: First floor, 38,926; second floor, 29,547; third
floor, 29,547; fourth floor, 21,759; total, 119,779 square feet.
BUILDING HEIGHT: First floor, 16'; second floor, 16';
third floor, 16'; fourth floor, 15'6"; total, 63'6".
BASIC CONSTRUCTION TYPE: New.
FOUNDATION: Concrete. **EXTERIOR WALLS:** Steel
structure, composite metal precast envelope.
ROOF: Modified bitumin. **FLOORS:** Concrete, steel deck.
INTERIOR WALLS: Metal stud, CMU, drywall.

FIRST FLOOR PLAN

ALACHUA COUNTY COURTHOUSE CRIMINAL JUSTICE CENTER
Construction Period: Nov 2001 to Sep 2003 • Total Square Feet: 119,779

C.S.I. Divisions (1 through 16)	COST	% OF COST	SQ.FT. COST		SPECIFICATIONS
BIDDING REQUIREMENTS	-	-	-		
1. GENERAL REQUIREMENTS	2,674,600	15.21	22.33	1	Summary of work, allowances, coordination, project meetings, submittals, construction facilities & temporary controls, contract closeout.
3. CONCRETE	1,789,400	10.18	14.94	3	Formwork, retaining walls, precast, pads.
4. MASONRY	1,265,200	7.20	10.56	4	Masonry & grout, accessories, unit, granite pavers.
5. METALS	1,095,400	6.23	9.15	5	Structural steel, joists, decking, miscellaneous, expansion joints.
6. WOOD & PLASTICS	1,015,600	5.78	8.48	6	Rough carpentry, architectural casework, wall paneling, finish carpentry.
7. THERMAL & MOIST. PROTECTION	447,200	2.54	3.73	7	Waterproofing, dampproofing, fireproofing, modified bituminous roof system, flashing.
8. DOORS AND WINDOWS	1,251,800	7.12	10.45	8	Hollow metal, wood doors, finish hardware, access doors, overhead doors, fire doors, aluminum curtainwall, storefront, glass & glazing.
9. FINISHES	2,012,500	11.45	16.80	9	Metal studs, drywall, insulation, stucco, hard tile, acoustical ceilings & wall panels, carpet, VCT, painting, wall coverings.
10. SPECIALTIES	183,700	1.04	1.53	10	Markerboards, toilet partitions, louvers, signage, flagpole, fire extinguishers, lockers, toilet accessories, inmate specialties, projection screens, dock bumpers, bike racks, miscellaneous.
12. FURNISHINGS	135,700	0.77	1.13	12	Courtroom bench seating.
14. CONVEYING SYSTEMS	654,400	3.72	5.46	14	Elevators (6).
15. MECHANICAL	2,779,600	15.81	23.21	15	Basic materials & methods, fire protection, plumbing, HVAC, insulation.
16. ELECTRICAL	2,278,800	12.95	19.03	16	Basic materials & methods, service & distribution, lighting, communications, controls, testing, special systems, fire alarm.
TOTAL BUILDING COST	**17,583,900**	**100%**	**$146.80**		
2. SITE WORK	735,100			2	Demolition, survey & Layout, improvements, earthwork, paving & surfacing, sewer & drainage, landscaping.
LANDSCAPING & OFFSITE WORK					
TOTAL PROJECT COST	**18,319,000**	*(Excluding architectural and engineering fees)*			

UPDATED ESTIMATE TO JAN 2011: $276.51 SF

Alachua County Courthouse
Criminal Justice Center
Gainesville, Florida

Architect: DLR Group, Inc. (A Florida Corp.)

The new Alachua County Courthouse Criminal Justice Center houses 11 new courtrooms and is situated on 6.2 acres. In an attempt to solve the county's operational issues of having a civil and a criminal court building, the new criminal courts facility brings with it the Clerk of the Court criminal operations, a portion of the court-reporting department, and staff from court administration. A new jury assembly space that is located in the criminal courthouse supports both buildings, making up for the lack of adequate assembly space in the existing courthouse.

The new criminal courthouse site encompasses 6 blocks a half mile west of the existing courthouse in downtown Gainesville, Florida. Since the 6-block site is more than adequate to support the new criminal courthouse, future growth beyond this project can be accommodated. The criminal courthouse with a ground floor of approximately 39,000 square feet, located at the northeast corner of the site, allowing for a future expansion to occur to the south. Up to as many as three future construction phases will make this site the new judicial complex for the downtown and County.

Photos Courtesy of George Cott of Chroma, Inc.

The development of the architecture has proceeded along the dual aims of creating a building, which would not only reflect the dignity and honor appropriate for a courthouse, but also would make a notable yet complementary civic addition to the fabric of the host downtown. The design draws from classical architecture with a vertical vocabulary, which is organized in a prototypical classical formula of base, middle and top. It is intended that this building have an architectural feel different from a commercial or religious structure. The strong solid ends and somewhat severe austerity of the elevations call attention to the seriousness of what this building represents and sets it apart from the more superficial aspects of our day-to-day commercial architecture.

The building mass is composed of 3 separate volumes and is a reflection of the functional aspects of the building. The first and largest is the main courts tower; levels 1-4 contain the courts and their direct support. The separation of the judicial chambers allowed for the creation of the low wing along southwest Second Avenue. The third component is the two-story entrance pavilion facing the north east of the low wing and tower.

This tri-part arrangement allows the building to have a more massive base and to attenuate as it reaches towards the sky, thereby creating a more visually pleasing and proportional silhouette. It also allowed the building to respond to different scales, creating both a gentle transition to the smaller size and scale of the buildings on the adjacent blocks as well as responding to the larger realm of the Gainesville skyline.

The asymmetrical 'Z' shaped plan relationship of tower to low wing and entry pavilion creates the opportunity to form a powerful urban space and entry forecourt. This plaza not only allows the tower to reinforce the urban edge of Main Street but also becomes a major component of the entry sequence in to the building.

LEED® PENDING
MANUFACTURERS/SUPPLIERS

DIV 08: *Curtainwall, Entrances & Storefronts:* Vistawall.

COST PER SQUARE FOOT $ 276.51

ARCHITECT
MHTN ARCHITECTS, INC.
420 East South Temple, #100
Salt Lake City, UT 84111
www.mhtn.com

CONSTRUCTION TEAM
STRUCTURAL ENGINEER: ABS Consulting, Inc.
310 South Main Street, #300, Salt Lake City, UT 84101
GENERAL CONTRACTOR: Layton Construction Co., Inc.
9090 South Sandy Parkway, Sandy, UT 84070
ELECTRICAL ENGINEER: BNA Consulting, Inc.
635 South State Street, Salt Lake City, UT 84111
MECHANICAL ENGINEER: Colvin Engineering Associates, Inc.
244 West 300 North, #200, Salt Lake City, UT 84103
ACOUSTICAL ENGINEER: Spectrum Engineers
175 South Main Street, #300, Salt Lake City, UT 84111

GENERAL DESCRIPTION
SITE: 2 acres.
NUMBER OF BUILDINGS: One.
BUILDING SIZE: First floor, 20,000; total, 20,000 square feet.
BUILDING HEIGHT: First floor, 33'8'; multiple roofs, highest point 33'8".
BASIC CONSTRUCTION TYPE: New/Design-Build.
FOUNDATION: Cast-in-place, reinforced, slab-on-grade.
EXTERIOR WALLS: CMU, curtainwall, brick, metal panels.
ROOF: Metal, membrane.
FLOORS: Concrete, access.
INTERIOR WALLS: Metal stud drywall.

Photos Courtesy of Paul Richer Photography

DRAPER LIBRARY
Construction Period: Nov 2004 to Oct 2005 • Total Square Feet: 20,000

C.S.I. Divisions (1 through 16)			COST	% OF COST	SQ.FT. COST	SPECIFICATIONS
		PROCUREMENT & CONT. REQ.	-		-	-
1.	1.	GENERAL REQUIREMENTS	453,414	15.81	22.67	-
3.	3.	CONCRETE	153,442	5.35	7.67	Forming & accessories, reinforcing, cast-in-place, cutting & boring.
4.	4.	MASONRY	190,000	6.63	9.50	Unit, faux stone assemblies.
5.	5.	METALS	298,438	10.41	14.92	Structural metal framing, joists, decking, cold-formed metal framing
6.	6.	WOOD/PLASTICS/COMPOSITE	97,712	3.41	4.89	Rough carpentry, finish carpentry, architectural woodwork.
7.	7.	THERMAL & MOIST. PROTECT	267,010	9.31	13.35	Dampproofing & waterproofing, thermal protection, roofing & siding panels, membrane roofing, flashing & sheet metal, roof & wall specialties & accessories, fire & smoke protection.
8.	8.	OPENINGS	150,156	5.24	7.51	Doors & frames, specialty doors & frames, entrances, storefronts & curtainwalls, windows, hardware, glazing, louvers & vents.
9.	9.	FINISHES	293,177	10.22	14.66	Plaster & gypsum, tiling, ceilings, flooring, acoustic treatment, painting & coating.
10.	10.	SPECIALTIES	126,621	4.42	6.33	Fireplaces & stoves.
11.	11.	EQUIPMENT	3,913	0.13	0.20	Residential.
12.	12.	FURNISHINGS	-	-	-	-
13.	13.	SPECIAL CONSTRUCTIONS	-	-	-	-
14.	14.	CONVEYING SYSTEMS	-	-	-	-
15.		MECHANICAL	496,459	17.31	24.82	Water based fire-suppression systems, piping & pumps, equipment, fixtures, air distribution, air cleaning devices, central heating equipment, central cooling equipment, central equipment.
15.	21.	FIRE SUPRESSION	-	-	-	-
15.	22.	PLUMBING	-	-	-	-
15.	23.	HVAC	-	-	-	-
16.	26.	ELECTRICAL	337,283	11.76	16.86	Medium-voltage distribution, lighting.
16.	27.	COMMUNICATIONS	-	-	-	-
TOTAL BUILDING COST			**2,867,625**	**100%**	**$143.38**	
2.	2.	SITEWORK	680,382			Sitework, excavation, utilities.
2.	31.	EARTHWORK	-		-	
2.	32.	EXTERIOR IMPROVEMENTS	-		-	
2.	33.	UTILITIES	-		-	
		LANDSCAPING & OFFSITE WORK	-		-	
TOTAL PROJECT COST			**3,269,703**	*(Excluding architectural and engineering fees)*		

UPDATED ESTIMATE TO JAN 2011: $242.62 SF

Draper Library
Draper, Utah

Architect: MHTN Architects, Inc.

The Draper Library design creates harmony with the surrounding farmland through timeless character and sustainable design elements. The design focuses on the efficient functionality of the library spaces while taking advantage of the wonderful mountain views to the east and south.

By using materials with recycled content, the Draper Library is a sustainable design and will qualify for LEED® certification.

The MHTN design team took the base program and enhanced it to optimize the relationships and fuctionality of the library. The result is a design with increased square footage, optimal adjacencies, circulation, and function, with several interior and exterior architectural features that make the building a learning tool for the community.

The soundness and quality of the design resonates in the materials used on the project including sustainable design principles such as the re-use of barn wood taken from an old Draper City barn. The interior experience of the library has been designed around the idea of connecting the visitor to their own landscape by framing large views of the mountains and Corner Canyon. The large east window is meant to orient and draw us through the library, with each area having its own connection to Draper's regional native landscape. For example, the Children's area windows are lower to encourage that connection to the teaching garden immediately outside, the adult reading area window is long and expansive to capture the surrounding panoramic landscape.

Photos Courtesy of David A. Harvey Photography

The library is designed to be a state-of-the-art facility organized around its functionality as a library. The large vaulted space recalls large barn structures of the past, and acknowledges Draper's agricultural history without literally trying to re-create any particular structure. The angular slope of the meeting room gives a nod to the slope and geometry of the mountains, and the IFA structure. Additional aspects give identity and scale to other parts of the library such as the staff area and young adult area.

Most materials in this structure have been selected with recycled content as one of the basic criteria. Many materials are not normally thought of as having recycled content, however in the Draper Library 40% of the carpet is made from recycled content; 40% of the steel structure; 80% of the roof deck; 57% of the ceiling tile; 25%

of the metal wall studs; 12% of the brick; 30% of the exterior metal wall panels; 32% of the aluminum window system; and 20% of the batt insulation. The circulation and reference desks are made of "wheatboard" board which is made of a composite of wheat hulls. An important part of the entry plaza is the reclaimed barnwood trellis which shades the meeting room windows. This trellis is made entirely from salvaged barnwood from local Draper history, donated by Mayor Smith from his family property.

By using sustainable design principles and sustainable materials, the Draper Library looks to the future while honoring the City's agricultural past.

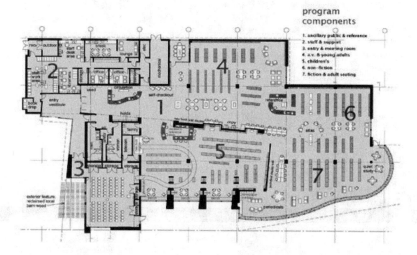

program components

1. ancillary public & reference
2. staff & support
3. entry & meeting room
4. a.v. & young adults
5. children's
6. non-fiction
7. fiction & adult seating

LEED® CERTIFICATION PENDING	
MANUFACTURERS/SUPPLIERS	
DIV 04:	*Brick:* **Interstate Brick.**
DIV 05:	*Metal Joists:* **Vulcraft.**
DIV 07:	*Manufactured Wall Panels:* **Omega®** Panels by **Laminators**; *Metal Roof:* **Berridge Manufacturing.**
DIV 08:	*Entrances & Storefronts, Windows:* **Vistawall.**
DIV 09:	*Carpet:* **Shaw**, **Lees**; *Gypsum:* **BPB**; *Paint:* **Columbia.**
DIV 16:	*Lighting:* **Lightolier**, **Autolux.**

COST PER SQUARE FOOT $ 242.62

ARCHITECT
McMillan Pazdan Smith
200 East Broad Street, #300,
Greenville, SC 29601
www.mcmillansmith.com

FILE UNDER
CIVIC
Greenville, South Carolina

CONSTRUCTION TEAM

STRUCTURAL ENGINEER:
Britt, Peters & Associates, Inc.
550 S. Main Street, #301, Greenville, SC 29601
GENERAL CONTRACTOR:
The Harper Corporation
35 West Court Street, #400, Greenville, SC 29601
LEED® COMMISSIONING AGENT:
KLG Jones
218 Trade Street, Greer, SC 29651

GENERAL DESCRIPTION

SITE: 7.21 Acres.
NUMBER OF BUILDINGS: One.
BUILDING SIZE: First floor, 14,769; second floor, 12,948; total, 27,717 square feet.
BUILDING HEIGHT: First floor, 14'; second floor, 27'8" min to 43'6" max; floor to floor, 14'; total, 27'8" min to 43'6" max.
BASIC CONSTRUCTION TYPE: Type: New/Steel Frame
FOUNDATION: Reinforced concrete, slab-on-grade.
EXTERIOR WALLS: Curtainwall, metal panels, stone veneer.
ROOF: Metal. **FLOOR::** Concrete.
INTERIOR WALLS: Metal stud drywall.

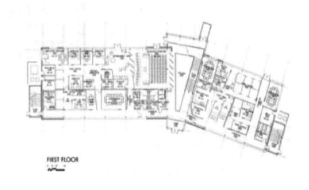

FIRST FLOOR

RENEWABLE WATER RESOURCES
Construction Period: May 2007 to May 2008 • Total Square Feet: 27,000

C.S.I. Divisions (1 through 16)			COST	% OF COST	SQ.FT. COST	SPECIFICATIONS
		PROCUREMENT & CONT. REQ.	-	-	-	-
1.	1.	GENERAL REQUIREMENTS	610,782	12.47	22.04	-
3.	3.	CONCRETE	296,359	6.05	10.69	Forming & accessories, reinforcing, cast-in-place. (Concrete Breakdown: cubic yards foundation, 222; cubic yards walls, 122; cubic yards floors, 220.)
4.	4.	MASONRY	152,853	3.12	5.51	Stone assemblies.
5.	5.	METALS	757,744	15.48	27.34	Structural metal framing, joists, decking, cold-formed metal framing, fabrications.
6.	6.	WOOD/PLASTICS/COMPOSITE	93,435	1.91	3.37	Rough carpentry, finish carpentry, architectural woodwork.
7.	7.	THERMAL & MOIST. PROTECT	316,217	6.46	11.41	Weather barriers, roofing & siding panels, joint protection.
8.	8.	OPENINGS	784,295	16.02	28.30	Doors & frames, entrances, storefronts, & curtainwalls, hardware, glazing, louvers & vents.
9.	9.	FINISHES	465,266	9.50	16.79	Plaster & gypsum board, ceilings, flooring, wall finishes, painting & coating.
10.	10.	SPECIALTIES	86,717	1.77	3.13	Fountain, other.
11.	11.	EQUIPMENT	24,913	0.51	0.90	Security, detention & banking, foodservice.
12.	12.	FURNISHINGS	5,500	0.11	0.20	-
14.	14.	CONVEYING SYSTEMS	51,100	1.04	1.84	Elevators (1).
15.	21.	FIRE SUPRESSION	57,353	1.17	2.07	Water-based fire-suppression systems.-
15.	22.	PLUMBING	92,916	1.90	3.35	Piping & pumps, equipment, fixtures.
15.	23.	HVAC	540,152	11.03	19.49	Air distribution, central heating, central cooling, central HVAC equipment.
16.	26.	ELECTRICAL	560,860	11.46	20.23	Medium-voltage distribution, lighting.
TOTAL BUILDING COST						
2.	2.	EXISTING CONDITIONS	36,577			Demolition & structure moving.
2.	31.	EARTHWORK	285,365			Site clearing, earth moving.
2.	32.	EXTERIOR IMPROVEMENTS	243,667			Bases, bollards, & paving.
2.	33.	UTILITIES	128,408			Water, sanitary sewer, storm.
TOTAL PROJECT COST			**5,590,479**	*(Excluding architectural and engineering fees)*		

UPDATED ESTIMATE TO JAN 2011: $ 255.08 SF

DCD Subscribers: Access this case study and hundreds more for instant date and location calculations at www.dcd.com.

Renewable Water Resources (ReWa) Administration Building

Greenville, South Carolina

Architect: McMillan Pazdan Smith

A new administrative building for ReWa consisted of demolishing and recycling the existing administration and engineering buildings and constructing a new two-story 27,700-square-foot Class A office building with associated parking. New construction was located within the existing building's footprints, resulting in minimal disturbance across the heavily wooded 7.21 acre site directly adjacent to the Reedy River.

Tasked with emulating the efficient, scientific processes of ReWa's water treatment processes, while engaging and respecting the existing pastoral riverside site, the Design Team created a building project to consolidate ReWa's corporate office functions as well as for community and educational events. This modern, organic aesthetic was created by utilizing simple, efficient forms, exposed structural steel framing, floor and roof-decking, along with exposed HVAC and other utilities, which also reflected the efficiency of ReWa's Treatment Process and created a showplace for their commitment to the community and sustainable development. Locally quarried stone was used minimally on the exterior, while metal paneling, curtainwall, roofing and sunshades dominated the facades and combined to reflect the modern organic nature of the development.

Intended programmatically to house departments which had been located in different buildings on and off the HQ site, the first floor houses the executive, human resources, customer service, finance and purchasing departments. A two-story glass atrium houses the Process Gallery, showcasing the water treatment process, which leads to an exterior patio area overlooking the scenic Reedy River. A board/multipurpose room expands into the Gallery (via) a custom wood pivoting wall system) for larger gatherings and also has a refreshment area for catered events. The second floor features the engineering and pretreatment departments, along with training/expansion space, seven conference rooms and an employee break room.

Photos Courtesy of Firewater Photography® 2009

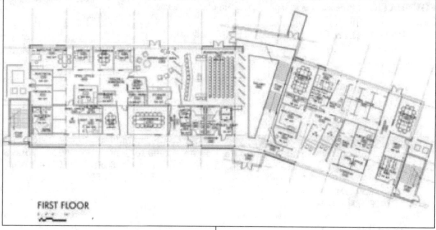

FIRST FLOOR

The entire project was intended to serve as a learning tool for the community about sustainability, which began with recycling the building materials of the existing Administration and Engineering buildings. Existing parking areas were reused, thereby maintaining the stands of mature white oak and pine trees. Free educational building tours and a custom graphics/signage system explains to visitors and occupants alike that the building's energy system utilizes daylighting, occupancy sensors and energy efficient light fixtures to minimize consumption, as well as a four-pipe HVAC system which results in energy efficiency over 20% percent better than ASHRAE 90.1 (2004 standards).

While 84% of non-hazardous materials were diverted from landfills, over 33% of the building materials (by cost) were made from post- and pre-consumer recycled content. Further reducing the Project's carbon footprint is that 45% of the building materials were extracted and manufactured regionally, within a 500-mile radius of Greenville, S.C.

A unique site irrigation system engineered by ReWa also uses the effluent water byproduct produced by a ReWa treatment plant located nearby, and plans are underway to partner with Greenville County to allow the "Greenville Greenways" system, a pedestrian/bicycle trails program, to cross the property along the banks of the Reedy River. The Owner and Design Team are currently pursuing LEED® Certification for the project, which was also designed to serve as a "Bird Habitat."

**LEED®
CERTIFICATION PENDING**

COST PER SQUARE FOOT $ 255.08

ARCHITECT OF RECORD	DESIGN ARCHITECT	FILE UNDER
Becker Morgan Group	GWWO, Inc./ Architects	*CIVIC*
309 South Governors Avenue, Dover, DE 19904	800 Wyman Park Drive, #300, Baltimore, MD 21211	*Wilmington, Delaware*
www.beckermorgan.com	www.gwwoinc.com	

CONSTRUCTION TEAM

STRUCTURAL ENGINEER: MacIntosh Engineering
300 Delaware Avenue, #820, Wilmington, DE 19801
CONSTRUCTION MANAGER: EDiS Company
110 South Poplar Street, #400, Wilmington, DE 19801
MECHANICAL & ELECTRICAL: Mahaffy & Associates, Inc.
800 Woodlawn Avenue, Wilmington, DE 19805
CIVIL ENGINEER: Rummel, Klepper & Kahl
81 W. Mosher Street, Baltimore, MD 21217

GENERAL DESCRIPTION

SITE: Site: 7 Acres in a 214 acre wetland reserve.
NUMBER OF BUILDINGS: One.
BUILDING SIZE: Size: 4-story with a total of 9,174 square feet.
BUILDING HEIGHT: 4-story with a total of 68'.
BASIC CONSTRUCTION TYPE: Heavy Timber Framing.
FOUNDATION: Pre-cast concrete piles w/cast-in-place concrete caps and concrete piers.
EXTERIOR WALLS: Cold form metal studs w/hand split cedar shake siding.

ROOF: Standing seam metal roof.
FLOORS: Heavy timber.
INTERIOR WALLS: Metal stud drywall, wheatboard.

Dupont Environmental Education Center
Construction Period: April 2008 to September 2009 • Total Square Feet: 9,174

C.S.I. Divisions (1 through 16)			COST	% OF COST	SQ.FT. COST	SPECIFICATIONS
		PROCUREMENT & CONT. REQ.	75,000	1.20	8.18	Solicitation, conditions of the contract, revisions, clarifications, and modifications.
1.	1.	GENERAL REQUIREMENTS	1,514,385	24.13	165.07	Price & payment procedures, administrative requirements, quality requirements, temporary facilities & controls, product requirements, execution & closeout, performance.
3.	3.	CONCRETE	486,369	7.75	53.02	Forming & accessories, reinforcing, cast-in-place, cast decks & underlayment, grouting (Concrete breakdown: cubic yards total, 1200).
4.	4.	MASONRY	151,754	2.42	16.54	Unit, manufactured.
5.	5.	METALS	1,531,519	24.40	166.94	Structural metal framing, joists, decking, metal fabrications.
6.	6.	WOOD/PLASTICS/COMPOSITE	163,865	2.61	17.86	Rough carpentry, finish carpentry, architectural woodwork, composite fabrications.
7.	7.	THERMAL & MOIST. PROTECT	102,394	1.63	11.16	Roofing & siding panels, flashing & sheet metal, roof & wall specialties & accessories, joint protection.
8.	8.	OPENINGS	132,780	2.12	14.47	Doors & frames, specialty doors & frames, entrances, storefronts & curtainwalls, windows, hardware, glazing.
9.	9.	FINISHES	354,577	5.65	38.65	Plaster & gypsum board, tiling, ceilings, flooring, wall fiishes, acoustic treatment, painting & coating.
10.	10.	SPECIALTIES	108,905	1.74	11.87	Other.
14.	14.	CONVEYING SYSTEMS	76,270	1.22	8.31	Elevators (1 passenger).
15.	21.	FIRE SUPRESSION	132,677	2.11	14.46	Water-based fire-suppression system, fire pumps.
15.	22.	PLUMBING	132,000	2.10	14.39	Piping & pumps, equipment, fixtures.
15.	23.	HVAC	360,443	5.74	39.29	Piping & pumps, air distribution, air cleaning devices, central heating equipment, central cooling equipment, central HVAC equipment. Also included Integrated Automatics: network equipment, automation instrumentation & terminal devices, facility controls, control sequences.
16.	26.	ELECTRICAL	952,658	15.18	103.84	Medium voltage electrical distribution, low-voltage electrical transmission, electrical & cathodic protection, lighting. Electronic Safety & Security: detection & alarm.
TOTAL BUILDING COST			**6,275,619**	**100%**	**$684.07**	
2.	2.	EXISTING CONDITIONS	50,000			Assessment, subsurface investigation, site remediation, water remediation, facility remediation.
2.	31.	EARTHWORK	1,907,822			Site clearing, earth moving, earthwork methods, shoring & underpinning, excavaton support & protection.
2.	32.	EXTERIOR IMPROVEMENTS	424,709			Bridges.
TOTAL PROJECT COST			**8,658,150**	*(Excluding architectural and engineering fees)*		

UPDATED ESTIMATE TO JAN 2011: $ 812.63 SF

DCD Subscribers: Access this case study and hundreds more for instant date and location calculations at www.dcd.com.

Dupont Environmental Education Center
At the Russell W. Peterson
Urban Wildlife Refuge
Wilmington, Delaware

Architect: Becker Morgan Group and GWWO, Inc./Architects

The new DuPont Environmental Education Center is part of efforts by the Riverfront Development Corporation of Delaware to restore marshlands along the Christina River, while creating economic vitality, enhancing the environment and promoting public access. The site for the new facility is an open expanse of reclaimed wetland located at the southern most edge of Wilmington, Delaware's urban environment. Housing exhibits, classrooms and offices for nature education and related recreation activities, as well as community meeting and gathering space, the building enhances and encourages the relationship between urban development along the waterfront and the natural environment.

The client's goal was "to create a symbiotic relationship between urban development along the waterfront and the natural environment." To fully realize the client's objectives it was important to locate the 4-story building within the 212-acre refuge. Elevated above the marshlands, the design responds to the dynamics of the natural systems - tidal river and wetlands - as well as infrastructural elements: an adjacent railway, high voltage power lines, and multiple utility rights of way. Keeping true to its setting, the building has a number of green innovations including infrastructure for both a wind turbine and a solar panel array. The site parking and walkway areas are lit by Photovoltaic technology. The building compliments include structural glue-laminated timber from local sources, low VOC paints, passive solar design, lighting controls and wheatboard paneling throughout.

The center is at once a sentinel standing guard over the refuge and a gateway between the urban environment and the last vestige of natural marshland. As they traverse the delicate bridge, visitors experience the tenuous connection between man and nature. Seen from afar, the building acts as a beacon marking the terminus of the city 's riverwalk. From within, windows on the north facade frame views to the city while the south opens toward the natural setting.

Photos Courtesy of Jay Greene

COST PER SQUARE FOOT $ 812.63

ARCHITECT
SCHMIDT COPELAND PARKER STEVENS, INC.
1220 West 6th, #300
Cleveland, OH 44113
www.scpsohio.com

CONSTRUCTION TEAM
STRUCTURAL ENGINEER: I.A. Lewin & Associates
4110 Mayfield Road, #B, South Euclid, OH 44121
GENERAL CONTRACTOR: Odyssey Builders, Inc.
1521 Lowell Avenue, Erie, PA 16505
ELECTRICAL & MECHANICAL ENGINEER:
Peters, Tschantz, & Bandwen, Inc.
275 Springside Drive, #300, Akron, OH 44333
LANDSCAPE ARCHITECT & COST ESTIMATOR:
Schmidt Copeland Parker Stevens, Inc.
1220 West 6th, #300, Cleveland, OH 44113

Photos Courtesy of Mabius Grey

GENERAL DESCRIPTION
SITE: 50 acres.
NUMBER OF BUILDINGS: One.
BUILDING SIZE: Basement, 650; first floor, 9,185; second floor, 855; total, 10,690 square feet.
BUILDING HEIGHT: Basement, 8'; first floor, 10'; second floor, 8'; total, 26'.
BASIC CONSTRUCTION TYPE: Addition, Renovation/5B.
FOUNDATION: Reinforced concrete, slab-on-grade.
EXTERIOR WALLS: Wood siding, fiber cement siding.
ROOF: Asphalt shingles, membrane, vegetated.
FLOORS: Concrete, wood.
INTERIOR WALLS: CMU, drywall

ASBURY WOODS NATURE CENTER
Construction Period: Oct 2004 to Nov 2005 • Total Square Feet: 10,690

C.S.I. Divisions (1 through 16)			COST	% OF COST	SQ.FT. COST	SPECIFICATIONS
1.	1.	GENERAL REQUIREMENTS	135,366	8.41	12.66	Price & payment procedures, administrative requirements, quality requirements, temporary facilities & controls, execution & closeout requirements.
3.	3.	CONCRETE	82,527	5.13	7.72	Formwork, reinforcing, cast-in-place.
4.	4.	MASONRY	39,886	2.48	3.73	Unit.
5.	5.	METALS	18,591	1.16	1.74	Fabrications.
6.	6.	WOOD/PLASTICS/COMPOSITE	403,598	25.08	37.75	Rough & finish carpentry, architectural woodwork.
7.	7.	THERMAL & MOIST. PROTECT	110,275	6.85	10.32	Dampproofing & waterproofing, thermal protection, weather barriers, steep slope roofing, roofing & siding panels, membrane roofing, flashing & sheet metal, fire & smoke protection.
8.	8.	OPENINGS	94,888	5.90	8.88	Doors & frames, entrances, storefronts, & curtainwalls, windows, roof windows & skylights, hardware, glazing.
9.	9.	FINISHES	128,312	7.97	12.00	Plaster & gypsum board, ceilings, flooring, wall finishes, acoustic treatment, painting & coating.
10.	10.	SPECIALTIES	8,948	0.56	0.84	Interior & exterior specialties.
11.	11.	EQUIPMENT	-	-	-	-
12.	12.	FURNISHINGS	6,280	0.39	0.59	Casework, other furnishings.
13.	13.	SPECIAL CONSTRUCTIONS	-	-	-	-
14.	14.	CONVEYING SYSTEMS	-	-	-	-
15.	21.	MECHANICAL	-	-	-	-
15.	22.	PLUMBING	99,874	6.20	9.34	Piping & pumps, equipment, fixtures, pool & fountain.
15.	23.	HVAC	239,200	14.86	22.38	Piping & pumps, air distribution, central equipment.
16.	26.	ELECTRICAL	241,619	15.01	22.60	Medium voltage distribution, low-voltage transmission, lighting, communications, electronic detection & alarm.
16.	27.	COMMUNICATIONS	-	-	-	-
TOTAL BUILDING COST			**1,609,364**	**100%**	**$ 150.55**	
2.	2.	EXISTING CONDITIONS				Assessment, demolition & structure moving, utilities, wells, planting.
TOTAL PROJECT COST			**1,866,979**	*(Excluding architectural and engineering fees)*		

UPDATED ESTIMATE TO JAN 2011: $224.62 SF

DCD Subscribers: Access this case study and hundreds more for instant date and location calculations at www.dcd.com.

Asbury Woods Nature Center
Erie, Pennsylvania

Photos Courtesy of Schmidtcopelandparkerstevens

In 2002, Schmidt Copeland Parker Stevens, Inc. was selected to serve as the lead architects and landscape architects for the improvements to Asbury Woods Nature Center. While the nature center was known as one of the premier environmental centers in northwestern Pennsylvania, it was operating out of a 70-year old cottage donated to the Millcreek Township School District by Dr. Otto Behrend, co-founder of the Hammermill Paper Company, in 1957. As a part of the master planning process performed by Schmidt Copeland Parker Stevens it was determined that the 2,000-square-foot cottage would need to be renovated, modernized and significantly enlarged to continue providing educational and recreational opportunities to the more than 130,000 visitors the center serves annually.

The completed Center now anchors a collage of buildings including a renovated pavilion, garage, new information kiosk, pergola and sugar shack. To compliment the much-needed accessible restrooms, office improvements, and classroom spaces for multiple concurrent school groups, the addition contains amenities ranging from a dedicated gift shop to an indoor turtle pond and waterfall.

Following the missions of both the Center and the design team, a wide variety of sustainable design and construction practices were used throughout the project and are supplemented by an extensive range of "green" materials and technologies. By balancing the educational experience with economically sound design solutions and the environment, the Nature Center has been crafted into a regional model of efficient and responsible use of resources to conserve water, energy, land and materials.

A geothermal heating exchange system consisting of fifteen 300-foot deep wells was installed beneath a grassy field adjacent to the center. The fluid circulating through the closed loop piping in these wells is pumped to the mechanical space in the basement of the original cottage where heat pumps extract heating or cooling as needed to satisfy the HVAC loads of the Center.

The 4,700-square-foot "green roof" was installed through funding secured from a Pennsylvania Energy Harvest grant. The vegetated roof moderates daily fluctuations in building temperature, reduces storm water runoff, improves air quality and mitigates urban heat island impacts.

The exterior walls of the addition are clad in fiber cement shingle panels, which effectively blend the new construction into the wood shingle of the historic cottage providing a long term, low maintenance siding alternative.

Operable windows allow views to the Center's surroundings as well as offering light and natural ventilation. Tubular skylights also bring natural light into several of the buildings spaces.

Interior finishes include sealed concrete floors, flooring composed of recycled rubber, low VOC paints, and carpet tiles with a number of green attributes. The wood wainscot used throughout the building was milled from trees harvested on site as part of a forestry management campaign.

The site work includes low maintenance native trees, wildflower seeding and wetland plantings. The design team minimized paved parking areas by providing overflow parking on stabilized lawn and much of the storm water piping was eliminated by using planted swales, which allow rain to naturally recharge groundwater.

In addition to recognition for the environmental accomplishments of the project, one of the greatest compliments that the design team received was the acknowledgment that even long-time visitors to Asbury Woods Nature Center had difficulty distinguishing where the architecture of the original much-loved structure ended and the new construction began. With the help of dedicated staff and a refreshed mission, the new building will see Asbury Woods Nature Center and its visitors through a long and sustainable future.

MANUFACTURERS/SUPPLIERS	
DIV 03:	*Concrete Stain:* Solomon Colors.
DIV 07:	*Fiber Cement Siding:* James Hardie
	Membrane: Sarnafail;
	Vegetated: Roof Meadow;
	Shingles: GAF
DIV 08:	*Entrances & Storefron:* Tubelite
	Windows: Integrity® by Marvin.
DIV 09:	*Acoustical Treatment:* USG Tectrum;
	Wall Carpet: Ozite Rib,
	Paint: ICI; *Sheet Flooring:* ECOsurfaces®
	By Dodge Regupol; *Carpet Tile:* Shaw
DIV 26:	*Lighting:* Lithonia, Hapco, Trend, Ruud,
	Lightolier, Linear, Bartco.

COST PER SQUARE FOOT $ 224.62

ARCHITECT
KKE Architects, Inc.
300 1ˢᵗ Avenue North, Minneapolis, MN 55401
www.kke.com

CONSTRUCTION TEAM

STRUCTURAL ENGINEER: Clark Engineering Corp.
621 Lilac Drive North, Minneapolis, MN 55422

GENERAL CONTRACTOR:

Merrimar Construction Corp.
18651 Buchanan Street NE, East Bethel, MN 55011

MECHANICAL AND ELECTRICAL ENGINEER:

Karges-Faulconebridge Inc.
670 West County Road B, St. Paul, MN 55113

COST ESTIMATOR: Faithful + Could
500 International Centre, 2ⁿᵈ Avenue South,
Minneapolis, MN 55402

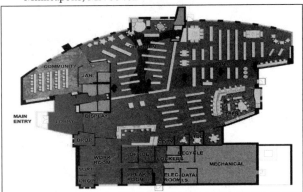

GENERAL DESCRIPTION

Site: 4 Acres.
Number of Buildings: One.
Building Size: First floor, 16,542; total 16,542 square feet.
Building Height: First floor, 16'; total 16'.
Basic Construction Type: New/Steel Frame.
Foundation: Cast-in place.
Exterior Walls: CMU, curtainwall.
Roof: Asphalt shingles.
Floors: Concrete.
Interior Walls: CMU, metal stud drywall.

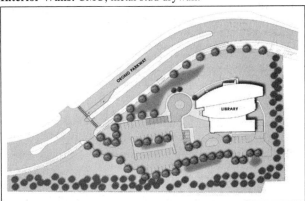

ELK RIVER LIBRARY
Construction Period: Nov 2006 to Sep 2007 • Total Square Feet: 16,542

C.S.I. Divisions (1 through 16)			COST	% OF COST	SQ.FT. COST	SPECIFICATIONS
		PROCUREMENT & CONT. REQ.	127,791	3.61	7.73	-
1.	1.	GENERAL REQUIREMENTS	75,816	2.14	4.58	Summary, price & payment procedures, administrative requirements, quality requirements, temporary facilities & controls, product requirements, execution & closeout requirements, performance requirements.
3.	3.	CONCRETE	200,000	5.64	12.09	Forming & accessories, reinforcing, cast-in-place.
4.	4.	MASONRY	325,000	9.17	19.65	Unit, manufactured masonry.
5.	5.	METALS	308,000	8.69	18.62	Structural metal framing, joists, decking, cold-formed metal framing, fabrications.
6.	6.	WOOD/PLASTICS/COMPOSITE	256,000	7.22	15.48	Rough carpentry, architectural woodwork.
7.	7.	THERMAL & MOIST. PROTECT	303,000	8.55	18.32	Dampproofing, membrane roofing, flashing & sheet metal, roof & wall specialties and accessories, fire & smoke protection, joint protection.
8.	8.	OPENINGS	271,000	7.65	16.38	Doors & frames, entrances, storefronts & curtain walls, windows, hardware, glazing.
9.	9.	FINISHES	383,000	10.81	23.15	Plaster & gypsum board, tiling, ceilings, flooring, wall finishes, acoustic treatment, painting & coating.
10.	10.	SPECIALTIES	31,900	0.90	1.93	Information, interior, safety, storage.
11.	11.	EQUIPMENT	83,000	2.34	5.02	Residential, educational & scientific.
12.	12.	FURNISHINGS	11,600	0.33	0.70	Furnishings & accessories, multiple seating, other.
15.	21.	FIRE SUPRESSION	28,000	0.79	1.69	Water-based fire-suppression systems.
15.	22.	PLUMBING	401,000	11.31	24.24	Piping & pumps, equipment, fixtures.
15.	23.	HVAC	381,000	10.75	23.03	Piping & pumps, air distribution, equipment, decentralized HVAC equipment.
16.	26.	ELECTRICAL	312,000	8.80	18.86	MediumElectronic detection & alarm.
16.	28.	ELECTRONIC SAFETY & SECURITY	46,000	1.30	2.78	Electronic detection & alarm.
TOTAL BUILDING COST			**3,544,107**	**100.00**	**$214.25**	
2.	2.	EXISTING CONDITIONS	15,000			Demolition.
2.	31.	EARTHWORK	130,500			-
2.	32.	EXTERIOR IMPROVEMENTS	158,100			-
2.	33.	UTILITIES	100,000			-
TOTAL PROJECT COST			**3,947,707**	*(Excluding architectural and engineering fees)*		

UPDATED ESTIMATE TO JAN 2011: $257.25 SF

DCD Subscribers: Access this case study and hundreds more for instant date and location calculations at www.dcd.com.

Elk River Library
Elk River, Minnesota

Architect: KKE Architects, Inc.

The Elk River Library is a LEED® registered project in the City of Elk River, Minnesota. The goals set by the building committee were to create a beautiful and lasting facility, create a civic campus with the adjacent city hall and public safety building, and take advantage of the views and connections with the adjacent Lake Orono and Orono Park.

The City of Elk River was committed to a high level of sustainable design in support of the City's leadership in and commitment to energy efficiency. Waste reduction, reduction in toxicity, energy conservation, and water conservation were all driving forces for the Elk River Library project. Sustainable design features included natural daylighting, a high efficiency mechanical system, and adjacent water gardens. Final certification will make this the first LEED® certified library in the state of Minnesota.

The architectural form of the library is inspired by daylight harvesting resulting in a modern, comfortable, and welcoming environment with the use of wood, warm colors, and various textural elements. Local environmental elements such as the two local rivers, and the library's proximity to Lake Orono, are the source of inspiration for the interior design concept. Those elements are represented by area rugs that have a pattern reminiscent of a sandy beach, interior window glazing with a water-like pattern, and two stylized "trees" in the children's area. Patrons can take advantage of various serene reading areas with a central fireplace, and picturesque views of Lake Orono. The teen

area is screened off from the adult area to provide a unique sense of place for teenage patrons. The color scheme is similar to that of the adult area, however the distinctive appearance derives from the rubber flooring, the deep window seating with its colorful stained glass, modern light fixtures, and study chairs made of aluminum. A large video screen will soon be incorporated for gaming, on-line meetings, and blogging for the purpose of creating a fun place for teens to hang out with friends and study.

The most notable accomplishment on this project is the energy efficiency. The system designed for the Elk River Library considered both occupant comfort and energy consumption as high priorities. While the goal established early in the project was to design a system that would reduce energy consumption by 40%, energy modeling resulted in a building that is 60% more efficient than other code-based designs. The geothermal well and the displacement ventilation system coupled with daylighting strategies helped exceed the goal.

Water use reduction is accomplished at the Elk River Library starting with the landscaping design. The design reduces the area of irrigation on site. Indigenous drought tolerant plantings, the water gardens and the high efficiency irrigation system all contribute to the water usage reduction.

LEED® CERTIFIED

MANUFACTURERS/SUPPLIERS	
DIV 07:	*Membrane:* Carlisle; *Skylight:* Kalwall.
DIV 08:	*Entrances & Storefront:* Tubelite; *Curtainwall:* Wausau; *Windows:* Traco.
DIV 09:	*Flooring:* Forbo.
DIV 26:	*Lighting:* Lithonia, Precision, Gotham.

COST PER SQUARE FOOT $257.25

<table>
<tr><td colspan="2">

ARCHITECT
Pate Design Group, Inc.
4168 Abbotts Bridge Road, Duluth, GA 30097
www.patedesigngroup.com

</td><td>

FILE UNDER
COMMERCIAL
Lawrenceville, Georgia

</td></tr>
</table>

CONSTRUCTION TEAM

STRUCTURAL ENGINEER: Bennett & Pless, Inc.
   3395 Northeast Expressway, Atlanta, GA 30341
GENERAL CONTRACTOR: Schoppman Company, Inc.
   1640 Powers Ferry Road, Building 7, #200, Marietta, GA 30067
ELECTRICAL & MECHANICAL ENGINEER:
   GTP Consulting Engineers
   2400 Pleasant Hill Road, #310, Duluth, GA 30096
LANDSCAPE ARCHITECT: Chattahoochee Home & Garden
   1101 Via Bayless, Marietta, GA 30066

GENERAL DESCRIPTION

SITE: .40 Acre.
NUMBER OF BUILDINGS: One.
BUILDING SIZE: First floor, 6,316; second floor, 9,645; total, 15,961 s.f.
BUILDING HEIGHT: First floor, 9'4"; second floor, 10'1"; total, 32'4".
BASIC CONSTRUCTION TYPE: Renovation/2A/Structural Steel, Wood.
FOUNDATION: Reinforced concrete, slab-on-grade.
EXTERIOR WALLS: CMU, brick. Roof: Asphalt shingles, metal.
ROOF: Asphalt shingles, metal.
FLOORS: Concrete. **INTERIOR WALLS:** CMU, metal stud drywall.

CROGAN STREET

Construction Period: Oct 2007 to Jan 2008 • Total Square Feet: 15,961

C.S.I. Divisions (1 through 16)			COST	% OF COST	SQ.FT. COST	SPECIFICATIONS
		PROCUREMENT & CONT. REQ.	-	-	-	-
1.	1.	GENERAL REQUIREMENTS	158,281	13.76	9.92	Price & payment procedures, administrative requirements, temporary facilities & controls, product requirements, execution & closeout requirements, performance requirements.
3.	3.	CONCRETE	73,876	6.42	4.63	Forming & accessories, reinforcing, cast-in-place, cast decks & underlayment, mass, cutting & boring.
4.	4.	MASONRY	148,734	12.93	9.32	Unit, stone assemblies, manufactured masonry.
5.	5.	METALS	198,755	17.28	12.45	Structural metal framing, joists, decking, cold-formed metal framing, fabrications, decorative.
6.	6.	WOOD/PLASTICS/COMPOSITE	54,503	4.74	3.41	Rough carpentry, finish carpentry.
7.	7.	THERMAL & MOIST. PROTECT	110,977	9.65	6.95	Dampproofing & waterproofing, thermal protection, weather barriers, roofing & siding panels, flashing & sheet metal.
8.	8.	OPENINGS	64,754	5.63	4.06	Doors & frames, specialty doors & frames, entrances, storefronts, & curtain walls, hardware, glazing, louvers & vents.
9.	9.	FINISHES	90,151	7.84	5.65	Plaster & gypsum board, tiling, ceilings, flooring, wall finishes, painting & coating, acoustic treatment.
10.	10.	SPECIALTIES	2,272	0.20	0.14	Exterior.
15.	21.	FIRE SUPRESSION	35,000	3.04	2.19	Water-based fire-suppression systems.
15.	22.	PLUMBING	30,000	2.61	1.88	Piping & pumps, equipment, fixtures.
15.	23.	HVAC	73,020	6.34	4.58	Piping & pumps, air distribution, central heating, central cooling, central HVAC equipment.
16.	26.	ELECTRICAL	110,080	9.56	6.90	Medium-voltage distribution, low-voltage transmission, facility power generating & storing equipment, lighting.
TOTAL BUILDING COST			**1,150,403**	**100%**	**$72.08**	
2.	2.	EXISTING CONDITIONS	138,607			Subsurface investigation, demolition & structure moving.
2.	31.	EARTHWORK	50,000			Site clearing, earth moving, shoring & underpinning, excavation support & protection, special foundations & load-bearing elements.
2.	33.	UTILITIES	15,000			Water, sanitary, storm drainage.
TOTAL PROJECT COST			**1,354,010**	*(Excluding architectural and engineering fees)*		

UPDATED ESTIMATE TO JAN 2011: $113.47 SF

Crogan Street
Lawrenceville, Georgia

Architect: Pate Design Group, Inc.

Photos Courtesy of Howard Doughty

This historic landmark located in the heart of Lawrenceville, Georgia has been a vital part of the city for over a century. It served as a horse and mule trade center for the county and was known as "Honest Alley". Later, it became a garage, a car dealership, a hardware store, a taxi office and a barber shop among other general retail establishments. These previous businesses brought people together for the purpose of trade and camaraderie. Honest Alley, LLC wanted to continue this tradition by creatingspaces that would attract people to the historic square. Pate Design Group, Inc. was chosen to coordinate the design for the renovation of this 16,000-square-foot building to house retail and commercial businesses. Among the goals of the renovation was to protect the structure, further enhance the activity and commerce-themed projects that have served to revitalize the city. Rather than to demolish the existing building, local investment partners Randy and Cindy Sutt were interested in a sustainable project to restore and enhance the site's traditional uses.

The original building was constructed of stone and wood timbers, as well as a wood slat floor. Years of roof leaks and water damage led this portion of the building to be refurbished by demolition of the existing roof and floor system. The roof was removed and replaced with new wood trusses, roof sheathing and a combination of metal standing seam roof material and shingles, which were placed on the side to tie the building into the neighboring building on the square. The wood floor had suffered irreparable damage and was removed. Concrete was poured in to replace it. Stone with grapevine mortar joints were kept intact, which made up the original building's facade that faces the nearby historic courthouse and was divided into two separate tenant spaces.

A two story CMU second addition to the building was added in the 1950's. This portion was open throughout and had a concrete floor and wood trusses with a shingled roof. The intent of the project was to shore up the existing structure with steel reinforcement and create a poured concrete floor that separated the later addition into two floors. This would allow for retail space above and future restaurant space below. The concrete floor was cut to allow for new foundation footers to be poured Steel columns were erected on these new footers to support a steel beam in the center. Bar joists that support the decking for the second floor is supported by this beam and mechanically attached steel angles through the existing CMU walls. A balcony and poured ramp were added to provide access to the second floor retail and commercial portion of the building.This would allow for retail space above and future restaurant space below. The concrete floor was cut to allow for new foundation footers to be poured. Steel columns were erected on these new footers to support a steel beam in the center.

Bar joists that support the decking for the second floor is supported bu this beam and mechanically attached steel angles through the existing CMU walls. A balcony and poured ramp were added to provide access to the second floor retail and commercial portion of the building.

An elaborate garden in the courtyard of the building is the centerpiece of the site. This garden was designed to use indigenous plants that would naturally attract butterflies to create a place where patrons and visitors to the town square could relax and enjoy the scenery. The garden serves as an invitation to visit Honest Alley and experience life as it has been for the past century.

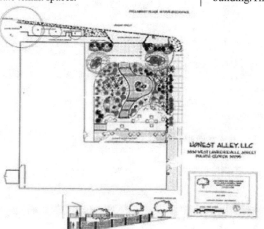

MANUFACTURERS/SUPPLIERS	
DIV 08:	*Entrances & Storefron: Windows:* **Vistawall**
DIV 09:	*Flooring:* **Dal-Tile**
DIV 26:	*Lighting:* **Daybrite**

COST PER SQUARE FOOT $ 113.47

ARCHITECT
Pryor & Morrow Architects, P. A.
5227 South Frontage Road, Columbus, MS 39701
www.pryor-morrow.com

FILE UNDER
COMMERCIAL
Starkville, Mississippi

CONSTRUCTION TEAM

STRUCTURAL & MECHANICAL ENGINEER:
Pryor & Morrow Architects, P. A.
5227 South Frontage Road, Columbus, MS 39701
GENERAL CONTRACTOR:
Yates Construction
P. O. Box 456, Philadelphia, MS 39350

GENERAL DESCRIPTION

SITE: . 1 Acre.
NUMBER OF BUILDINGS: One.
BUILDING SIZE: First floor, 5,200; total, 5,200 square feet.
BUILDING HEIGHT: First floor, 36'; total, 36'.
BASIC CONSTRUCTION TYPE: Wood Frame/New.
FOUNDATION: Cast-in-place, slab-on-grade.
EXTERIOR WALLS: CMU, brick, wood stud backup.
ROOF: Asphalt shingles.
FLOORS: Wood. **INTERIOR WALLS:** Drywall over wood studs.

THE CITIZENS BANK
Construction Period: Feb 2007 to Sep 2007 • Total Square Feet: 5,200

C.S.I. Divisions (1 through 16)			COST	% OF COST	SQ.FT. COST	SPECIFICATIONS
		PROCUREMENT & CONT. REQ.	172,794	13.25	33.23	General conditions.
1.	1.	GENERAL REQUIREMENTS	64,451	4.94	12.39	Tax, modification procedures.
3.	3.	CONCRETE	262,939	20.17	50.57	Forming & accessories, reinforcing, cast-in-place, grouting (197 cubic yards concrete for foundation).
4.	4.	MASONRY	133,323	10.22	25.64	Unit, stone assemblies, manufactured masonry.
5.	5.	METALS	50,118	3.85	9.64	Structural metal framing, fabrications.
6.	6.	WOOD/PLASTICS/COMPOSITE	140,721	10.79	27.06	Rough carpentry, finish carpentry, architectural woodwork.
7.	7.	THERMAL & MOIST. PROTECT	59,910	4.60	11.52	Dampproofing & waterproofing, thermal protection, steep slope roofing, roofing & siding panels, flashing & sheet metal, fire & smoke protection, joint protection.
8.	8.	OPENINGS	111,115	8.52	21.37	Doors & frames, speciality doors & frames, entrances, storefronts, & curtain walls, windows, roof windows & skylights, hardware, glazing, louvers & vents.
9.	9.	FINISHES	124,382	9.54	23.92	Plaster & gypsum board, tiling, flooring, painting & coating.
10.	10.	SPECIALTIES	17,221	1.32	3.31	Other.
15.	22.	PLUMBING	18,000	1.38	3.46	Fixtures.
15.	23.	HVAC	57,500	4.41	11.06	Piping & pumps, central heating, central cooling, central HVAC equipment.
16.	26.	ELECTRICAL	91,424	7.01	17.58	Medium voltage distribution, electrical & cathodic protection.
TOTAL BUILDING COST			**1,303,898**	**100%**	**$250.75**	
2.	31.	EARTHWORK	73,592			Earth moving.
TOTAL PROJECT COST			**1,377,490**	*(Excluding architectural and engineering fees)*		

UPDATED ESTIMATE TO JAN 2011: $371.22 SF

DCD Subscribers: Access this case study and hundreds more for instant date and location calculations at www.dcd.com.

The Citizens Bank
Starkville, Mississippi

Architect: Pryor & Morrow Architects, P.A.

Photos Courtesy of Don Beard

The 4,000-square-foot branch bank in Starkville, Mississippi represents The Citizens Bank's first unit in the Golden Triangle Area. The Bank decided to make a strong statement about its commitment to sustainability with the project. Also, the notion of respecting the vernacular design traditions of the area strongly influenced the outcome of the project. The combination of sustainability within the context of a traditional design vocabulary offered interesting challenges to the design team and rendered a unique solution to the Bank's project program.

Traditional design features include solid masonry walls, historic detailing and a clock tower that is similar to Starkville's historic Cotton Mill Tower. The Citizens Bank building has become a landmark in Starkville, Mississippi and has established a strong presence in the Golden Triangle Community.

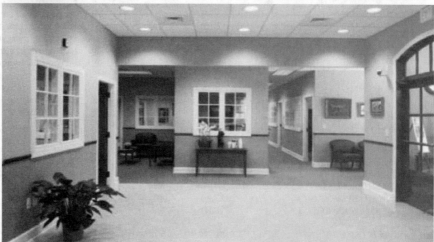

SUSTAINABLE

Sustainability features included:

- Super Insulation
- High Peformance HVAC systems
- Day-lighting
- High output fluorescent lighting
- Ideal solar orientation
- Pervious paving

MANUFACTURERS/SUPPLIERS

| DIV 07: | *Metal Roofing:* **Berridge.** |
| DIV 08: | *Windows:* **Pella.** |

COST PER SQUARE FOOT $ 371.22

ARCHITECT
Moshier Studio
201 South Highland Avenue, #203, Pittsburgh, PA 15206
www.moshierstudio.com

FILE UNDER
COMMERCIAL
Pittsburg, Pennsylvania

CONSTRUCTION TEAM

ELECTRICAL ENGINEER:
Carlins Consulting
8005 Broadlawn Dr., Pittsburgh, PA 15237-4152
MECHANICAL ENGINEER & COMMISSIONING AGENT:
BDA Engineering
217 W. 8th Ave., West Homestead, PA 15120
GENERAL CONTRACTOR:
Jendoco Construction Company
2000 Lincoln Road, Pittsburgh, PA 15235
LEED® CONSULTANT:
Moshier Studio
201 South Highland Avenue, #203, Pittsburgh, PA 15206

GENERAL DESCRIPTION

SITE: .75 Acre.
NUMBER OF BUILDINGS: One; 16 classrooms, seating capacity, 170.
BUILDING SIZE: Basement, 10,000; first floor, 10,000; second floor, 10,000; third floor, 10,000; total, 40,000 square feet.
BUILDING HEIGHT: Basement, 12'; first floor, 16'; second floor, 16'; third floor, 32'; total, 76'.
BASIC CONSTRUCTION TYPE: Renovation/Brick.
FOUNDATION: Reinforced concrete.
EXTERIOR WALLS: CMU, brick.
ROOF: Membrane, slate.
FLOORS: Concrete, wood.
INTERIOR WALLS: CMU, metal stud drywall.

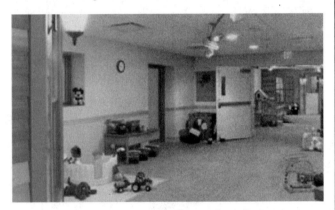

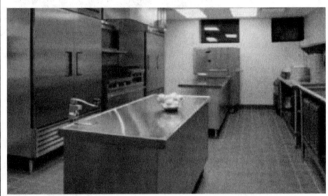

CARRIAGE HOUSE CHILDREN'S CENTER
Construction Period: May 2006 to Apr 2007 • Total Square Feet: 40,000

C.S.I. Divisions (1 through 16)			COST	% OF COST	SQ.FT. COST	SPECIFICATIONS
		PROCUREMENT & CONT. REQ.	52,503	2.88	2.63	Permits, fees.
1.	1.	GENERAL REQUIREMENTS	289,974	15.92	14.50	Summary, price & payment procedures, administrative requirements, quality requirements, performance requirements.
3.	3.	CONCRETE	51,345	2.82	2.57	Forming & accessories, reinforcing, cutting & boring.
4.	4.	MASONRY	37,300	2.05	1.87	Unit, manufactured masonry.
5.	5.	METALS	37,180	2.04	1.86	Structural metal framing, fabrications, decorative metal.
6.	6.	WOOD/PLASTICS/COMPOSITE	133,562	7.33	6.68	Rough carpentry, finish carpentry, architectural woodwork.
7.	7.	THERMAL & MOIST. PROTECT	60,608	3.33	3.03	Dampproofing & waterproofing, thermal protection, weather barriers, roof & wall specialties & accessories.
8.	8.	OPENINGS	66,929	3.68	3.35	Doors & frames, specialty doors & frames, windows, hardware.
9.	9.	FINISHES	185,227	10.17	9.26	Plaster & gypsum board, tiling, ceilings, flooring, painting & coating.
10.	10.	SPECIALTIES	5,850	0.32	0.29	Information, exterior, other.
11.	11.	EQUIPMENT	55,901	3.07	2.79	Foodservice.
12.	12.	FURNISHINGS	1,559	0.09	0.08	—
15.	22.	PLUMBING	146,300	8.03	7.31	Piping & pumps, equipment, fixtures.
15.	23.	HVAC	504,000	27.69	25.20	HVAC equipment, decentralized HVAC equipment.
16.	26.	ELECTRICAL	187,725	10.31	9.38	Medium voltage distribution, lighting.
		ELECTRONIC SAFETY & SECURITY	5,000	0.27	0.25	
TOTAL BUILDING COST			**1,820,963**	**100%**	**$91.05**	
1.	30.	EXISTING CONDITIONS	111,111			Demolition.
2.	31.	EARTHWORK	11,720			—
TOTAL PROJECT COST			**1,943,794**	*(Excluding architectural and engineering fees)*		

UPDATED ESTIMATE TO JAN 2011: $122.12 SF

Carriage House Children's Center
Pittsburg, Pennsylvania

Architect: Moshier Studio

Carriage House Children's Center is a pioneer in the field of early childhood education, with Southwestern Pennsylvania's first infant and toddler program in 1974. The organization's entrepreneurial and risk-taking spirit has resulted in several major accomplishments over the last 35 years.

In 1985 Carriage House purchased the 113-year-old, 40,000-square-foot Wightman School Community Building and relocated in 1986. Since that time, Carriage House has overseen seven major renovations to the Wightman School Community Building transforming the building from a dilapidated, public school building that did not even meet fire and safety codes and was once considered for demolition, to one that is now gold certified by the U.S. Green Building Council and is registered with the National Registry of Historic Places.

The Wightman School Community Building is a multi-use facility where over 17,000 square feet of multipurpose and office space is leased to local non-profits, small businesses and other family-oriented groups. CHCC manages the entire building and operates the CHCC childcare program, which consists of ten classrooms between the lower level and the first floor. Administrative offices, conference rooms, and multiple tenant office spaces are located on the second floor. The third floor is home to a gymnasium and five music/dance studios.

In 2006 and 2007 Carriage House renovated the entire lower level of the building (9,720 square feet), which houses the infant and toddler programs, the kitchen, the boiler, and other storage and multi-use rooms. Carriage House also retrofitted many of the building's lighting and plumbing fixtures to meet LEED® criteria. Air conditioning was added to the first floor.

Extensive demolition of non-load bearing and bearing walls located all four classrooms (infants, young toddlers, toddlers, and one extra classroom) at the front of the building, where they each benefit from more natural light. Spaces not used by children, such as the kitchen and the laundry room, have taken their place where the windows struggle to receive light. The center of the lower level has been opened up to create a large indoor play area that is subdivided for two groups to use at the same time access to daylight

Photos Courtesy Linda Jeub

and views have been opened to increase the amount of natural light in the center of the lower level. Two classrooms have unique entrances to this central space- two enclosed porches for the young toddlers and older toddlers. Preschool sized bathrooms have been located in two of the classrooms for children who are making this transition.

The new "green" kitchen has reach-in coolers, food storage areas, and the commercial dishwasher adjacent to the elevator for good delivery access, and for ease of sending prepared food up to the other classrooms. The kitchen is designed to comply with Energy Star ratings at 60 or above. The entire building is now rated as Energy Star 81.

All plumbing faucets and toilets are low-flow fixtures. With the exception of one teacher sink per classroom, all children's faucets are hands-free to promote healthy handwashing methods.

The new, lower-level classrooms have a suspended ceiling for acoustical control, as high as the bottom of the original beams, and not blocking the windows. Heated and cooled air is delivered from the individual heat pumps via exposed round ducts below the ceiling. All classrooms are equipped with individual thermostats.

On the first and lower-level indirect lighting illuminates the ceiling, giving off an impression of openness and more light output with less energy. T-8, daylighting systems have been installed throughout the lower-level. The lights are dimmable in all classrooms and are programmed to coordinate with the children's napping schedules.

LEED®
EXISTING BUILDING GOLD

COST PER SQUARE FOOT $ 122.12

ARCHITECT
Faulkner Architects a.p.c.
12242 Business Park Drive, #18, Truckee, CA 96161
www.faulknerarchitects.com

FILE UNDER
COMMERCIAL
Truckee, California

CONSTRUCTION TEAM

STRUCTURAL ENGINEER:
Gabbart & Woods Structural Engineers
877 Tahoe Boulevard, Incline Village, NV 89451
GENERAL CONTRACTOR & COST ESTIMATOR:
Robert Marr Construction, Inc.
10363 High Street, Truckee, CA 96161
MECHANICAL & ELECTRICAL ENGINEER:
MSA Engineering
4599 Longely Lane, Reno, NV 89502

GENERAL DESCRIPTION

SITE: . 6.5 Acres.
NUMBER OF BUILDINGS: One; 16 classrooms, seating capacity, 170.
BUILDING SIZE: First floor, 6,309; total, 6,309 square feet.
BUILDING HEIGHT: First floor, 14'; total, 14'.
BASIC CONSTRUCTION TYPE: New/VNR/Tilt Up.
FOUNDATION: Reinforced concrete.
EXTERIOR WALLS: Tilt up.
ROOF: Built up.
FLOORS: Concrete.
INTERIOR WALLS: Wood stud drywall.

GRAY'S CROSSING CART BARN
Construction Period: Sep 2006 to June 2007 • Total Square Feet: 6,309

C.S.I. Divisions (1 through 16)			COST	% OF COST	SQ.FT. COST	SPECIFICATIONS
1.	1.	GENERAL REQUIREMENTS	—	—	—	—
3.	3.	CONCRETE	262,317	24.99	41.58	Forming & accessories, reinforcing, cast-in-place, grouting, cutting & boring (Concrete breakdown: cubic yards foundation, 66; cubic yards walls, 110; cubic yards floors, 150).
5.	5.	METALS	82,538	7.86	13.08	Structural metal framing, joists, decking, fabrications.
6.	6.	WOOD/PLASTICS/COMPOSITE	247,661	23.60	39.26	Rough carpentry, finish carpentry.
7.	7.	THERMAL & MOIST. PROTECT	66,332	6.32	10.51	Dampproofing & waterproofing, membrane roofing, flashing & sheet metal.
8.	8.	OPENINGS	35,912	3.42	5.69	Doors & frames, specialty doors & frames, windows, hardware.
9.	9.	FINISHES	70,685	6.74	11.20	Plaster & gypsum, flooring, painting & coating.
10.	10.	SPECIALTIES	14,780	1.41	2.34	Storage, other.
15.	21.	FIRE SUPPRESSION	21,000	2.00	3.33	Water-based fire-suppression systems.
15.	22.	PLUMBING	69,197	6.59	10.97	Piping & pumps, equipment, fixtures.
15.	23.	HVAC	59,780	5.70	9.48	Piping & pumps, air distribution, central heating, central cooling, central HVAC equipment.
16.	26.	ELECTRICAL	119,326	11.37	18.91	Medium-voltage distribution, low-voltage transmission, lighting.
TOTAL BUILDING COST			**1,049,528**	**100%**	**$166.35**	
2.	31.	EARTHWORK	153,058			Site clearing, earth moving, earthwork methods, excavation support & protection.
TOTAL PROJECT COST			**1,202,586**	*(Excluding architectural and engineering fees)*		

UPDATED ESTIMATE TO JAN 2011: $216.29 SF

DCD Subscribers: Access this case study and hundreds more for instant date and location calculations at www.dcd.com.

Gray's Crossing Cart Barn
Truckee, California

Architect: Faulkner Architects, a.p.c.

Photos Courtesy Faulkner Architects a.p.c.

A private golf course community in a northern Sierra Nevada mountain environment selected Faulkner Architects for a new structure to house their golf carts. The goal was to provide an aesthetically pleasing, environmentally-friendly and economical golf cart storage facility which blended well with the natural landscape of the Tahoe region, yet could withstand the harsh climate. The project sits at an elevation of 6,000 feet where freezing temperatures, heavy snow loads and high UV exposure are of great concern.

The roof structure incorporates a LEED® (R)-approved EPDM roofing system in an effort to avoid a "hot spot" created by buildings in forest ecosystems. Hot-rolled steel columns and beams support pitched roofs, the steel allowing for less bulky structural elements that withstand heavy snow loads without dominating the landscape. Exterior concrete tilt-up panels were acid-stained to match the color and shade of surrounding white pine tree bark, while dark-bronze anodized windows complement the natural stain and help complete the rustic unobtrusive character of the building.

Numerous building materials were incorporated for their green characteristics and renewable properties: concrete with fly ash, soy-based insulation, steel and sheet rock with recycled content, LEED®approved EPDM roofing system. Environmentally conscious materials combined with the natural stain and texture of the concrete allow the building to not only blend well with the Lake Tahoe environment, but to exist harmoniously within it.

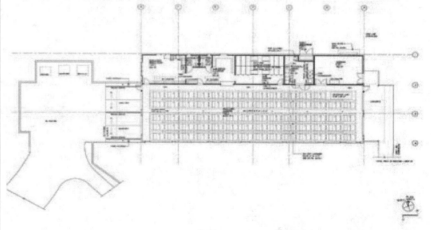

COST PER SQUARE FOOT $ 216.29

ARCHITECT
McGill, Smith, Punshon
3700 Park 42nd Drive, #190B, Cincinnati, OH 45241
www.mcgillsmithpunshon.com

CONSTRUCTION TEAM

STRUCTURAL ENGINEER: Britt, Peters & Associates, Inc.
550 S. Main Street, #301, Greenville, SC 29601
GENERAL CONTRACTOR: The Harper Corporation
35 West Court Street, #400, Greenville, SC 29601
LEED® COMMISSIONING AGENT: KLG Jones
218 Trade Street, Greer, SC 29651

GENERAL DESCRIPTION

SITE: .25 Acres. **NUMBER OF BUILDINGS:** One.
BUILDING SIZE: First floor, 2,300; total, 2,300 square feet.
BUILDING HEIGHT: First floor, 21'; total, 21'.
BASIC CONSTRUCTION TYPE: New
FOUNDATION: Cast-in-place, reinforced, pier & grade beam.
EXTERIOR WALLS: Fiber cement board siding, curtainwall.
ROOF: Reflective. **FLOOR:** Concrete
INTERIOR WALLS: CMU, wood stud drywall.

WINTON WOODS CAMPGROUND OFFICE & RETAIL BUILDING
Construction Period: June 2008 to Nov 2008 • Total Square Feet: 2,300

C.S.I. Divisions (1 through 16)			COST	% OF COST	SQ.FT. COST	SPECIFICATIONS
1.	1.	GENERAL REQUIREMENTS	63,902	11.59	27.78	—
3.	3.	CONCRETE	46,449	8.42	20.20	—
4.	4.	MASONRY	45,000	8.16	19.57	Unit, stone assemblies, manufactured masonry.
5.	5.	METALS	29,250	5.30	12.72	Structural metal framing, fabrications.
6.	6.	WOOD/PLASTICS/COMPOSITE	89,000	16.14	38.70	Rough carpentry, finish carpentry, architectural woodwork.
7.	7.	THERMAL & MOIST. PROTECT	48,070	8.72	20.90	Dampproofing & waterproofing, thermal protection, weather barriers, roofing & siding panels, flashing & sheet metal, joint protection.
8.	8.	OPENINGS	39,000	7.07	16.96	Doors & frames, specialty doors & frames, entrances, storefronts, curtainwalls, windows, roof windows & skylights, hardware, glazing, louvers & vents.
9.	9.	FINISHES	24,243	4.40	10.54	Plaster & gypsum board, ceilings, flooring, wall finishes, acoustic treatment, painting & coating.
10.	10.	SPECIALTIES	3,294	0.60	1.43	Interior.
15.	22.	PLUMBING	31,950	5.78	13.89	—
15.	23.	HVAC	70,375	12.76	30.60	—
16.	26.	ELECTRICAL	61,000	11.06	26.51	—
TOTAL BUILDING COST			**551,533**	**100%**	**$239.80**	
2.	31.	EARTHWORK	4,690			—
2.	32.	EXTERIOR IMPROVEMENTS	24,000			—
2.	33.	UTILITIES	2,800			Storm.
TOTAL PROJECT COST			**583,023**	*(Excluding architectural and engineering fees)*		

UPDATED ESTIMATE TO JAN 2011: $ 291.96 SF

Winton Woods Campground Office and Retail Building
Cincinnati, Ohio

Architect: McGill, Smith, Punshon

The intent on this building was to be a practice round for the Hamilton County Park District (Ohio) at building green. They decided to take a shot at environmentally conscious design by targeting green products, design, management and intent to better understand the thought process that goes into a LEED®-certified building. Park planners engaged Dan Montgomery (AIA, LEED®) of McGill Smith Punshon Architects in the design and drawing of the building. This was to be the monument that welcomed their campers into the Winton Woods Campground, providing them a place to check in, gear up and use as a resource during their camping experience. Todd Palmeter, Park Planner and Tim Hendrixson, PE, Park Engineer were the park planners tapped to manage this project and were involved from the beginning in working with Dan Montgomery on the design. Todd, a landscape architect, and Tim, a civil engineer, both influenced the selection of many of the green finish products as well as worked to design parts of the project that synergized with their competencies.

Green products that were used range from cement board siding, solar-reflecting roofing, green certified lumber, cast stone veneer, geothermal HVAC, waterless urinals, energy-efficient lighting fixtures, and natural lighting just to name a few. The building oriented itself to face the western exposure with a large Low-E curtain wall to a gather up all the sunlight possible for natural lighting. Additionally, there was a construction waste management plan in effect on site. This meant that all construction materials were to be separated according to pure waste, recyclable materials, renewable materials, or reusable materials. All products shipped to site were requested to be shipped using as little paper, cardboard, plastic, etc. as possible so as to minimize waste. Approximately twenty cubic yards of actual non-renewable waste was created on the job, a very minimal amount. Prime contractors were required to submit estimates on their proposed waste and recyclables, as well as submit records

Photos Courtesy of Oliver Roe, DER Development

of their actual numbers at the end of the project. Estimates were beaten across the board.

Altogether the project was completed successfully, on time, and to the owner's great satisfaction. It was amazing to see a building comprised of lumber, structural steel, fiber cement siding, masonry, and glass go up with the intent on short-term paybacks on utilities and overall low-impact on the environment.

Everyone involved felt that it was sending the right message in a place where the great outdoors are a priority.

COST PER SQUARE FOOT $ 291.96

ARCHITECT
Zimmer Gunsul Frasca Architects LLP
1800 K Street, N.W., #200, Washington, DC 20006
www.zgf.com

FILE UNDER
EDUCATIONAL
Carlisle, Pennsylvania

CONSTRUCTION TEAM

STRUCTURAL ENGINEER:
LeMessurier Consultants
675 Massachusetts Avenue, Cambridge, MA 02139
GENERAL CONTRACTOR:
Reynolds Construction Management
3300 North Third Street, Harrisburg, PA 17110
ELECTRICAL & MECHANICAL ENGINEER:
BR+A Consulting Engineers, LLC
311 Arsenal Street, Watertown, MA 02472
LANDSCAPE ARCHITECT:
Chattahoochee Home & Garden
1101 Via Bayless, Marietta, GA 30066

GENERAL DESCRIPTION

SITE: 2.88 Acres.
NUMBER OF BUILDINGS: One; 3 classrooms with seating capacity for 30; auditorium seating 80.
BUILDING SIZE: Basement, 13,613; first floor, 32,175; second floor, 26,454; third floor, 14,953; total, 87,195 square feet.
BUILDING HEIGHT: Basement, 15'; first floor, 15'3"; second floor, 15'; floor to floor, 12'; penthouse, 21'4"; total, 66'7".
BASIC CONSTRUCTION TYPE: Addition/Steel Frame.

FOUNDATION: Cast-in-place, pier & grade beam, reinforced concrete, slab-on-grade.
EXTERIOR WALLS: CMU, stone, metal panels.
ROOF: Metal, membrane, vegetated.
FLOORS: Concrete.
INTERIOR WALLS: CMU, metal stud drywall.

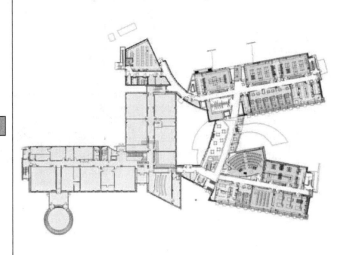

RECTOR SCIENCE COMPLEX
Construction Period: July 2006 to May 2008 • Total Square Feet: 87,195

C.S.I. Divisions (1 through 16)			COST	% OF COST	SQ.FT. COST	SPECIFICATIONS
		PROCUREMENT & CONT. REQ.	170,000	0.71	1.95	-
1.	1.	GENERAL REQUIREMENTS	1,322,041	5.51	15.16	-
3.	3.	CONCRETE	2,143,129	8.94	24.58	Forming & accessories, reinforcing, cast-in-place (Concrete breakdown: 12,978 cubic yards foundation, 1,490 cubic yards walls, 5,850 cubic yards basement, 2,780 cubic yards first floor).
4.	4.	MASONRY	1,242,300	5.18	14.25	Unit, stone.
5.	5.	METALS	2,071,661	8.64	23.76	Steel erection, decking, fabrication, freight.
6.	6.	WOOD/PLASTICS/COMPOSITE	451,000	1.88	5.17	Rough carpentry, finish carpentry, architectural woodwork.
7.	7.	THERMAL & MOIST. PROTECT	3,056,521	12.75	35.05	Waterproofing & dampproofing, insulation, roof & wall panels, membrane roofing, metal shingles, roof hatches, (vegetated roof donated).
8.	8.	OPENINGS	1,271,888	5.31	14.59	Doors & frames, entrances, storefronts, hardware, glazing.
9.	9.	FINISHES	2,553,882	10.65	29.29	Plaster & gypsum board, ceilings, flooring, wall finishes, painting & coating.
10.	10.	SPECIALTIES	178,495	0.74	2.05	Screens, specialties.
12.	12.	FURNISHINGS	1,658,314	6.92	19.02	Casework.
14.	14.	CONVEYING SYSTEMS	122,300	0.51	1.40	Elevators (2: 1 passenger, 1 freight).
15.	21.	FIRE SUPRESSION	383,690	1.60	4.40	Pumps, controls, piping.
15.	22.	PLUMBING	1,342,534	5.60	15.40	Piping & pumps, equipment, fixtures.
15.	23.	HVAC	3,217,316	13.42	36.90	Submittals, piping, ductwork, equipment, instrumentation & controls, testing, adjusting & balancing.
16.	26.	ELECTRICAL	2,788,650	11.64	31.97	-
TOTAL BUILDING COST			**23,973,721**	**100%**	**$274.94**	-
2.	2.	EXISTING CONDITIONS	348,583			Demolition.
2.	31.	EARTHWORK	629,783			
2.	32.	EXTERIOR IMPROVEMETNS	522,337			Paving, sidewalks, landscaping.
2.	33.	UTILITIES	112,849			Water, sanitary sewer, storm.
TOTAL PROJECT COST			**25,587,273**	*(Excluding architectural and engineering fees)*		

UPDATED ESTIMATE TO JAN 2011: $362.17 SF

DCD Subscribers: Access this case study and hundreds more for instant date and location calculations at www.dcd.com.

Rector Science Complex Stuart Hall and James Hall, Dickinson College
Carlisle, Pennsylvania

Architect: Zimmer Gunsul Frasca Architects LLP

Photos Courtesy of Reynolds Construction Management

The building's skeleton is comprised of poured concrete foundations with a steel composite superstructure and concrete slabs on deck. Some exterior features of the building, besides the limestone and metal panel facade, include a fluid applied vapor barrier, zinc-coated copper standing seam, vegetated and white thermoplastic roofs, and an architecturally appealing sunscreen system fabricated using high percent recycled content aluminum by Industrial Louvers, Inc. Carlisle SynTec, located in Carlisle, Penn., donated the vegetated roof. The surrounding landscape was restored with native plants that are watered via a storm water retention pond/bioswale. There is also preferred parking for carpooling.

Some interior features of the building include scored and integrally colored concrete floors, perforated, wood paneled ceilings, a passenger elevator with a hanging pit, and wall panels for informational postings. To reduce water consumption, waterless urinals and electric eye sensors on faucets and lavatories were installed.

For the LEED® Energy Recovery credit, four enthalpy heat recovery wheels and one plate-to-plate heat exchanger were installed to provide proper ventilation and clean air throughout the building.

The new Stuart Hall and James Hall addition to the Rector Science Complex has achieved LEED® Gold certification from the U.S. Green Building Council.

There is a capstone phase for the Rector Science Complex that is currently in the planning stage. It consists of a third wing or hall that will provide 70,000 square feet of teaching labs and research space.

LEED®
NEW BUILDING GOLD
MANUFACTURERS/SUPPLIERS

DIV 07:	*Vegetated Roof:* Carlisle Syntec Roofing
DIV 10:	*Louvers:* Industrial Louvers

COST PER SQUARE FOOT $ 362.17

ARCHITECT
Nacht & Lewis Architects
600 Q Street, #100, Sacramento, CA 95811
www.nlarch.com

FILE UNDER
EDUCATIONAL
Yuba City, California

CONSTRUCTION TEAM

OWNER: Yuba City School Unified District
750 Palora Avenue, Yuba City, CA 95991
STRUCTURAL ENGINEER:
Buehler & Buehler Structural Engineers
600 Q Street, #200, Sacramento, CA 95811
MECHANICAL ENGINEER & COMMISSIONING AGENT:
Capital Engineering Consultants, Inc.
11020 Sun Center Drive, Rancho Cordova, CA 95670
ELECTRICAL ENGINEER: The Engineering Enterprise
853 Lincoln Way, #105, Auburn, CA 95603
GENERAL CONTRACTOR: Sundt Construction
2860 Gateway Oaks Drive, #300, Sacramento, CA 95833

GENERAL DESCRIPTION

SITE: 21 Acres. **NUMBER OF BUILDINGS:** 13 – 3 site built, 10 modular. 46 classrooms seating 1,300 students; Auditorium, 4,946 sq. ft. seating 733 occupants; Gym, 5,968 sq. ft. seating 771 occupants.
BUILDING SIZE: First floor, 86,000; total, 86,000 square feet.
BUILDING HEIGHT: First floor, 38'6"; total, 38'6".
BASIC CONSTRUCTION TYPE: New/Structural steel braced frame (CHPS Certified).
FOUNDATION: Cast-in-place, slab-on-grade.
EXTERIOR WALLS: CMU, storefront, cement plaster.
ROOF: Metal, modified bitumen.
FLOORS: Concrete. **INTERIOR WALLS:** Metal stud drywall.

RIVERBEND ELEMENTARY SCHOOL
Construction Period: Apr 2006 to July 2007 • Total Square Feet: 86,000

C.S.I. Divisions (1 through 16)			COST	% OF COST	SQ.FT. COST	SPECIFICATIONS
		PROCUREMENT & CONT. REQ.	-	-	-	
1.	1.	GENERAL REQUIREMENTS	4,469,155	27.27	51.97	Summary, price & payment procedures, temporary facilities & controls, product requirements, execution & closeout requirements.
3.	3.	CONCRETE	1,977,230	12.06	22.99	Forming & accessories, reinforcing, cast-in-place, cast decks & underlayment, grouting.
4.	4.	MASONRY	239,688	1.46	2.79	Unit.
5.	5.	METALS	1,984,420	12.11	23.07	Structural metal framing, joists, decking, cold-formed metal framing, fabrications, decorative metal.
6.	6.	WOOD/PLASTICS/COMPOSITE	107,995	0.66	1.26	Rough carpentry, finish carpentry, architectural woodwork, plastic fabrications.
7.	7.	THERMAL & MOIST. PROTECT	531,653	3.24	6.18	Dampproofing & waterproofing, thermal protection, weather barriers, roofing & siding panels, modified bitumen built-up roofing, flashing & sheet metal, roof & wall specialties
8.	8.	OPENINGS	545,827	3.33	6.35	Doors & frames, specialty doors & frames, entrances, storefronts, & curtain walls, windows, hardware, glazing.
9.	9.	FINISHES	1,734,885	10.58	20.17	Plaster & gypsum board, tiling, ceilings, flooring, wall finishes, acoustical treatment, painting & coating.
10.	10.	SPECIALTIES	181,396	1.11	2.11	Information, interior, safety, storage.
11.	11.	EQUIPMENT	489,450	2.99	5.69	Vehicle & pedestrian, commercial, foodservice, educational & scientific, athletic & recreational, collection & disposal.
12.	12.	FURNISHINGS	66,806	.41	.78	Casework, furniture, multiple seating.
13.	13.	SPECIAL CONSTRUCTION	48,918	0.30	0.57	Special purpose rooms.
14.	14.	CONVEYING SYSTEMS	8,675	0.05	0.10	Lifts.
15.	21.	FIRE SUPRESSION	216,000	1.32	2.51	Water-based fire-suppression systems, fire-extinguishing systems.
15.	22.	PLUMBING	-	-	-	Included in HVAC: Piping & pumps, equipment, fixtures, gas & vacuum systems for laboratory & healthcare.
15.	23.	HVAC	1,711,774	10.44	19.90	Air distribution, air cleaning devices, decentralized HVAC equipment.
16.	26.	ELECTRICAL	2,076,682	12.67	24.15	Medium-voltage distribution, facility power generating & storing equipment, electrical & cathodic protection, lighting.
TOTAL BUILDING COST			**16,390,554**	**100%**	**$190.59**	
2.	2.	EARTHWORK	3,554,167			
2.	32.	EXTERIOR IMPROVEMENTS	856,396			
TOTAL PROJECT COST			**20,801,117**	*(Excluding architectural and engineering fees)*		

UPDATED ESTIMATE TO JAN 2011: $221.95 SF

DCD Subscribers: Access this case study and hundreds more for instant date and location calculations at www.dcd.com.

Riverbend Elementary School
Yuba City, California

Architect: Nacht & Lewis Architects

Photos Courtesy of Donald Satterlee Photography

The Riverbend Elementary School is modeled around a K-8 curriculum model. This program presents unique challenges in dealing with such diverse age groups. Special attention was required to address the unique needs of younger kindergarten children and the advanced curriculum needs of 7th & 8th graders.

The school is organized in small grade-level clusters around a central courtyard. Facilities include an administrative and counseling office, library, multi-purpose room with performing arts capabilities and a full-size independent gymnasium. Site amenities include generous turf playfields, a running track and equipment areas for the individual grade levels.

Sustainable design elements were also fundamental in the projects development. Natural daylight is abundant in virtually all spaces on the campus. High efficiency mechanical systems, low-water use plumbing fixtures, and automatic lighting controls contribute to a facility that exceeds the energy requirement of California Title 24 by better than 30%. The school is also recognized by the Collaborative of High Performance Schools for its energy conscious design and sustainable features.

The design goals were to create a new model for an elementary school campus that would focus on sustainability, provide a sense of community and that promote student achievement; and to create a facility that would reflect the School District's commitment to the community and to planning for the future.

The school site is located several hundred yards west of the Feather River and on the edge of new suburban development. The Feather River, its landforms and vegetation, became the design inspiration for the hardscape and landscape design of the campus. There is a symbolic levee in the center of the campus and the selection of trees, ground covers, and paving patterns were selected specifically to relate to the river environment giving the school and the students a unique sense of place.

The sustainability components of the project were developed using the Collaborative for High Performing Schools' Best Practices Manual and from the U.S. Green Building Council LEED® program. The building forms and materials are derived from the local agricultural vernacular and are reinterpreted looking towards the future. The roof forms and materials become the integrated support structure for more than 300 kVA of thin film photovoltaic panels producing enough power to lower the utility costs to run the school by over 30%.

The Gymnasium, Administration Building and the Library Building were site built and maintained the campus's focus on green design with an emphasis on energy efficiency and day-lighting. All three buildings incorporate large amounts of translucent insulated window panels, which let diffused light in and lowered the energy loss typical of traditional windows. The building forms were designed based on the layout of thin film photovoltaic panels that are integrated directly into the metal roofing system without the need for additional structural supports. The PV system is designed to provide up to 100% of the peak electrical demand of the campus on a bright day and more than 30% of the campus's annual energy needs. Based on the success of the PV system the district is looking into opportunities to expand the system at other schools.

This project was designed as a model of Green Design and sustainability for public schools in California and scored 38 points in the Collaborative for High Performing Schools rating system. The buildings on campus combine community wide centralized planning, high efficiency equipment and environmental controls, translucent glazing, and rooftop photovoltaic panels and the elimination of potable water for landscape irrigation. The campus is an active laboratory of sustainable design ideas and is used in the educational curriculum of the students.

This elementary school campus is based on a "super sized" Kindergarten thru 8th grade model.. As our society becomes more fragmented and transient, students have been looking to their schools as a form of stability and continuity.

MANUFACTURERS/SUPPLIERS	
DIV 07:	*Metal Roofing:* **Garland.**
DIV 08	*Glass Low E:* **PPG;** *Entrances & Storefronts, Windows, Curtainwall:* **Kawneer;** *Daylighting:* **Kalwall.**
DIV 09:	*Carpet:* **Collins & Aikman;** *VCT:* **Armstrong.**

COST PER SQUARE FOOT $ 221.95

ARCHITECT
Du Bose Associates, Inc. Architects
49 Woodland Street, Hartford, CT 06105
www.dbarch.com

FILE UNDER
EDUCATIONAL
Chicopee, Massachusetts

CONSTRUCTION TEAM

STRUCTURAL ENGINEER:
Santo Domingo Engineering, LLC
2074 Park Street, Hartford, CT 06106

GENERAL CONTRACTOR:
P & S Construction, Inc.
35 John Street, Lowell, MA 01852

MECHANICAL & ELECTRICAL ENGINEER:
Loureiro Engineering Associates, Inc.
100 Northwest Drive, Plainville, CT 06062

COST ESTIMATOR:
Leach Consulting Company, LLC
1010 Wethersfield Avenue, #306, Hartford, CT 06114

CIVIL ENGINEER:
URS Corporation AES
500 Enterprise Drive, Rocky Hill, CT 06067

GENERAL DESCRIPTION

SITE: 2.5 Acres.
NUMBER OF BUILDINGS: One; 22 classrooms each seating 20 to 25.
BUILDING SIZE: First floor, 26,000; total, 26,000 square feet.
BUILDING HEIGHT: First floor, 21'; total, 21'.
BASIC CONSTRUCTION TYPE: New/2B Unprotected Non-Combustible.
FOUNDATION: Slab-on-grade.
EXTERIOR WALLS: Brick, cement composite panel rain screen system.
ROOF: Membrane. **FLOOR:** Concrete
INTERIOR WALLS: Metal stud drywall.

WESTOVER JOB CORPS CENTER
Construction Period: Oct 2006 to Feb 2008 • Total Square Feet: 26,000

	C.S.I. Divisions (1 through 16)		COST	% OF COST	SQ.FT. COST	SPECIFICATIONS
		PROCUREMENT & CONT. REQ.	50,189	1.01	1.93	Contract Modifications
1.	1.	GENERAL REQUIREMENTS	333,043	6.72	12.81	General conditions, fees, testing, contingency, allowances, temporary facilities, superintendance.
3.	3.	CONCRETE	319,000	6.43	12.27	Forming & accessories, reinforcing, cast-in-place, precast, cast decks & underlayment, grouting (Concrete breakdown: Foundation: 140 cubic yards; Floors: 400 cubic yards; Walls: n/a).
4.	4.	MASONRY	85,000	1.71	3.27	Unit.
5.	5.	METALS	711,250	14.35	27.36	Structural metal framing, joists, decking, cold formed metal framing, fabrications.
6.	6.	WOOD/PLASTICS/COMPOSITE	92,000	1.86	3.54	Rough carpentry, finish carpentry, architectural woodwork.
7.	7.	THERMAL & MOIST. PROTECT	611,000	12.32	23.50	Dampproofing & waterproofing, thermal protection, weather barriers, roofing & siding panels, membrane roofing, flashing & sheet metal, roof & wall specialties & accessories, fire & smoke protection, joint protection.
8.	8.	OPENINGS	358,000	7.22	13.77	Doors & frames, specialty doors & frames, entrances, storefronts, & curtainwalls, windows, hardware.
9.	9.	FINISHES	652,900	13.17	25.11	Plaster & gypsum board, tiling, ceilings, flooring, wall finishes, acoustic treatment, painting & coating.
10.	10.	SPECIALTIES	29,500	0.60	1.13	Information, interior, safety, storage.
12.	12.	FURNISHINGS	16,000	0.32	0.62	—
15.	22.	PLUMBING	225,000	4.54	8.65	Piping & pumps, equipment, fixtures.
15.	23.	HVAC	293,007	5.91	11.27	Piping & pumps, ductwork, fans, fire dampers, HVAC, insulation, louvers, testing & balancing.
16.	26.	ELECTRICAL	1,147,000	23.13	44.12	Distribution, fixtures, telephone & data, transformers, lightning protection, coordination.
16.	28.	COMMUNICATIONS	35,000	0.70	1.35	Fire alarm.
TOTAL BUILDING COST			**4,957,889**	**100%**	**$190.69**	
2.	2.	EXISTING CONDITIONS	133,000			Assessment, demolition & structure moving.
2.	31.	EARTHWORK	549,600			Site clearing, earth moving.
2.	32.	EXTERIOR IMPROVEMENTS	95,200			Bases, bollards, & paving, improvements.
TOTAL PROJECT COST			**5,735,689**	*(Excluding architectural and engineering fees)*		

UPDATED ESTIMATE TO JAN 2011: $ 233.40 SF

Westover Job Corps Center
Chicopee, Massachusetts

Architect: Du Bose Associates, Inc. Architects

Photos Courtesy of
Donald Satterlee Photography

The new 26,000-square-foot classroom building for the Westover Job Corps Center is master-planned to become the first building of a new vocational education quadrangle. The building's most public functions are designed to engage an outdoor entrance plaza in order to enhance the sense of "campus", to greet pedestrian traffic from the residence halls on the main quadrangle on campus and to allow ease of nighttime use. The Job Corps concept of creating a "Community of Learners" was achieved through a campus setting that fosters independence and individual growth. The use of bold interior color gives the building an inviting presence and conveys an energetic attitude to the Job Corps Center students.

The building exceeds the stringent Massachusetts energy code by using a high R-Value cement composite panel "rain screen wall system". Other "Green Design" features include the use of natural light via clerestories and large windows in public areas and classrooms, waterless urinals and occupancy sensors for lighting. Energy STAR® rated equipment was used throughout.

The planning process involved working with the Department of Labor (DOL) prototypes in review sessions with DOL and P.B. Dewberry (DOL's Program Manager), as well as with the center's operator. The prototypes evolved into a layout that is unique to the Westover Center's needs.

The Job Corps program targets students that either did not complete high school and/or completed high school but do not have a career goals.

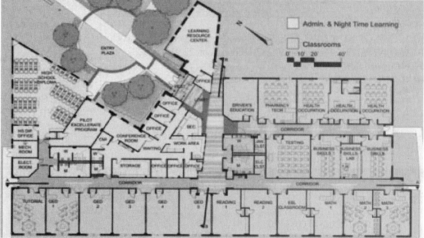

The program teaches vocational skills that focus on the local area's economy and job placement needs. The center also offers GED programs and encourages all students to complete their high school diploma. Nighttime classes and nighttime availability of study and library spaces are also intended to encourage study skills.

The existing campus reuses a portion of the Westover Air Force Base. The original Westover buildings are typically three stories with exposed concrete frames and brick and glass infill. The new building uses light and dark colored brick stripes at the main entrance to relate to the original concrete and brick buildings on campus.

Michael O'Malley, Head Architect for the Department of Labor (DOL) summed up DOL's satisfaction with the completed classroom building

by saying, "The use of light and color identity were the primary focus of the design, which was well achieved and spectacular. It looks well thought out and inviting."

MANUFACTURERS/SUPPLIERS	
DIV 04	*Brick:* Redland; *Cement Composite Panel:* Cement Board Fabricators.
DIV 07:	*Roof:* Firestone.
DIV 08	*Windows:* **Winco;** *Entrances & Storefronts:* **EFCO Corporation.** **Kawneer;** *Daylighting:* **Kalwall.**
DIV 09:	*Tile:* **Dal-Tile;** *Carpet:* **Monterrey;** *VCT:* **Armstrong.**
DIV 26:	*Lighting:* **Lithonia,** *Lite Control,* **Ellipitar, Cooper.**

COST PER SQUARE FOOT $ 233.40

ARCHITECT

RB+B Architects/Hutton Architecture Studio
315 E. Mountain Avenue, #100, Fort Collins, CO 80524
www.rbbarchitects.com

CONSTRUCTION TEAM

STRUCTURAL ENGINEER:
The Sheflin Group, Inc.
6638 West Ottawa Avenue, #230, Littleton, CO 80128

GENERAL CONTRACTOR:
W. O. Danielson Construction
2970 South Fox Street, Englewood, CO 80110

ELECTRICAL ENGINEER:
The RMH Group
12600 West Colfax Avenue, #A-400, Lakewood, CO 80215

MECHANICAL ENGINEER:
Shaffer + Baucom Engineering Consultants
7333 West Jefferson Avenue, #230, Lakewood, CO 80235

KITCHEN CONSULTANT:
William Caruso & Associates
9200 E. Mineral Avenue, #320, Englewood, CO 80112

GENERAL DESCRIPTION

SITE: 10.02 Acres.
NUMBER OF BUILDINGS: 1; 32 classrooms seating capacity 30 each.
BUILDING SIZE: First floor, 50,232; second floor, 22,914; total, 73,146 square feet.
BUILDING HEIGHT: First floor, 21'; total, 21'.

SECOND FLOOR

FIRST FLOOR

FOUNDATION: Cast-in-place, reinforced concrete, drilled piers.
EXTERIOR WALLS: CMU, curtainwall, metal siding.
ROOF: Membrane.
FLOOR: Concrete, precast (first floor).
INTERIOR WALLS: CMU, metal stud drywall.

DOUGLAS COUNTY ELEMENTARY SCHOOL

Construction Period: Oct 2006 to Nov 2007 Total Square Feet: 73,146

C.S.I. Divisions (1 through 16)			COST	% OF COST	SQ.FT. COST	SPECIFICATIONS
		PROCUREMENT & CONT. REQ.	322,755	3.08	4.41	General conditions.
1.	1.	GENERAL REQUIREMENTS	1,341,627	12.80	18.34	Summary, price & payment procedures, administrative requirements, execution & closeout requirements, performance requirements.
3.	3.	CONCRETE	1,193,412	11.39	16.32	Forms & accessories, reinforcement, cast-in-place.
4.	4.	MASONRY	543,960	5.19	7.44	Unit, cleaning.
5.	5.	METALS	1,152,045	10.99	15.75	Structural framing, miscellaneous, stairs, structural steel erection, expansion control.
6.	6.	WOOD/PLASTICS/COMPOSITE	281,400	2.69	3.85	Rough carpentry, architectural woodwork.
7.	7.	THERMAL & MOIST. PROTECT	563,706	5.38	7.71	Waterproofing, dampproofing, water repellants, insulation, membrane roofing, flashing & trim.
8.	8.	OPENINGS	569,313	5.43	7.78	Aluminum storefront, overhead doors, wood doors, steel doors & frames, curtainwall, hardware, glazing.
9.	9.	FINISHES	848,351	8.10	11.60	Gypsum wall systems, acoustical, resilient flooring, carpet, seamless flooring, paint, vinyl wallcovering.
10.	10.	SPECIALTIES	143,896	1.37	1.97	Interior, visual display board, lockers, fire extinguisher & cabinets, toilet accessories, partitions.
11.	11.	EQUIPMENT	199,621	1.90	2.73	Projection screens, food service, residential appliances, gym.
12.	12.	FURNISHINGS	26,038	0.25	0.36	Window treatment.
14.	14.	CONVEYING SYSTEMS	73,943	0.71	1.01	Elevators (1).
15.	21.	FIRE SUPRESSION	97,850	0.93	1.34	—
15.	22.	PLUMBING	822,717	7.85	11.24	Equipment, piping, insulation.
15.	23.	HVAC	1,351,085	12.90	18.47	AHUs, boiler, chillers, ice storage, piping & pumps, temperature controls.
16.	26.	ELECTRICAL	869,933	8.30	11.89	Mobilization, basic materials & methods, service & distribution, lighting.
16.	28.	COMMUNICATIONS	50,738	0.48	0.69	—
16.	28.	ELECTRONIC SAFETY & SECURITY	27,399	0.26	0.37	Fire alarm.
TOTAL BUILDING COST			**10,479,789**	**100.00%**	**$143.27**	
2.	31.	EARTHWORK	260,084			Grading, excavation.
2.	32.	EXTERIOR IMPROVEMENTS	769,449			Traffic controls, pavement, playground surface system, irrigation system, fence, playground equipment,
2.	33.	UTILITIES	212,803			
TOTAL PROJECT COST			**11,722,125**	*(Excluding architectural and engineering fees)*		

UPDATED ESTIMATE TO JAN 2011: $ 189.01

DCD Subscribers: Access this case study and hundreds more for instant date and location calculations at www.dcd.com.

Douglas County Elementary School
Parker, Colorado

Architect: RB+B Architects/Hutton Architecture Studio

As the result of a design competition, this Elementary School was designed as an energy efficient prototype, allowing for academic flexibility and adaptation to various sites in Douglas County. The high performance design includes a high-efficiency building envelope coupled with state-of-the-art daylighting strategies. The facility provides significant savings to the Owner by performing at less than half the energy cost of existing schools. Increased occupant comfort, higher student performance, lower energy costs (including 30% smaller than average boilers), and unique design features all add to the success of this innovative educational project.

The challenge of the project was to design a prototype that allowed the most flexibility to accommodate varying sites without compromising the unique attributes of each building. Incorporating sustainable design features added to the challenge but rewarded in a state-of-the-art high-performance school.

Classrooms filled with natural light have been shown repeatedly to not only improve student performance but also enhance student and staff attitudes and reduce absenteeism. Over half the classrooms in this prototype are able to achieve required light levels without any artificial lighting for most of the school day (the remainder only require about 30% supplemental lighting). Bright, inviting, vibrant spaces are spaces where students want to be.

Marie Unger, Principal of Douglas County Elementary School No. 43 stated, "Our kids have been coming home at the end of their school day with a new enthusiasm for taking care of our environment. It has been reported to me by several families that their children are demanding a home recycling program. We even have families who do not live within walking distance (so they have to find their own transportation) and their children have requested a carpool to get them to school!"

Sustainable design features include:

- Daylighting - Natural light enhances student learning and creates invigorating indoor spaces.
- Sun Shades - Shading the south facing facade reduces heat gain from direct sunlight penetrating through windows and reduces glare inside classrooms.
- Exterior Insulation - Upgraded insulation levels in exterior walls and the roof reduce heat gain in summer and heat loss in winter, resulting in lower utility costs.
- Reduced Cooling Load - Less electrical lighting and shaded southern elevations reduce the annual cooling load. Less air conditioning is needed to maintain the comfort level and energy use is reduced.
- Reduced HVAC Equipment - A lower cooling load allows for the installation of smaller HVAC cooling systems.
- Ice Storage and Peak Demand Reduction - Ice is made at night, taking advantage of off-peak energy rates and is stored in large insulated ice storage tanks. The system then uses this stored ice during the day for cooling.
- Displacement Ventilation - Displacement ventilation supplies air to a space near the floor level with very low velocity. This configuration improves air quality, helps to minimize air movement noise levels and saves energy.

"Furthermore, (just last week) a mom shared with me that her daughter wants to eat dinner by candlelight more often, to save energy. She said that her daughter likes it, that we often have plenty of light in the cafeteria and that we don't need to turn on any lights at all."

MANUFACTURERS/SUPPLIERS	
DIV 07:	*Membrane:* Carlisle.
DIV 08	*Curtainwall, Entrances & Storefronts:* **United States Aluminum;** *Daylighting:* **Solatube;** *Hardware:* **Ingersoll Rand.**
DIV 14:	*Elevators:* **KONE.**
DIV 26:	*Lighting:* **Hubbell, Prescolite, KIM, Columbia, Pinnacle, Kewall.**

COST PER SQUARE FOOT $ 189.01

ARCHITECT
McKissick Associates
317 N. Front Street, Harrisburg, PA 17101
www.mckissickassociates.com

FILE UNDER
EDUCATIONAL
Havertown, Pennsylvania

CONSTRUCTION TEAM

STRUCTURAL ENGINEER:
Baker Ingram & Associates, Inc.
1547 Oregon Pike, Lancaster, PA 17601
MECHANICAL & ELECTRICAL ENGINEER:
H. F. Lenz
1407 Scalp Avenue, Johnstown, PA 15904
GENERAL CONTRACTOR:
John S. McManus, Inc.
9 Smithbridge Road, Chester Heights, PA 19017
CIVIL ENGINEER: H. F. Lenz
1407 Scalp Avenue, Johnstown, PA 15904

GENERAL DESCRIPTION

SITE: 3.8 Acres.
NUMBER OF BUILDINGS: One; 37 classrooms, 850 pupil seating
BUILDING SIZE: First floor, 57,755; second floor, 13,800; mechanical mezzanine, 13,800; total, 85,355 square feet.
BUILDING HEIGHT: First floor, 12'8"; second floor, 12'8"; mechanical mezzanine, 11'6"; total, 36'10".
BASIC CONSTRUCTION TYPE: New/Type 2 Non Combustible.
FOUNDATION: Cast-in-place.
EXTERIOR WALLS: Brick, metal panel.
ROOF: Modified bitumen. **FLOORS:** Concrete, concrete plank.
INTERIOR WALLS: CMU, metal stud drywall.

MANOA ELEMENTARY SCHOOL
Construction Period: May 2007 to Nov 2008 • Total Square Feet: 85,355

C.S.I. Divisions (1 through 16)			COST	% OF COST	SQ.FT. COST	SPECIFICATIONS
		PROCUREMENT & CONT. REQ.	301,400	1.81	3.53	Mobilization, supervision, general conditions, bond.
1.	1.	GENERAL REQUIREMENTS	46,200	0.28	0.54	Price & payment procedures, administrative requirements, quality requirements, temporary facilities & controls, product requirements, execution & closeout requirements.
3.	3.	CONCRETE	1,600,000	9.63	18.75	Forming & accessories, reinforcing, cast-in-place, precast, cast decks & underlayment, grouting, cutting & boring (concrete breakdown: 473 cubic yards foundation, 11 cubic yards walls, 1,866 cubic yards).
4.	4.	MASONRY	2,176,000	13.10	25.49	Unit, manufactured masonry.
5.	5.	METALS	1,300,000	7.83	15.23	Structural metal framing, joists, decking, fabrications.
6.	6.	WOOD/PLASTICS/COMPOSITE	130,000	0.78	1.52	Rough carpentry, finish carpentry, architectural woodwork, structural plastics, plastic fabrications.
7.	7.	THERMAL & MOIST. PROTECT	1,600,000	9.63	18.75	Dampproofing, membrane roofing, flashing & sheet metal, roof & wall specialties & accessories.
8.	8.	OPENINGS	600,000	3.61	7.03	Doors & frames, entrances, storefronts, & curtainwalls, windows, hardware.
9.	9.	FINISHES	1,050,000	6.32	12.30	Plaster & gypsum board, tiling, ceilings, flooring, wall finishes, painting & coating.
10.	10.	SPECIALTIES	200,000	1.20	2.34	Information, interior, metal building assembly.
11.	11.	EQUIPMENT	350,000	2.11	4.10	Food service, athletic & recreational, other.
12.	12.	FURNISHINGS	470,000	2.83	5.51	Casework, other.
14.	14.	CONVEYING SYSTEMS	70,000	0.42	0.82	Elevators (1).
15.	21.	FIRE SUPRESSION	256,000	1.55	3.00	Water-based fire-suppression systems, fire-extinguishing systems.
15.	22.	PLUMBING	762,000	4.59	8.93	Piping & pumps, equipment, fixtures.
15.	23.	HVAC	2,724,000	16.40	31.91	Piping & pumps, air distribution, air cleaning devices, central heating equipment, central HVAC.
16.	26.	ELECTRICAL	2,002,542	12.06	23.46	Medium-voltage distribution, low-voltage transmission, facility electrical power generating & storing equipment, electrical & cathodic protection, lighting.
16.	27.	COMMUNICATIONS	607,597	3.66	7.12	Structured cabling, data, voice, audio-video, distributed communications & monitoring systems.
16.	28.	ELECTRONIC SAFETY & SECURITY	362,861	2.19	4.25	—
TOTAL BUILDING COST			**16,608,600**	**100%**	**$194.58**	
2.	2.	EXISTING CONDITIONS	83,500			—
2.	31.	EARTHWORK	321,000			Earthmoving.
2.	32.	EXTERIOR IMPROVEMENTS	796,800			Bases, bollards, & paving, improvements, planting.
2.	33.	UTILITIES	412,500			Water, sanitary sewerage, storm drainage, electrical.
TOTAL PROJECT COST			**18,222,400**	*(Excluding architectural and engineering fees)*		

UPDATED ESTIMATE TO JAN 2011: $247.99 SF

DCD Subscribers: Access this case study and hundreds more for instant date and location calculations at www.dcd.com.

Manoa Elementary School
Havertown, Pennsylvania

Architect: McKissick Associates

The Manoa Elementary School can accommodate 4 classrooms per grade level with a total capacity of 850 pupils. With limited open land in this Philadelphia streetcar suburb, the only available site was 3.1 acres of a 10-acre community sports complex. This necessitated a 3-story building plan. The building and site circulation pattern was intentionally compact in design to maximize the amount of athletic field space that remains available to the community, including fields for lacrosse, soccer, field hockey and football. The use of subsurface storm water detention and a hard surface playground for event overflow parking assisted in retaining green field space.

A sub divisible gymnasium room, cafeteria, 2-music rooms, art room, and 14-flexible learning spaces provide support for the District's enrichment and special needs programs. Spatial efficiency is maximized by allowing the use of the cafeteria as both stage and sub divisible large group instructional area. A reinforced masonry bearing and pre-cast concrete plank structural system was utilized for the classroom wing to substantially reduce construction time and to limit the overall height of the building to 30-feet to meet local zoning requirements. To further reduce the building's apparent mass within this residential neighborhood, the exterior skin utilizes a mixture of reflective zinc colored metal panels, allowing the classroom and gymnasium wings to assume the color of the surrounding environment.

Internally the building features an integrated data fiber optic backbone with wireless networking and a complete modern "voice over" IP communication system. In addition to portable wireless labs, each classroom has 4 computers, as well as a mounted LCD projector and smart board capability.

Green components of the building are cost-effective and protective of the environment. Sustainable features include occupancy sensors for lighting, insulated glass windows and doors, high efficiency indirect/direct lighting and daylighting. The acid-etched and sealed concrete floors in circulation areas require minimal maintenance while avoiding the use of manmade products.

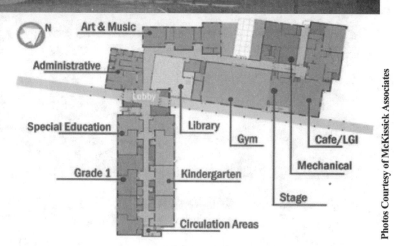

Photos Courtesy of McKissick Associates

Corridors have a wainscot of bamboo wood - a renewable resource. The high-tech heating and cooling system provides dehumidification capability, controlling mold and allowing for superior heating and cooling recovery with the use of energy recovery ventilators.

The emphasis on sustainability was reinforced by development of interior design themes based upon the five Greek elements. Floors became part of this integral design throughout the school, and each of the 3-classroom floors has a different visual theme based upon the elements of Water (1st Floor classrooms), Earth (2nd Floor classrooms) and Air (3rd Floor classrooms). The elements of Ether (main entry) and Fire (library and 1st Floor administration & gymnasium wing) were also incorporated. These themes are reinforced by the choice of flooring, wall colors, cabinets,

vinyl tiles and stair treads which allow the children to wayfind their floor by color. The multipurpose space in the Cafeteria and Stage incorporate all of the themes. The multi-colored earth-toned floor has a star on the stage level and a sun on the cafeteria level. An acid etched hand-print border using the handprints of students and staff imparts immediate ownership.

MANUFACTURERS/SUPPLIERS	
DIV03:	*Precast:* High Concrete Structures, Inc. *Concrete Stain:* Lithocrome®
DIV07:	*Metal Panel:* **Kingspan ASI;** *Modified Bituminous Membrane roofing:* Tremco.
DIV08	*Wood Windows:* Pella; *Aluminum Windows:* **EFCO Corporation;** *Rolling Shutter & Overhead Security Rills:* **Cornell;** *Wood Doors:* **Graham.**
DIV09:	*Paint:* **Sherwin Williams;** *Ceilings:* **Armstrong.**
DIV10:	*Folding Partitions:* **Modernfold.**

COST PER SQUARE FOOT $ 247.99

ARCHITECT	FILE UNDER
KKE Architects, Inc.	*EDUCATIONAL*
300 First Avenue North, Minneapolis, MN 55401	*New Prague, Minnesota*
www.kke.com	

CONSTRUCTION TEAM

STRUCTURAL ENGINEER:

Ericksen Roed & Associates, Inc.

2550 University Avenue West, #201S, St. Paul, MN 55114

ELECTRICAL & MECHANICAL ENGINEER:

Hallberg Engineering

1750 Commerce Court, White Bear Lake, MN 55110

LANDSCAPE ARCHITECT:

Anderson-Johnson Associates, Inc.

7575 Golden Valley Road, #200, Minneapolis, MN 55427

CONSTRUCTION MANAGER & COST ESTIMATOR

AMCON Construction

1715 Yankee Doodle Road, #200, Eagan, MN 55121

GENERAL DESCRIPTION

SITE: 20 Acres.

NUMBER OF BUILDINGS: One.

BUILDING SIZE: First floor, 53,481; second floor, 36,676; total, 90,157 square feet.

BUILDING HEIGHT: First floor, 14'8"; second floor, 14'8"; total, 29'6".

BASIC CONSTRUCTION TYPE: New

FOUNDATION: Cast-in-place, reinforced concrete, slab-on-grade.

EXTERIOR WALLS: Precast.

ROOF: Membrane.

FLOORS: Precast.

INTERIOR WALLS: CMU, metal stud drywall.

RAVEN STREAM ELEMENTARY SCHOOL

Construction Period: Apr 2005 to Aug 2006 • Total Square Feet: 90,157

C.S.I. Divisions (1 through 16)			COST	% OF COST	SQ.FT. COST	SPECIFICATIONS
		PROCUREMENT & CONT. REQ.	—	—	—	
1.	1.	GENERAL REQUIREMENTS	806,000	6.50	8.94	—
3.	3.	CONCRETE	1,698,400	13.69	18.84	Cast-in-place, forming & accessories, precast.
4.	4.	MASONRY	256,125	2.06	2.84	Unit.
5.	5.	METALS	603,075	4.86	6.69	—
6.	6.	WOOD/PLASTICS/COMPOSITE	144,813	1.17	1.61	Rough carpentry & finish carpentry.
7.	7.	THERMAL & MOIST. PROTECT	594,342	4.79	6.59	Fire & smoke protection, joint protection, membrane roofing, roof & wall specialties & accessories.
8.	8.	OPENINGS	586,640	4.73	6.51	Doors, frames, hardware, specialty doors & frames, windows.
9.	9.	FINISHES	1,403,265	11.31	15.56	Ceilings, flooring, painting & coating, plaster & gypsum board, tiling.
10.	10.	SPECIALTIES	87,500	0.71	0.97	Flagpole, information, interior, marker boards.
11.	11.	EQUIPMENT	334,722	2.70	3.71	Athletic & recreational, food service, student lockers.
12.	12.	FURNISHINGS	810,750	6.54	8.99	Casework, furniture.
14.	14.	CONVEYING SYSTEMS	45,750	0.37	0.51	Elevator (1).
15.	21.	FIRE SUPRESSION	160,100	1.29	1.78	Water-based fire-suppression systems.
15.	22.	PLUMBING	1,628,250	13.13	18.06	—
15.	23.	HVAC	1,422,335	11.47	15.78	—
16.	26.	ELECTRICAL	1,501,422	12.10	16.65	—
16.	27.	COMMUNICATIONS	320,000	2.58	3.55	—
TOTAL BUILDING COST			**12,403,489**	**100.00%**	**$137.58**	
2.	31.	EARTHWORK	422,840			Earthwork.
2.	32.	EXTERIOR IMPROVEMENTS	515,003			Bases, ballasts, & paving, irrigation, improvements.
2.	33.	UTILITIES	246,448			
TOTAL PROJECT COST			**13,587,780**	*(Excluding architectural and engineering fees)*		

UPDATED ESTIMATE TO JAN 2011: $194.34 SF

DCD Subscribers: Access this case study and hundreds more for instant date and location calculations at www.dcd.com.

Raven Stream Elementary School
New Prague, Minnesota

Architect: KKE Associates, Inc.

Photos Courtesy of McKissick Associates

The New Prague Area School District was planning to build two new Elementary Schools for the first time in 30 years. The District chose a process to complete the schematic design of the new elementary schools prior to the referendum passage and simultaneously with the search for land.

The schematic design process involved a committee of 40 people from all parts of the District as well as selected educators. The goals of the committee were as follows: Design a building that could be used on different sites and maintain an East/West orientation for the classrooms; Develop separate zones for education and community use; and apply the sustainable design concepts of the LEED® program.

The building responded to the lack of a site with classroom wings that could flip in different directions to maintain the East/West orientation while still providing separation between the site elements of bus, car and play/athletic zones. The East/West orientation is important for energy efficiency by providing consistent sunlight with ease of control in the classroom. The selection of materials and heating cooling systems also responded to the LEED® criteria.

The two-story cafeteria is the center of activity that provides orientation within the building to which the entry, gymnasium, administration, arts and media center are directly connected for community use and separate for the classroom areas.

The building's design utilizes multiple sustainable design principles while meeting the District's needs for a prototypical and economical design to be built in two locations. Daylighting opportunities were maximized through proper site orientation, which partnered with variable electrical controls and allow for reduced energy consumption. Classroom windows are sized to approximately 17% of the floor area that results in an average 35 foot-candle level. Mechanical systems were selected based on their high efficiency, low maintenance and ability to provide a high level of indoor air quality. Classroom windows are sized to approximately 17% of the floor area that results in an average 35 foot-candle level. Mechanical systems were selected based on their high efficiency, low maintenance and ability to provide a high level of indoor air quality. The cost-effective building materials were selected to utilize regional resources, for their reuse potential and to include recycled and renewable products.

The structural system of structural steel columns and beams, precast plank floors, and precast walls and metal panel exterior walls provided the owner with a cost efficient building system while meeting LEED® criteria.

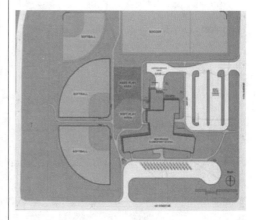

MANUFACTURERS/SUPPLIERS	
DIV03:	*Precast:* Hanson.
DIV07:	*Membrane Roofing:* Firestone
DIV09:	*Vinyl:* **Mannington**; *Carpet:* **Mohawk**

COST PER SQUARE FOOT $ 194.34

ARCHITECT
NEAL PRINCE + PARTNERS ARCHITECTS
110 West North Street, #300
Greenville, Sc 29601
www.neal-prince.com

CONSTRUCTION TEAM

STRUCTURAL ENGINEER:
Professional Engineering Associates
110 Edgeworth Street, Greenville, SC 29607
ELECTRICAL & MECHANICAL ENGINEER:
Mccracken & Lopez
1300 Baxter Street, #350, Charlotte, NC 28204
GENERAL CONTRACTOR:
Triangle Construction
P. O. Box 6266, Greenville, SC 29606
COST ESTIMATOR
Triangle Construction
P. O. Box 6266, Greenville, SC 29606

GENERAL DESCRIPTION

SITE: 4 Acres.
NUMBER OF BUILDINGS: One.
BUILDING SIZE: First floor, 34,000; second floor, 34,000; total, 68,000 square feet.
BUILDING HEIGHT: First floor, 11'10"; second floor, 10'; total 28'.
BASIC CONSTRUCTION TYPE: Renovation.
FOUNDATION: Shallow spread footings.
EXTERIOR WALLS: Brick.
ROOF: Existing.
FLOORS: Carpet, slate, tile, rubber, linoleum.
INTERIOR WALLS: Gypsum, wood paneling.

FURMAN HALL, FURMAN UNIVERSITY
Construction Period: Dec 2003 to Aug 2005 • Total Square Feet: 68,000

C.S.I. Divisions (1 through 16)			COST	% OF COST	SQ.FT. COST	SPECIFICATIONS
		PROCUREMENT & CONT. REQ.	9,445	0.20	0.14	Permits.
1.	1.	GENERAL REQUIREMENTS	657,565	14.22	9.67	Bond, fee, overhead.
3.	3.	CONCRETE	237,781	5.14	3.50	Cast-in-place, forming, beams.
4.	4.	MASONRY	127,977	2.77	1.88	Masonry, cast stone, FRP cornice.
5.	5.	METALS	182,960	3.96	2.69	Miscellaneous steel, structural steel.
6.	6.	WOOD/PLASTICS/COMPOSITE	196,000	4.24	2.88	Carpentry, cabinets, trim.
7.	7.	THERMAL & MOIST. PROTECT	149,882	3.24	2.20	Aluminum composite panels, insulation, roofing, spray on fireproofing & firestop.
8.	8.	OPENINGS	311,857	6.75	4.59	Automatic door openers, curtainwall, storefront, FRP doors, skylights, wood windows.
9.	9.	FINISHES	846,502	18.31	12.45	Acoustical ceilings, carpet, VCT, tile, drywall & framing, EIFS, painting, stone flooring.
10.	10.	SPECIALTIES	38,862	0.84	0.57	Cabinets, markerboards, toilet accessories & partitions, louvers.
11.	11.	EQUIPMENT	31,490	0.68	0.46	Projection screens.
12.	12.	FURNISHINGS	3,860	0.08	0.06	Entrance mats.
15.	21.	FIRE SUPRESSION	140,000	3.03	2.06	Fire protection.
15.	22.	PLUMBING	167,160	3.62	2.46	Plumbing.
15.	23.	HVAC	755,154	16.34	11.11	HVAC.
16.	26.	ELECTRICAL	766,336	16.58	11.26	
TOTAL BUILDING COST			**4,622,831**	**100%**	**$67.98**	
2.	2.	EXISTING CONDITIONS	159,800			Demolition.
2.	31.	EARTHWORK	4,160			Soil poison.
2.	32.	EXTERIOR IMPROVEMENTS	94,135			Asphalt paving, clean & repair canopy, sidewalk, sitework for outdoor classroom.
2.	33.	UTILITIES	29,554			Water, storm drain, fire.
TOTAL PROJECT COST			**4,910,480**	*(Excluding architectural and engineering fees)*		

UPDATED ESTIMATE TO JAN 2011: $121.12 SF

DCD Subscribers: Access this case study and hundreds more for instant date and location calculations at www.dcd.com.

Furman Hall, Furman University
Greenville, South Carolina

Architect: Neal Prince + Partners Architects

Photos Courtesy of Marc Lamkin Photography

Furman Hall was the first academic building constructed when Furman University relocated to its present campus and has been used continuously for more than 40 years. In an effort to bring its facilities into compliance with current academic standards, Furman Hall selected Neal Prince + Partners to undertake a complete renovation, including reconfiguration of its interior spaces, adding new energy efficient windows, updating HVAC and electrical systems, and providing new computer network systems and new finishes. The 68,000 square foot building houses nine academic departments, with a capacity of more than one thousand occupants. The renovated Furman Hall blends state-of-the-art technology and building systems within the context of the "traditional" Furman University architecture.

Neal Prince + Partners worked closely with the Furman administration and staff to develop design criteria and a programmatic layout for the available existing building footprint. The University did not want to consider external additions to the existing building that would encroach upon the open spaces in the heart of campus surrounding Furman Hall; therefore, making efficient use of space became an organizing concept. The Furman campus is typically stylized Georgian architecture, but often classical detailing was absent from building facades constructed in the late 1950's. The design solution needed to address the image of the building as well as the functionality. New entry elements,

complete with proportional detailing, canopies, lighting, and signage add presence and elegance to the prominent edifice.

Interior spaces were reconfigured to promote easy circulation and to concentrate classroom areas in the middle section of the linear building. Major educational departments were clustered in suites, arranged at the ends of the buildings, while smaller suites were located with easy access to main circulation paths. A sweeping two story monumental stair connecting an enlarged entry foyer and central corridors adds function and beauty to the renovated spaces.

Each classroom was upfitted with current technology, including motorized projection screens, ceiling mounted digital projectors, centralized lighting and HVAC controls, and a document station for each instructor. A working broadcast studio, electronic language lab, multi-media lecture hall, and computer labs were each integrated into the existing structure.

Plain painted concrete block walls were replaced with new partitions and stained wood clad column enclosures. New varied ceiling treatments, such as curved vaults interspersed with groined ceilings, help break up long corridors.

Existing construction was replaced with new energy efficient windows and HVAC units, recyclable carpets and tiles, and high efficiency lighting. As part of the waste management program, over seventy-five percent of construction and demolition waste was diverted from the waste stream. Furman University has adopted a policy of sustainable design, and accordingly, Furman Hall is registered as a LEED® project. The building is currently being reviewed for conformance with a Silver Certification rating.

A coordinated effort by architect, owner, and contractor resulted in minimum disruptions, particularly during periods of exam preparation and administration. The end result is a handsome addition to the campus fabric and a satisfied client.

LEED® SILVER PENDING	
MANUFACTURERS/SUPPLIERS	
DIV 04:	*Brick:* **Hanson Brick.**
DIV 08:	*Entrances & Storefronts:* **YKK America;** *Windows:* **Traco Historic Series;** *FRP Doors:* **Special Lite;** *Glazing:* **AFDG Glass** supplied by Tri-State Glass; *Skylights:* **Solatube.**
DIV 09:	*Carpet:* **Shaw,** *Porcelain Tile::* *Crossville; Paint:* **Sherwin Williams;** *Slate:* **Vermont Structural Slate;** *Linoleum:* **Forbo.**
DIV 10:	*Louvers:* **American Warming & Ventilating**

COST PER SQUARE FOOT $ 121.12

ARCHITECT
LAUER-MANGUSO& ASSOCIATES ARCHITECTS
4080 Ridge Lea Road Buffalo, NY 14228
www.lauer-manguso.com

· FILE UNDER
EDUCATIONAL
Amherst, New York

CONSTRUCTION TEAM

STRUCTURAL ENGINEER:
   Lauer-Manguso & Associates Architects
   4080 Ridge Lea Road, Buffalo, NY 14228
ELECTRICAL & MECHANICAL ENGINEER:
   RobsonWoese, Inc.
   501 John James Audubon Parkway, Amherst, NY 14228
LANDSCAPE ARCHITECT:
   Frank T. Brzezinski, RLA
   69 Foxboro Lane, East Amherst, NY 14051
GENERAL CONTRACTOR & COST ESTIMATOR
   ADF Construction Corp.
   455 Commerce Drive, Amherst, NY 14228

GENERAL DESCRIPTION

SITE: 1 Acre.
NUMBER OF BUILDINGS: One.
BUILDING SIZE: First floor, 3,462; total, 3,462 square feet.
BUILDING HEIGHT: First floor, 10', total, 18'.
BASIC CONSTRUCTION TYPE: New/VI.
FOUNDATION: Concrete.
EXTERIOR WALLS: Siding.
ROOF: Asphalt shingle.
FLOORS: Carpet, tile, VCT.
INTERIOR WALLS: Drywall.

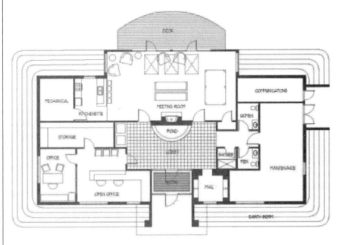

COMMUNITY BUILDING FOR STUDENT HOUSING NEIGHBORHOOD

CREEKSIDE VILLAGE COMMUNITY CENTER
Construction Period: Mar 2002 to Aug 2002 • Total Square Feet: 3,462

C.S.I. Divisions (1 through 16)			COST	% OF COST	SQ.FT. COST	SPECIFICATIONS
		PROCUREMENT & CONT. REQ.	—	—	—	—
1.	1.	GENERAL REQUIREMENTS	23,358	7.85	6.75	General requirements, overhead & profit, fees.
3.	3.	CONCRETE	24,464	8.22	7.07	Formwork, cast-in-place, reinforcing.
4.	4.	MASONRY	9,154	3.08	2.64	Masonry & grout.
5.	5.	METALS	—	—	—	—
6.	6.	WOOD/PLASTICS/COMPOSITE	28,544	9.60	8.24	Rough carpentry, wood trusses, wall panels, finish carpentry.
7.	7.	THERMAL & MOIST. PROTECT	24,894	8.37	7.19	Insulation, roofing, siding, trim, accessories, sealants & caulking.
8.	8.	OPENINGS	23,659	7.95	6.83	Metal doors & frames, wood & plastic doors, hardware, windows.
9.	9.	FINISHES	34,629	11.64	10.00	Drywall, carpet, VCT, tile, painting.
10.	10.	SPECIALTIES	4,540	1.53	1.31	mailboxes.
11.	11.	EQUIPMENT	—	—	—	—
12.	12.	FURNISHINGS	10,205	3.43	2.95	Cabinets, window treatments, mats.
14.	14.	CONVEYING SYSTEMS	—	—	—	—
15.	15.	MECHANICAL	67,706	22.76	19.56	Basic materials & methods, plumbing, HVAC.
16.	26.	ELECTRICAL	46,304	15.57	13.38	Basic materials & methods.
TOTAL BUILDING COST			**297,457**	**100%**	**$85.92**	
2.	2.	SITEWORK	19,975			Preparation, earthwork, sewerage & drainage.
2.		LANDSCAPING & OFFSITE WORK	62,892			Landscaping.
TOTAL PROJECT COST			**380,324**	*(Excluding architectural and engineering fees)*		

UPDATED ESTIMATE TO JAN 2011: $138.94 SF

Creekside Village Community Center
Amherst,, New York

Architect: Lauer-Manguso & Associates Architects

Photos Courtesy of Marc Lamkin Photography

The University at Buffalo with the desire to provide quality upper-class student housing required a new community center for the residents. The University also required some important design criteria they wanted to incorporate into the project. The site is environmentally significant and to demonstrate their concern, the University wanted the building to promote the concepts of "Green Building Design". In addition, the character of the structure should reflect the "arts/craft" style, which is prevalent to Western New York's architectural heritage.

This building is the "Commons Building", which serves as the social and administrative hub of the neighborhood (232 bed townhouse complex). The interior consists of office spaces, central mail pick-up, maintenance room and large meeting/social room. Site amenities include a pedestrian bridge over an existing creek, outdoor deck space, natural wildflower lawn, bike racks and special parking spaces for car pooling and electric car charging station.

The building uses forms and colors that "remind" us of the arts/craft style, while constructed to save approximately 20% more energy than a conventionally built structure.

To validate the green building concept, the building has been " LEED® Certified". The LEED® (TM) (Leadership in Energy and Environmental Design) green building rating system is a voluntary, consensus based national standard for developing high-performance sustainable buildings.

Currently this building is one of only two LEED®Certified structures in New York State. To achieve this required a great effort by all team members (Owner, Architect, consultants and contractors) to understand the strict requirements for construction and documentation of same.

The building uses numerous environmentally friendly materials, including building items with recycled content such as flooring, decking, countertops, cabinets, etc. and also low-voc coatings for all walls and floors.

Partially bermed walls provide passive insulation while solar powered deck lighting illuminate outdoor spaces and contribute to lower energy consumption. Special lighting controls not only turn the lights off when a room is unoccupied, they also reduce artificial lighting levels in proportion to the available natural light to further reduce the energy needs. A decorative indoor pond helps to humidify the building's interior while reducing comfort-heating requirements. Heat recovery from the ventilation system, high efficiency HVAC equipment, and DDC controls ensure the building uses the smallest amount of energy while keeping everyone comfortable.

MANUFACTURERS/SUPPLIERS	
DIV 03:	*Foundation Insulation:* **Owens Corning Foamular 250.**
DIV 07:	*Asphalt Shingles:* **Owens Corning** *Cementitious Siding:* **James Hardie;** *Housewrap Moisture Barrier:* **Owens Corning;** *Skylights:* **Wasco** E-Class Skywindow
DIV 08:	*Entrances & Storefronts:* **Vistawall,** *Closers: Vinyl Windows:* **Marquee 800** Series by Philips Products; *Hollow Metal Doors & Frames:* **Benchmark;** *Wood Doors:* Craftmaster Avalon®
DIV 09:	*Gypsum:* Georgia Pacific Tough Rock® *Carpet:* Bentley; *VCT:* **Armstrong;** *Rubber Base:* **Roppe;** Hardwood: Hartco; *Floor Finish:* **L.M. Scofield "A-50"**
DIV 10:	*Postal Specialties:* **Salsbury Industries.**
DIV 15:	*Fan & Ventilator:* **Greenheck;** *Louvers:* **Ruskin.**

COST PER SQUARE FOOT $ 138.94

ARCHITECT
KKE ARCHITECTS, INC.
300 First Avenue North
Minneapolis, MN 55401
www.kke.com

FILE UNDER
EDUCATIONAL
Chaska, Minnesota

CONSTRUCTION TEAM

STRUCTURAL ENGINEER:
Ericksen Roed & Associates
2550 University Avenue, #201-S, St. Paul, MN 55114
ELECTRICAL & MECHANICAL ENGINEER:
Dunham Associates
8200 Normandale Blvd., #500, Minneapolis, MN 55437
LANDSCAPE ARCHITECT:
Dahlgren, Shardlow & Uban, Inc.
300 First Avenue North, #210, Minneapolis, MN 55401
GENERAL CONTRACTOR:
CM Construction Company, Inc.
12215 Nicollet Avenue South, Burnsville, MN 55337
COST ESTIMATOR:
PPM
1858 East Shore Drive, St. Paul, MN 55109

GENERAL DESCRIPTION

SITE: 27 Acres.
NUMBER OF BUILDINGS: One; 9th grade center, 35 classrooms with a maximum of 32 seating capacity each; gymnasium seating capacity of 604.
BUILDING SIZE: Lower level, 68,588; main level, 55,525; penthouse, 5,171; total, 129,284 square feet.
BUILDING HEIGHT: First floor, 15'4"; second floor, 15'4"; total, 32'.

BASIC CONSTRUCTION TYPE: New.
FOUNDATION: Concrete.
EXTERIOR WALLS: Block, metal. ROOF: Membrane.
ROOF: Membrane.
FLOORS: Wood, resilient, carpet.
NTERIOR WALLS: Metal studs, gypsum.

PIONEER RIDGE CENTER
Construction Period: Mar 2001 to Sep 2002 • Total Square Feet: 129,284

C.S.I. Divisions (1 through 16)			COST	% OF COST	SQ.FT. COST	SPECIFICATIONS
		PROCUREMENT & CONT. REQ.	481,915	3.60	3.73	—
1.	1.	GENERAL REQUIREMENTS	70,000	0.52	0.54	—
3.	3.	CONCRETE	1,188,725	8.88	9.19	Formwork, reinforcement, cast-in-place, curing, precast, cementitious decks & toppings, grout.
4.	4.	MASONRY	696,430	5.21	5.39	Masonry & grout, accessories, unit, simulated masonry.
5.	5.	METALS	1,788,410	13.37	13.83	Materials, fastening, structural metal framing, joists, decking, cold formed metal framing, fabrications, ornamental, expansion control.
6.	6.	WOOD/PLASTICS/COMPOSITE	236,000	1.76	1.83	Fasteners & adhesives, rough carpentry, finish carpentry, architectural woodwork.
7.	7.	THERMAL & MOIST. PROTECT	788,100	5.89	6.10	Waterproofing, vapor retarders, air barriers, insulation, fireproofing, firestopping, manufactured roofing & siding, membrane roofing, flashing & sheet metal, roof specialties & accessories.
8.	8.	OPENINGS	1,010,975	7.56	7.82	Metal doors & frames, wood & plastic doors, door opening assemblies, special doors, entrances & storefronts, metal windows, wood & plastic windows, special windows, hardware, glazing, glazed curtainwalls.
9.	9.	FINISHES	1,554,320	11.62	12.02	Gypsum, tile, acoustical treatment, wood flooring, resilient flooring, carpet, special coatings, painting.
10.	10.	SPECIALTIES	218,835	1.64	1.69	Visual display board, louvers & vents, grilles & screens, wall & corner guards, flagpoles, pedestrian control devices, lockers, fire protection, protective covers, operable partitions, storage shelving, toilet & bath accessories.
11.	11.	EQUIPMENT	464,150	3.47	3.59	Audio-visual, loading dock, food service, athletic, recreational & therapeutic.
12.	12.	FURNISHINGS	379,260	2.83	2.93	Window treatment, furniture & accessories, rugs & mats, multiple seating.
14.	14.	CONVEYING SYSTEMS	37,400	0.28	0.29	Elevators (1).
15.	15.	MECHANICAL	3,304,770	24.70	25.56	Basic materials & methods, insulation, fire protection, plumbing, HVAC, heat generation, refrigeration, heat transfer, air distribution, controls, testing, adjusting & balancing.
16.	26.	ELECTRICAL	1,160,585	8.67	8.98	Basic materials & methods, medium voltage distribution, service & distribution, lighting, communications, controls, testing.
TOTAL BUILDING COST			**13,379,875**	**100%**	**$103.49**	
2.	2.	SITEWORK	1,627,125			Preparation, earthwork, paving & surfacing, utility piping materials, water distribution, sewerage & drainage, landscaping.
2.		LANDSCAPING & OFFSITE WORK	—			Included in Site Work.
TOTAL PROJECT COST			**15,007,000**	*(Excluding architectural and engineering fees)*		

UPDATED ESTIMATE TO JAN 2011: $172.82 SF

DCD Subscribers: Access this case study and hundreds more for instant date and location calculations at www.dcd.com.

Pioneer Ridge Center
Chaska, Minnesota

Architect: KKE Architects, Inc.

Photos Courtesy of KKE Architects

Bringing together students from diverse middle schools, Pioneer Ridge Freshmen Center fosters a strong social and academic bond among all district 9th graders allowing the students to nurture relationships and grow together for one significant, transitional year.

The school is designed as a cool place that freshmen would want to spend time. Its distinctive colors, textures and shapes create a memorable learning context, especially the elliptically shaped media center that serves as the eye of knowledge with classrooms spiraling from it.

Another gathering place, The Commons, offers a vortex of social interaction. The openness and connectedness of The Commons allows for a constant sense of community in both sight and sound. It also offers an outdoor courtyard overlooking a grove of trees and adjacent wetland.

Sensitive to the environment, the project was constructed using sustainable practices such as sourcing the majority of materials from local quarries, and building with long-lasting brick, precast concrete, metal panels and glass.

The building is situated tightly against the site's western edge to make the most of its natural landscape. Through careful siting and approach, the center is presented as a gateway to the community both in structure and in spirit.

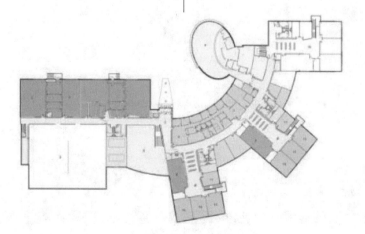

MANUFACTURERS/SUPPLIERS	
DIV 03:	*Block:* **Anchor Block.**
DIV 07:	*Metal Panels:* **Centria**
	Membrane Roofings: **Firestone.**
DIV 08:	*Curtainwall:* **EFCO Corporation;**
	Doors: **Curries Manufacturing;**
	Windows: **Wasau Window & Wall**
	Systems; *Glass:* **Oldcastle Glass.**
DIV 09:	*Carpet:* **Collins & Aikman;**
	Acoustical Treatment: **Armstrong;**
	Drywall: **National Gypsum;**
	Paint: **Pratt & Lambert.**
DIV 10:	*Toilet Partitions:* **Santana;**
	Baby Changers: **Koala.**

COST PER SQUARE FOOT $ 172.82

ARCHITECT
RB+B Architects, Inc.
315 E. Mountain Avenue, #100, Fort Collins, CO 80524
www.rbbarchitects.com

CONSTRUCTION TEAM

STRUCTURAL ENGINEER:
JVA, Inc.
25 Old Town Square, #200, Fort Collins, CO 80524

GENERAL CONTRACTOR:
Dohn Construction
2642 Midpoint Drive, #A, Fort Collins, CO 80525

ELECTRICAL & MECHANICAL ENGINEER:
Shaffer Baucom Engineering Consultants
7333 West Jefferson Avenue, #230, Lakewood, CO 80235

LEED® CONSULTANT:
Institute for the Built Environmnet
Spruce Hall, Colorado State University, Fort Collins, CO 80523

GENERAL DESCRIPTION

SITE: 13.63 Acres.
NUMBER OF BUILDINGS: One; 26 classrooms.
BUILDING SIZE: First floor, 46,430; second floor, 16,261; total, 62,691 square feet.
BUILDING HEIGHT: First floor, 9'4"; second floor, 9'4"; floor to floor, 13'4"; total, 28'.
BASIC CONSTRUCTION TYPE: New.
FOUNDATION: Cast-in-place, slab-on-grade.

EXTERIOR WALLS: CMU, curtainwall.
ROOF: Built-up, metal, membrane.
FLOOR: Concrete.
INTERIOR WALLS: CMU, metal stud drywall.

BETHKE ELEMENTARY SCHOOL

Construction Period: Apr 2007 to June 2008   Total Square Feet: 62,691

C.S.I. Divisions (1 through 16)			COST	% OF COST	SQ.FT. COST	SPECIFICATIONS
		PROCUREMENT & CONT. REQ.	446,499	5.24	7.12	Solicitation, instructions for procurement, available information, procurement forms & supplements, contracting forms & supplements, project forms, condition of the contract, revisions, clarifications, & modifications.
1.	1.	GENERAL REQUIREMENTS	667,469	7.83	10.65	Summary, price & payment, administrative, quality requirements, temporary facilities & controls, product requirements, execution & closeout, performance.
3.	3.	CONCRETE	412,283	4.84	6.58	Forming & accessories, reinforcing, cast-in-place, precast.
4.	4.	MASONRY	624,315	7.32	9.96	Unit, stone assemblies.
5.	5.	METALS	944,553	11.08	15.07	Structural metal framing, joists, decking, cold-formed metal framing, fabrications, decorative.
6.	6.	WOOD/PLASTICS/COMPOSITE	370,000	4.34	5.90	Rough carpentry, finish carpentry, architectural woodwork.
7.	7.	THERMAL & MOIST. PROTECT	578,474	6.79	9.23	Thermal protection, weather barriers, steep slope roofing, roofing & siding panels, membrane roofing, flashing & sheet metal, roof & wall specialties & accessories, joint protection.
8.	8.	OPENINGS	631,643	7.41	10.08	Doors & frames, specialty doors & frames, entrances, storefronts, curtainwalls, windows, roof windows & skylights, hardware, glazing, louvers & vents.
9.	9.	FINISHES	515,681	6.05	8.23	Plaster & gypsum board, tiling, ceilings, flooring, wall finishes, acoustic treatment, painting & coating.
10.	10.	SPECIALTIES	158,250	1.86	2.52	Information, interior, safety, exterior.
11.	11.	EQUIPMENT	232,991	2.73	3.72	Residential, foodservice, athletic & recreational.
12.	12.	FURNISHINGS	85,293	1.00	1.36	Casework, furnishings & accessories.
14.	14.	CONVEYING SYSTEMS	74,328	0.87	1.19	Elevators (1).
15.	21.	FIRE SUPRESSION	108,952	1.28	1.74	Water-based fire suppression system.
15.	22.	PLUMBING	612,900	7.19	9.78	Piping & pumps, equipment, fixtures, gas & vacuum systems for laboratory & healthcare facilities.
15.	23.	HVAC	969,200	11.37	15.46	Piping & pumps, air distribution, central heating, central cooling, central HVAC.
16.	24.	INTEGRATED AUTOMATION	183,900	2.16	2.93	Controls.
16.	26.	ELECTRICAL	552,191	6.47	8.80	Power generating & storing equipment, lighting.
16.	28.	COMMUNICATIONS	285,495	3.35	4.55	Structured cabling, data, voice, audio-video.
16.	28.	ELECTRONIC SAFETY & SECURITY	70,044	0.82	1.11	Access control & intrusion detection, surveillance, detection & alarm.
TOTAL BUILDING COST			**8,524,461**	**100%**	**$135.98**	
2.	31.	EARTHWORK	120,932			Earth moving, earthwork methods, excavation support & protection.
2.	32.	EXTERIOR IMPROVEMENTS	1,126,095			Bases, bollards, & paving, improvements, irrigation, planting.
2.	33.	UTILITIES	142,355			Water, sanitary, storm, electrical, communications.
TOTAL PROJECT COST			**9,913,843**	*(Excluding architectural and engineering fees)*		

UPDATED ESTIMATE TO JAN 2011: $ 176.70 SF

Bethke Elementary School
Timnath, Colorado

Architect: RB+B Architects, Inc.

Photos Courtesy of Paul Brokering Photography

I n May of 2000, the Poudre School District (PSD) sponsored a design competition for a Prototype three round K-6 elementary school. Along with traditional Education and Technical Specifications, the three competition finalists were given PSD's "Sustainable Design Guidelines". These Guidelines were the result of 18 months of a comprehensive participatory process involving facility personnel, educators, and administrators. Thus "total District buy-in" was paramount in a fundamental shift away from conventional design. Due to meticulous and prudent management of the Bond funds, PSD was able to fund a 4th elementary school in 2006. The result was Bethke Elementary School, opening in the Fall of 2008.

The school is designed on an East-West axis. Classrooms face North and South and employ "cool-daylighting(R)", a natural and artificial lighting strategy developed by the team's daylighting consultant. Indirect, evaporative cooling and heat-recovery wheels, along with a very tight building envelope will drop the building's energy use down to well below 1/3 that of a typical Colorado school.

An array of building products made from recycled materials are used throughout the building. Outside, landscape features were strategically placed along with bio-swales, low water use plantings, and alternative paving. All of this was accomplished within a very sustainable cost.

Classrooms filled with natural light have been shown repeatedly to not only improve student performance but also enhance student and staff attitudes and reduce absenteeism.

Sustainable design features include:

- Daylighting - Natural light enhances student learning and creates invigorating indoor spaces.
- Exterior Insulation - Upgraded insulation levels in exterior walls and the roof reduce heat gain in summer and heat loss in winter, resulting in lower utility costs. A closed-cell polyurethane spray insulation was applied to all exterior walls to provide insulation, as well as act as the vapor retarder and rain screen.
- Reduced Cooling Load - Less electrical lighting reduces the annual cooling load. Less air conditioning is needed to maintain the comfort level and energy use is reduced.
- Reduced HVAC Equipment - A lower cooling load allows for the installation of smaller HVAC cooling systems.
- Displacement Ventilation - Displacement ventilation supplies air to a space near the floor level with very low velocity. This configuration improves air quality, helps to minimize air movement noise levels and saves energy.

Over half the classrooms in this prototype are able to achieve required light levels without any artificial lighting for most of the school day (the remainder only require about 30% supplemental lighting). Bright, inviting, vibrant spaces are spaces where students want to be.

LEED® FOR SCHOOLS - GOLD

COST PER SQUARE FOOT $ 176.70

ARCHITECT
Cornerstone Design-Architects
48-50 West Chestnut Street, #400, Lancaster, PA 17603
www.cornerstonedesign.com

FILE UNDER
EDUCATIONAL
Lancaster, Pennsylvania

CONSTRUCTION TEAM

STRUCTURAL ENGINEER:
    Greenebaum Structures, P.C.
    26-28 Market Square, #2, Manheim, PA 17545

MECHANICAL ENGINEER:
    Accu-Aire Mechanical Services, Inc.
    3545 Marietta Avenue, Lancaster, PA 17601

PLUMBING ENGINEER:
    Haller Enterprises, Inc.
    212 Bucky Drive, Lititz, PA 17543

ELECTRICAL ENGINEER:
    Mast Electrics, Inc.
    16 Holly Drive, Leola, PA 17540

GENERAL CONTRACTOR & COST ESTIMATOR
    High Construction Company
    1853 William Penn Way, Lancaster, PA 17601

GENERAL DESCRIPTION

SITE: 4.53 Acres.
NUMBER OF BUILDINGS: One; school with 15 classrooms.
BUILDING SIZE: First floor, 13,480; second floor, 12,324; third floor, 12,149; total, 37,953 square feet.
BUILDING HEIGHT: First floor, 13'4"; second floor, 13'4"; third floor, 13'4"; total, 42'6".
BASIC CONSTRUCTION TYPE: New/Steel Frame/IIB.
FOUNDATION: Cast-in-place.
EXTERIOR WALLS: CMU, brick, curtainwall, EIFS.
ROOF: Metal, membrane.
FLOORS: Concrete.
INTERIOR WALLS: CMU, metal stud drywall.

RUTT ACADEMY CENTER
Construction Period: Oct 2007 to Oct 2008 • Total Square Feet: 37,953

	C.S.I. Divisions (1 through 16)	COST	% OF COST	SQ.FT. COST	SPECIFICATIONS
1.	1. GENERAL REQUIREMENTS	968,928	14.45	25.53	Summary, price & payment procedures, administrative requirements, quality requirements, temporary facilities & controls, product requirements, execution & closeout requirements, performance requirements.
3.	3. CONCRETE	281,437	4.20	7.42	Forming & accessories, reinforcing, cast-in-place, cast decks & underlayment, grouting, cutting & boring.
4.	4. MASONRY	982,671	14.65	25.89	Unit, stone assemblies.
5.	5. METALS	753,966	11.24	19.87	Structural metal framing, joists, decking, cold-formed metal framing, fabrications, decorative metal.
6.	6. WOOD/PLASTICS/COMPOSITE	193,564	2.89	5.10	Rough carpentry, finish carpentry, architectural woodwork.
7.	7. THERMAL & MOIST. PROTECT	302,635	4.51	7.97	Dampproofing & waterproofing, thermal protection, weather barriers, roofing & siding panels, membrane roofing, fire & smoke protection, joint protection.
8.	8. OPENINGS	615,918	9.19	16.23	Doors & frames, specialty doors & frames, entrances, storefronts, & curtainwalls, windows, hardware, glazing, louvers & vents.
9.	9. FINISHES	488,408	7.28	12.87	Plaster & gypsum board, tiling, ceilings, flooring, painting & coating.
10.	10. SPECIALTIES	37,985	0.57	1.00	Interior, storage, other.
14.	14. CONVEYING SYSTEMS	61,100	0.91	1.61	Elevators (1).
	FIRE SUPRESSION	120,237	1.79	3.17	Water-based fire-suppression systems, fire-extinguishing systems, fire pumps.
	PLUMBING	251,155	3.75	6.62	Piping & pumps, equipment, fixtures, pool & fountain plumbing systems, gas & vacuum systems for laboratory & healthcare facilities.
15.	15. HVAC	894,030	13.33	23.56	Piping & pumps, air distribution, air cleaning devices, central heating, central cooling, central HVAC.
16.	26. ELECTRICAL	753,526	11.24	19.84	Medium-voltage distribution, low-voltage transmission, electrical & cathodic protection.
	TOTAL BUILDING COST	**6,705,560**	**100%**	**$176.68**	
2.	2. EXISTING CONDITIONS	730,700			Assessment, subsurface investigation, demolition & structure moving, site remediation.
	TOTAL PROJECT COST	**7,436,260**	*(Excluding architectural and engineering fees)*		

UPDATED ESTIMATE TO JAN 2011: $219.42 SF

DCD Subscribers: Access this case study and hundreds more for instant date and location calculations at www.dcd.com.

Rutt Academic Center, Lancaster Mennonite School
Lancaster, Pennsylvania

Architect: Cornerstone Design-Architects

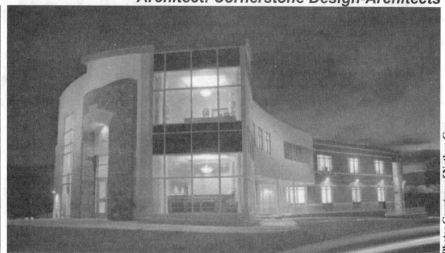

Photos Courtesy of Nathan Cox

Completed in October of 2008, the Lancaster Mennonite School Rutt Academic Center is a project that evolved from its original inception in January of 2002. The six year process of planning, strategizing, and fund-raising brought to fruition a building that integrates both classroom spaces for students and the central offices location for the Lancaster Mennonite School System. This facility contains fifteen classrooms including areas of study such as family and consumer sciences, business, math, and science.

Designed by Cornerstone Design-Architects based in Lancaster, Pennsylvania, this facility was intended as the gateway to campus and serves as a security barrier for the school. The main entrance to the building is located on the second floor and a long, winding entrance ramp guides visitors from the main northern parking lot through the academic center to the existing Fine Arts Center located just south of the new facility. This ramp guides occupants through an atrium space flooded with natural light and also serves as a community space for both students and visitors.

The exterior of the building was designed to respond to the existing architectural style on campus, but yet was designed to look towards the future of building construction on campus. Materials such as red brick, limestone, and decorative concrete masonry retain the traditional elements of the campus, but the curved and flowing facades of lightweight materials and glass curtain walls look toward the future.

While this project was not in pursuit of LEED® certification, many green and sustainable attributes were adopted within the design of the facility. Lancaster Mennonite School wants to instill a sense of environmental responsibility and education into their students for future generations within the community. This facility will serve as an educational tool for many students for decades to come. Some of the sustainable features include a geothermal heating and cooling system coupled with forty-two wells which are drilled to a depth of three-hundred fifty feet each. Energy-recovery units were installed to maximize the efficiency of outside

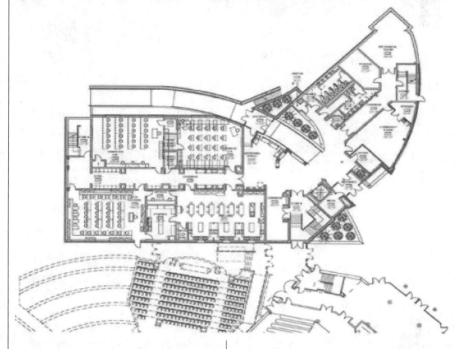

air intake and exhaust. A green roof is installed on a third floor patio area adjacent to the biology science classrooms. Students are able to learn more about plant species and sustainability in this learning environment. Along with increased insulation values within the floor, roof, and wall structures, radiant heating is installed in the concrete floor of the atrium space.

The project has been considered a success by Lancaster Mennonite School and their staff, board members, and students. Students now have more inspiring spaces to learn, communicate, socialize, and grow both spiritually and educationally

COST PER SQUARE FOOT $ 219.42

ARCHITECT
Turner Partners Architecture, LP
333 Cypress Run, #350, Houston, TX 77094
www.turnerpartnersarchitecture.com

CONSTRUCTION TEAM

STRUCTURAL ENGINEER:
Haynes Whaley Associates, Inc.
2000 W. Sam Houston Parkway South, #1800, Houston, TX 77042
GENERAL CONTRACTOR:
Fretz Construction Company
6301 Long Drive, Houston, TX 77087
ELECTRICAL & MECHANICAL ENGINEER:
Holste & Associates, Inc.
6671 Southwest Freeway, #850, Houston, TX 77074 Project
LEED® CONSULTANT:
Institute for the Built Environmnet
Spruce Hall, Colorado State University, Fort Collins, CO 80523

GENERAL DESCRIPTION

SITE: 16 Acres.
NUMBER OF BUILDINGS: One.
BUILDING SIZE: First floor, 12,000; second floor, 12,000; each additional floor, 12,000; total, 36,000 square feet.
BUILDING HEIGHT: First floor, 13'4"; second floor, 13'4"; each additional floor, 13'4"; total, 40'.
BASIC CONSTRUCTION TYPE: Structural Steel/New.
FOUNDATION: Slab-on-grade.
EXTERIOR WALLS: Brick, curtainwall, cast stone, metal panels.
ROOF: Built-up. **FLOORS:** Concrete, steel composite.
INTERIOR WALLS: Metal stud drywall, brick.

1st Floor Plan

Site Plan

CENTER FOR THE SCIENCES
Construction Period: June 2006 to Aug 2007 Total Square Feet: 36,000

C.S.I. Divisions (1 through 16)			COST	% OF COST	SQ.FT. COST	SPECIFICATIONS
		PROCUREMENT & CONT. REQ.	710,408	9.11	19.73	Conditions of the contract.
1.	1.	GENERAL REQUIREMENTS	4,306	0.06	0.12	—
3.	3.	CONCRETE	340,494	4.37	9.46	Forming & accessories, reinforcing, cast-in-place, grouting.
4.	4.	MASONRY	283,850	3.64	7.88	Unit, stone assemblies.
5.	5.	METALS	823,460	10.56	22.87	Structural metal framing, decking, cold-formed metal framing, fabrications, decorative metal.
6.	6.	WOOD/PLASTICS/COMPOSITE	440,937	5.65	12.25	Architectural woodwork.
7.	7.	THERMAL & MOIST. PROTECT	529,726	6.79	14.72	Dampproofing & waterproofing, thermal protection, roofing & siding panels, membrane roofing, flashing & sheet metal, roof & wall specialties & accessories, fire & smoke protection, joint protection.
8.	8.	OPENINGS	879,405	11.28	24.43	Doors & frames, specialty doors & frames, entrances, storefronts, & curtainwalls, windows, roof windows & skylights, hardware, glazing.
9.	9.	FINISHES	920,911	11.81	25.58	Plaster & gypsum board, tiling, ceilings, flooring, wall finishes, acoustic treatment, painting & coating.
10.	10.	SPECIALTIES	101,522	1.30	2.82	Information, interior.
11.	11.	EQUIPMENT	165,000	2.12	4.58	Educational & scientific.
12.	12.	FURNISHINGS	6,414	0.08	0.18	Casework.
14.	14.	CONVEYING SYSTEMS	46,705	0.60	1.30	Elevators (1).
15.	21.	FIRE SUPRESSION	67,380	0.86	1.87	Water-based fire-suppression systems.
15.	22.	PLUMBING	767,235	9.84	21.31	Piping & pumps, equipment, fixtures, gas & vacuum systems for laboratory & healthcare facilities.
15.	23.	HVAC	1,030,250	13.20	28.62	Piping & pumps, air distribution, air cleaning devices, central heating HVAC equipment, central cooling equipment, central HVAC equipment, decentralized HVAC equipment.
16.	26.	ELECTRICAL	681,100	8.73	18.92	Medium-voltage distribution, low-voltage transmission, electrical & cathodic protection, lighting.
TOTAL BUILDING COST			**7,799,103**	**100%**	**$216.64**	
2.	2.	EXISTING CONDITIONS	49,823			Demolition & structure moving, site remediation.
2.	31.	EARTHWORK	108,386			Site clearing, earthmoving, earthwork methods.
2.	32.	EXTERIOR IMPROVEMENTS	37,008			Bases, bollards, & paving, improvements, irrigation, planting.
TOTAL PROJECT COST			**7,994,320**	*(Excluding architectural and engineering fees)*		

UPDATED ESTIMATE TO JAN 2011: $ 299.12 SF

St. Agnes Academy Center for the Sciences and Student Services
Houston, Texas

Architect: Turner Partners Architecture, LP

Located in Houston, Texas, St. Agnes Academy ranks as one of the city's most respected college preparatory schools, devoted exclusively to educating young women.

The new Center for the Sciences and Student Services is the first phase of a comprehensive Master Plan for upgrading the campus instructional spaces and clarifying the overall circulation pattern. This facility not only includes new laboratory and lecture facilities for students; student services and conference/meeting spaces are included to support the entire campus. All are focused on a new student entrance, the three-story rotunda shaped to evoke a lantern signifying St. Agnes Academy's role as a light for the spiritual and intellectual growth of its students.

The building is a three level structure including a 12,300-square-foot ground floor on which new offices for Student Services, as well as conference facilities for up to 250 persons, are situated. Stacked above are two additional floors housing eight state-of-the-art science laboratory and lecture rooms encompassing 23,700 square feet of space.

The St. Agnes Campus is located on approximately 16 acres of land in an older urban neighborhood that is experiencing significant redevelopment. Although facilities are shared with a boys' college preparatory school on an adjacent site, land is at a premium for this 800-student institution.

A primary objective for the structure was to contribute to the campus as a compatible and integrated component, yet establish enough of an individual identity to provide orientation for visitors and students. Although the exterior materials selected are similar or identical to the existing buildings, the composition of form and mass for the Center varies as an expression of the functions within, and of its role as the new main entrance for students. Toward this end, the exterior elevations of the building are composed as layers of opaque brick and cast stone panels; glass planes are also incorporated in the overall composition, providing approximately 45% building transparency.

The rotunda "lantern" element was built to contrast with the orthogonal regularity of the remainder of the existing campus. Its highly visible entry is a focal point to pedestrian traffic, as well as to vehicles approaching by means of the main entrance drive.

The campus is interlaced with a network of landscaped courtyards. This provided an opportunity to create a learning environment rich in natural light, as well as in views of nature and the daily interactive activities around the building. Thus, the building steps back at the ground level, increasing view opportunities from interior spaces and bringing daylight deep into the office floor plates. Above, the laboratories are awash in natural light that nonetheless is controlled through use of exterior sunshades and interior blinds.

Photos Courtesy of Aker/Zvonkovic Photography

MANUFACTURERS/SUPPLIERS	
DIV 07:	*Aluminum Composite Panels:* **Alucobond**®; *Column Covers:* **Fry Reglet;** *Built-Up Roofings:* **Johns Manville.**
DIV 08:	*Curtainwall, Entrances & Storefronts:* **Vistawall.** *Glazing:* **Guardian.**
DIV 09:	*Flooring:* **Armstrong;** *Design Weave, Base:* **Roppe;** *Acoustical Treatment:* **Armstrong;**
DIV 26:	*Lighting:* **Cooper, HADCO, Hydrel, Atlantic.**

COST PER SQUARE FOOT $ 299.12

ARCHITECT

RDG Planning & Design
301 Grand Avenue, Des Moines, IA 50309
www.rdgusa.com

FILE UNDER

EDUCATIONAL
Ames, Iowa

CONSTRUCTION TEAM

STRUCTURAL ENGINEER:
 Charles Saul Engineering
 4308 University Avenue, Des Moines, IA 50311
GENERAL CONTRACTOR:
 Miron Construction Company, Inc.
 9440 Atlantic Drive S.W. #3, Cedar Rapids, IA 52404
ELECTRICAL & MECHANICAL ENGINEER:
 MEP Associates, LLC
 2900 43rd Street, #100, Rochester, MN 55901
COST ESTIMATOR:
 Stecker-Harmsen, Inc.
 510 S. 17th Street, #110, Ames, IA 50010

GENERAL DESCRIPTION

SITE: .9 Acres. **NUMBER OF BUILDINGS:** One.
BUILDING SIZE: First floor, 11,000; second floor, 11,000; total, 22,000 square feet.
BUILDING HEIGHT: First floor, 15'; second floor, 15'; light monitor, 6'; total, 36'.
BASIC CONSTRUCTION TYPE: Addition/Structural steel/Cast-in-place.
FOUNDATION: Cast-in-place, slab-on-grade, re-used existing concrete site walls.

EXTERIOR WALLS: CMU, curtainwall, metal panels.
ROOF: Membrane, vegetated. **FLOORS:** Concrete, polished concrete.
INTERIOR WALLS: CMU, metal stud drywall, cast concrete exposed.

KING PAVILION IOWA STATE UNIVERSITY, COLLEGE OF DESIGN
Construction Period: Aug 2008 to July 2009 Total Square Feet: 22,000

C.S.I. Divisions (1 through 16)			COST	% OF COST	SQ.FT. COST	SPECIFICATIONS
		PROCUREMENT & CONT. REQ.	44,324	1.11	2.01	Instructions for procurement, available information, procurement forms & supplements, contracting forms & supplements, project forms, conditions of the contract, revisions, clarifications, & modifications.
1.	1.	GENERAL REQUIREMENTS	209,607	5.23	9.53	Summary, administrative requirements, quality requirements, temporary facilities & controls, product requirements, execution & closeout, performance.
3.	3.	CONCRETE	421,825	10.53	19.17	Forming & accessories, reinforcing, cast-in-place, precast, cast decks & underlayment, grouting, cutting & boring (Concrete breakdown: cubic yards foundation, 90; cubic yards walls, 190; cubic yards floors, 230).
4.	4.	MASONRY	14,124	0.35	0.64	Unit.
5.	5.	METALS	711,400	17.76	32.34	Structural metal framing, joists, decking, cold-formed metal framing, fabrications.
6.	6.	WOOD/PLASTICS/COMPOSITE	55,000	1.37	2.50	Rough carpentry, finish carpentry, architectural woodwork.
7.	7.	THERMAL & MOIST. PROTECT	658,196	16.43	29.92	Dampproofing & waterproofing, thermal protection, weather barriers, roofing & siding panels, membrane roofing, flashing & sheet metal, roof & wall specialties & accessories, joint protection.
8.	8.	OPENINGS	302,802	7.56	13.76	Doors & frames, entrances, storefronts, & curtainwalls, windows, hardware, glazing.
9.	9.	FINISHES	168,059	4.20	7.64	Plaster & gypsum board, tiling, ceilings, flooring, painting & coating.
10.	10.	SPECIALTIES	28,420	0.71	1.29	Interior.
15.	21.	FIRE SUPRESSION	5,500	0.14	0.25	Water-based fire-suppression systems.
15.	22.	PLUMBING	314,107	7.84	14.28	Piping & pumps, equipment, fixtures.
15.	23.	HVAC	424,725	10.60	19.31	Piping & pumps, air distribution, air cleaning devices, decentralized HVAC equipment.
16.	25.	INTEGRATED AUTOMATION	97,000	2.42	4.41	Instrumentation & terminal devices, facility controls, control sequences.
16.	26.	ELECTRICAL	376,032	9.39	17.09	Medium-voltage distribution, low-voltage transmission, lighting.
16.	27.	COMMUNICATIONS	26,127	0.65	1.19	Structured cabling, data, voice, audio-video.
16.	28.	ELECTRONIC SAFETY & SECURITY	148,841	3.71	6.76	Access control & intrusion detection, detection & alarm, monitoring & control.
TOTAL BUILDING COST			**4,006,089**	**100%**	**$182.09**	
2.	2.	EXISTING CONDITIONS	75,000			Demolition & structure moving.
2.	31.	EARTHWORK	118,100			Site clearing, earth moving, earthwork methods, excavation support & protection.
2.	32.	EXTERIOR IMPROVEMENTS	48,211			Bases, bollards, & paving, improvements.
2.	33.	UTILITIES	185,000			Water, sanitary sewer, storm drainage.
TOTAL PROJECT COST			**4,432,400**	*(Excluding architectural and engineering fees)*		

UPDATED ESTIMATE TO JAN 2011: $ 235.30 SF

DCD Subscribers: Access this case study and hundreds more for instant date and location calculations at www.dcd.com.

King Pavilion
Iowa State University, College of Design
Ames, Iowa

Architect: RDG Planning & Design

The project challenge was to create a relatively small addition to the rear of the existing six-story College of Design facility. The building program includes housing Iowa State University's Freshman Core Design program and second year Architecture, Landscape Architecture, and Interior Design studio classrooms that would be able to facilitate student interaction, and exploration of ideas.

The project is sustainably designed and has achieved a LEED® Platinum certification. One of the project's main goals is to create a "living laboratory" that will demonstrate sustainability to the students learning in the facility.

The design solution presented is of a "Pavilion" design, pulled slightly away from the main building and organized as a two-story form. Open studio classroom environments are efficiently organized around a central core space that function as flexible experimentation space. The center volume allows natural daylight to penetrate deep into the center of the building to the lower level. Clerestory and full height corner perimeter windows captures daylight into each studio classroom.

The building will essentially require no electric lighting during daytime hours. The use of a vegetated roof on the facility reduces heat island affect and storm water management needs on the property.

The building was built on existing concrete site walls that became the foundation walls for the addition. Concrete foundation walls, footings, and floor slab are insulated with 2-inches of rigid insulation. As the building is depressed into the site, a soil berm is also providing additional passive insulation for two sides of the building. Concrete was chosen as a material because it is a regional material, and it also provided recycled content. Floor slabs were polished and left exposed as the interior finishes for increased life cycle value to the owner, lower maintenance requirements, and no VOC's to impact IAQ.

Photos Courtesy of Cameron Campbell, Photographer; RDG Planning & Design

FIRST FLOOR PLAN

GROUND FLOOR PLAN

A ceramic frit and metal sunscreen were implemented to protect glazing with a west exposure helping to reduce solar heat gain. The rest of the facade was constructed of 2-inch insulated metal panels over continuous rigid insulation with building wrap, fastened to 6-inch metal studs. The stud cavity was filled with 6-inch cotton bat insulation. The cotton bat insulation is 100% recycled cotton material and is also a rapidly renewable material. The building's envelope had a significant contribution to daylight and view, energy efficiency, as well as recycled content percentage.

LEED® PLATINUM		
MANUFACTURERS/SUPPLIERS		
DIV 07:	*Roofing:* **Sarnafil.**	
DIV 08:	*Entrances & Storefronts:* **Kawneer;**	
	Windows, Daylighting: **Wausau;**	
	Glazing: **PPG Solarban® 60.**	
DIV 09:	*Polished Concrete:* **Retroplate;**	
	Tile: **American Olean;**	
DIV 16:	*Architectural Lighting:* **Edison Price.**	

COST PER SQUARE FOOT $ 235.30

ARCHITECT
HOLABIRD & ROOT LLC
300 West Adams Street
Chicago, IL 60606
www.holabird.com

FILE UNDER
EDUCATIONAL
PELLA, Iowa

CONSTRUCTION TEAM

GENERAL CONTRACTOR: The Weitz Company
5901 Thorton Avenue, Des Moines, IA 50321
STRUCTURAL ENGINEER: Holabird & Root LLC
300 West Adams Street, Chicago, IL 60606
ELECTRICAL & MECHANICAL ENGINEER:
Holabird & Root LLC
300 West Adams Street, Chicago, IL 60606
LANDSCAPE ARCHITECT: RGD Planning & Design
301 Grand Avenue, Des Moines, IA 50309
COST ESTIMATOR: The Weitz Company
5901 Thorton Avenue, Des Moines, IA 50321

GENERAL DESCRIPTION

SITE: 3 acres. **NUMBER OF BUILDINGS:** One.
BUILDING SIZE: First floor, 34,497; second floor, 29,446;
each additional floor, 7,253; total, 71,196 square feet.
BUILDING HEIGHT: First floor, 12'; second floor, 14'-8";
Penthouse, 18'4"; total, 45'.
BASIC CONSTRUCTION TYPE: Addition/Renovation.
FOUNDATION: Concrete.
EXTERIOR WALLS: Masonry.
ROOF: Single Ply TPO membrane.
FLOORS: Tile, terrazzo, resilient, carpet.
INTERIOR WALLS: Gypsum, metal studs.

VERMER SCIENCE CENTER, CENTRAL COLLEGE
Construction Period: June 2001 to Aug 2003 • Total Square Feet: 71,196

C.S.I. Divisions (1 through 16)	COST	% OF COST	SQ.FT. COST		SPECIFICATIONS
BIDDING REQUIREMENTS	ó	ó	ó		Included in Div. 1.
1. GENERAL REQUIREMENTS	2,237,815	13.44	31.43	1	Summary of work, allowances, measurement & payment,alternates/alternatives, modification procedures, coordination,field engineering, regulatory requirements, identification systems,references, special project procedures, project meetings,submittals, quality control, construction facilities & temporarycontrols, material & equipment, contract closeout.3.
3. CONCRETE	1,784,805	10.72	25.07	3	Formwork, reinforcement, accessories, cast-in-place, curing,cementitious decks & toppings, grout.
4. MASONRY	491,604	2.95	6.90	4	Masonry & grout, accessories, unit, stone, masonry restoration& cleaning.
5. METALS	882,475	5.30	12.40	5	Materials, coatings, fastening, structural metal framing, joists,decking, cold formed metal framing, fabrications, ornamental.
6. WOOD & PLASTICS	586,082	3.52	8.23	6	Fasteners & adhesives, rough carpentry, finished carpentry,wood treatment, architectural woodwork.
7. THERMAL & MOIST. PROTECTION	271,615	1.63	3.82	7	Waterproofing, dampproofing, insulation, EIFS, firestopping,membrane roofing, traffic coatings, flashing & sheet metal, roofspecialties & accessories, joint sealers.
8. DOORS AND WINDOWS	1,655,957	9.95	23.26	8	Metal doors & frames, wood & plastic doors, special doors,entrances & storefronts, metal windows, hardware, glazing,glazed curtainwalls.
9. FINISHES	1,634,760	9.82	22.96	9	Metal support systems, gypsum, tile, terrazzo, acousticaltreatment, special wall surfaces, resilient flooring, carpet, specialflooring, painting, wall coverings.
10. SPECIALTIES	110,879	0.67	1.56	10	Visual display board, compartments & cubicles, wall & cornerguards, fire protection, toilet & bath accessories.
11. EQUIPMENT	727,804	4.37	10.22	11	Lab casework, fume hood, equipment.
12. FURNISHINGS	15,650	0.09	0.22	12	Window treatment, rugs & mats.
13. SPECIAL CONSTRUCTIONS	338,026	2.03	4.75	13	Greenhouse, fountain.
14. CONVEYING SYSTEMS	51,250	0.31	0.72	14	Elevators (1).
15. MECHANICAL	3,774,106	22.67	53.01	15	Basic materials & methods, insulation, fire protection, plumbing, HVAC, heat transfer, air distribution controls, chiller plant.
16. ELECTRICAL	2,086,221	12.53	29.03	16	Basic materials & methods, medium voltage distribution, service& distribution, lighting, special systems, communications.
TOTAL BUILDING COST	**16,649,049**	**100%**	**$233.85**		
2. SITE WORK LANDSCAPING & OFFSITE WORK	786,923			2	Demolition, preparation, dewatering, shoring & underpinning,excavation support systems, earthwork, paving & surfacing,improvements, landscaping.
TOTAL PROJECT COST	**17,435,972**	*(Excluding architectural and engineering fees)*			

UPDATED ESTIMATE TO JAN 2011: $435.91

Vermeer Science Center, Central College
Pella, Iowa

Architect: Holabird & Root LLC

Holabird & Root renovated and expanded the Vermeer Science Center. Built in 1978, the Vermeer Science Center needed significant upgrades to mechanical systems, computer network connections, classrooms and laboratory spaces. The new wraparound addition houses biology, chemistry, physics, pre-engineering, environmental studies, mathematics and computer science disciplines. Advanced student and faculty research labs are grouped together by physical requirements such as the concentration of fume hoods or the need for proximity to service areas or research greenhouse. All of the faculty offices are clustered together on the north side of the addition to encourage faculty interaction. The prominent corners of the addition feature common spaces used by all of the departments such as study rooms, computer labs and a multipurpose, 100-seat auditorium. Other spaces include a new greenhouse, museum display areas, and a special collections science library and reading room.

Holabird & Root regards environmental stewardship as an inherent part of design. The architect's team of Leadership in Energy and Environmental Design™ (LEED®) accredited architects and Certified Indoor Air Quality Professional (CIAQP) engineer worked to establish the Vermeer Science Center as the first LEED®-certified facility in Iowa. The project received Silver level of certification through collaboration among all parties involved.

Fifty percent of the Vermeer Science Center's site area has been restored from impervious surfaces to adaptive vegetation, including the conversion of an unused tennis court into a grass area.

A photovoltaic array, which was donated to the College, powers a fountain. The water level of the fountain fluctuates with the intensity of the sun. A concrete cistern near the greenhouse collects rainwater from the existing building roof surfaces that is in turn used to water plants in the greenhouse.

Ninety percent of the regularly occupied spaces permit views to the outdoors.

The building is 60% more efficient than the ASHRAE 90.1 base building model. Due to the efficiency of the system, there will be an annual reduction of 2,262,800 lbs. of CO_2.

Photos Courtesy of Dale Photographics

Next the fresh air is mixed with the non-lab space air and ventilates the lab spaces. This method allows the laboratory ventilation air to be used to ventilate the entire building, saving the ventilation air for the offices, lecture halls, and classrooms.

The fume hood exhaust system is manifolded, allowing energy recovery coils in the exhaust air stream to reclaim energy used to precondition the makeup air for the entire building. A payback analysis on the heat recovery system determined a payback of 3.5 years. A specialized control sequence allows for decommissioning of fume hoods and further reduces the minimum requirement per hood during the summer.

A specialized fan coil unit cools the sensible solar load projected into the west labs. This resulted in a reduction in chiller tonnage consumed while allowing natural light to enhance the indoor lab environment.

A controller receives a signal from the space occupancy sensor to ensure ventilation occurs only when the space is occupied. Carbon dioxide monitors located in the return ducting and in the most remote classroom help ensure the building is performing above the minimum ventilation requirements.

LEED® PLATINUM MANUFACTURERS/SUPPLIERS	
DIV 07:	*Membrane:* Firestone.
DIV 08:	*Sectional Overhead Door:* Richard Wilcox; *Counter Fire Shutter:* Raynor *Steel Doors & Frames:* Curries; *Wood Doors:* Marshfield; *Hinges:* McKinney; *Door Closers:* LCN; *Exit Devices:* Von Duprin.
DIV 09:	*Paint:* Diamond Vogel Paints.
DIV 14:	*Elevator:* Otis.

COST PER SQUARE FOOT $ 435.91

ARCHITECT
HOLLIS + MILLER ARCHITECTS
8205 W. 108th Terrace, #200, Overland Park, KS 66210
www.hollisandmiller.com

CONSTRUCTION TEAM

STRUCTURAL ENGINEER: Bob D. Campbell and Company
4338 Belleview, Kansas City, MO 64111
CONSTRUCTION MANAGER:
JE Dunn Construction Company
901 Charlotte, Kansas City, MO 64106
MECHANICAL, ELECTRICAL & TECHNOLOGY:
Henderson Engineers, Inc.
8325 Lenexa Drive, Lenexa, KS 66214
COST ESTIMATOR:Construction Management Resources
5201 Johnson Drive, #201, Mission, KS 66205

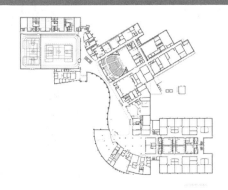

GENERAL DESCRIPTION

SITE: 80 acres. **NUMBER OF BUILDINGS:** One.
BUILDING SIZE: First floor, 167,471; second floor, 157,905; total, 325,376 square feet.
BUILDING HEIGHT: First floor, 14'; second floor, 15'4"; total, 29'4".
BASIC CONSTRUCTION TTYPE: Structural Steel/IIB.
FOUNDATION: Cast-in-place. **EXTERIOR WALLS:** CMU, brick, curtainwall.
ROOF: Membrane. **FLOORS:** Concrete.
INTERIOR WALLS: CMU, metal stud drywall.

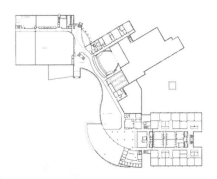

STALEY HIGH SCHOOL
Construction Period: May 2006 to May 2008 • Total Square Feet: 325,376

C.S.I. Divisions (1 through 16)			COST	% OF COST	SQ.FT. COST	SPECIFICATIONS
		PROCUREMENT & CONT. REQ.	2,870,065	4.71	8.82	Solicitation, instructions for procurement, available information, procurement forms & supplements, contracting forms & supplements, project forms, conditions of the contract, revisions, clarifications, & modifications.
1.	1.	GENERAL REQUIREMENTS	4,848,245	7.96	14.90	Summary, price & payment procedures, administrative requirements, quality requirements, temporary facilities & controls, product requirements, execution & closeout requirements, performance requirements.
3.	3.	CONCRETE	5,704,442	9.36	17.53	Forming & accessories, reinforcing, cast-in-place.
4.	4.	MASONRY	8,178,238	13.42	25.13	Unit.
5.	5.	METALS	5,443,702	8.93	16.73	Structural metal framing, joists, decking, cold-formed metal framing, fabrications.
6.	6.	WOOD/PLASTICS/COMPOSITE	2,418,404	3.97	7.43	Rough carpentry, finish carpentry.
7.	7.	THERMAL & MOIST. PROTECT	3,393,285	5.57	10.43	Membrane roofing, flashing & sheet metal, joint protection.
8.	8.	OPENINGS	2,371,960	3.89	7.29	Doors & frames, specialty doors & frames, windows, roof windows & skylights, hardware, glazing.
9.	9.	FINISHES	4,561,788	7.49	14.02	Plaster & gypsum board, ceilings, flooring, painting & coating.
10.	10.	SPECIALTIES	645,611	1.06	1.98	Information, interior, other.
11.	11.	EQUIPMENT	1,594,752	2.62	4.90	Foodservice, educational & scientific, athletic & recreational, other.
13.	13.	FURNISHINGS	760,255	1.25	2.34	Casework, multiple seating, other.
14.	14.	CONVEYING SYSTEMS	176,909	0.29	0.54	Elevator (1).
15.	21.	FIRE SUPRESSION	624,534	1.02	1.92	Water-based fire-suppression system.
15.	22.	PLUMBING	2,293,352	3.76	7.05	Piping & pumps, fixtures, gas & vacuum systems for laboratory & healthcare facilities.
15.	23.	HVAC	6,808,505	11.17	20.93	Piping & pumps, HVAC air distribution, decentralized HVAC equipment.
16.	26.	ELECTRICAL	7,128,135	11.70	21.91	Medium-voltage distribution, lighting.
16.	27.	COMMUNICATIONS	1,118,074	1.83	3.44	Data, audio-video.
TOTAL BUILDING COST			**60,940,256**	**100.00**	**$187.29**	
2.	2.	EXISTING CONDITIONS	40,000			Subsurface investigation.
2.	31.	EARTHWORK	4,652,640			Site clearing, earth moving, earthwork methods.
2.	32.	EXTERIOR IMPROVMENTS	4,132,536			Improvements.
2.	33.	UTILITIES	3,541,335			Wells, sanitary sewerage utilities, electrical.
TOTAL PROJECT COST			**73,306,767**	*(Excluding architectural and engineering fees)*		

*Cost of High School only. Total project cost $85,185,269 included a District Athletic Complex, roadway construction and site preparation for an adjacent elementary school.

UPDATED ESTIMATE TO JAN 2011: $249.42 SF

DCD Subscribers: Access this case study and hundreds more for instant date and location calculations at www.dcd.com.

Staley High School
Kansas City, Missouri

Architect: Hollis + Miller Architects

Photos Courtesy of Alistar Tutton Photography and Hollis + Miller Architects

S taley High School is the newest facility for the North Kansas City School District and is also Missouri's first school to be awarded LEED® Silver. The project team was led by Hollis + Miller Architects in collaboration with a variety of team members, including: JE Dunn Construction Company, construction manager; Henderson Engineers, mechanical, electrical and technology engineers; and, the North Kansas City School District, teachers, students and community. Hollis + Miller Architects' team was led by Partner-in-Charge Kirk Horner, AIA and Project Manager Larry Jordan, AIA.

This is the first new high school for the district in more than 35 years and it needed to accommodate significant changes in educational philosophies. The school board selected Hollis + Miller Architects to lead the effort and the design of the new high school facility. There were an array of challenging educational requirements: It needed to be a facility that would be more than just a high school for teenagers it needed to be a continuing education center for adult education accommodating satellite college courses; it needed to be a facility that would be responsive to its environment and was designed as a sustainable building; and flexible able to expand as the student base continues to grow.

Staley High School is sited on 80 acres - 48 contain "green space" with three ponds. Strategic site planning allowed minimal disruption or changes to the existing site features and topography. Except for one pond, all of the wetland areas were retained. A new larger pond was added. All of these features are planned as active functions with the high school's educational spaces, including: biology and environmental labs, walking/running/cross country trails and natural storm water filtration features. The building and paving components were carefully planned to maximize the percentage of undisturbed open space on the site and to retain the original drainage patterns.

Initially, the building has a 1,500 student capacity with the potential to be expanded in the future to 2,000. The facility includes a 750-seat auditorium, a 300-seat flex-theatre, a gymnasium capable of hosting four simultaneous basketball games and a 70,000 square foot great hall. The great hall is the unifying space for the entire facility and features a dynamically-shaped two-story

curvelinear space that streches from one end of the building to the other. Virtually all students and staff use this space several times daily, passing through between classes, breaking for lunch or hanging out before and after school. It's everyone's space.

The school opened in August 2008 and was immediately awarded LEED® Silver status. Through early discussions with the school board, the concept of a sustainable facility quickly became one of the fundamental goals of the project. The preparation and planning for LEED® certification involved the entire building team and all phases of design. In addition to involvement with the building and site design, decisions for sustainability also included construction sequences and methods as well as the development of the school's curriculum and educational programming. Sustainable design concepts are employed throughout the building. The most obvious direction was the effort to bring sunlight into the building. Shaded windows and curtainwalls are used throughout the exterior walls. Materials used in the building were selected for their recycled content, VOC content and the distance from the site that they were manufactured. Undoubtedly, the most energy efficient feature of the facility is the geo-thermal loop system that provides tempered water for a hydronic heating and cooling system. This system is calculated to use 42 percent less energy than a conventional system.

One of the significant results of the partnership led to the district's decision concerning green concepts. In the process of determining costs and expenses over the lifetime of the building, the district determined that the costs will provide a payback to the community and district in less than ten years.

The district now embraces sustainability concepts district-wide. The issue of sustainable practices is now even a part of each high school's required curriculum. The project was completed on-time and was approximately two percent under budget.

LEED® SILVER MANUFACTURERS/SUPPLIERS	
DIV 07:	*Membrane:* GAF.
DIV 08:	*Window, Curtainwall, Entrances & Storefronts::* YKK America.
DIV 09:	*Flooring:* Everlast, Desco, Armstrong, Mannington, Miliken.
DIV 14:	*Elevator::* Thyssen Krupp

COST PER SQUARE FOOT $249.42

ARCHITECT
Blackney Hayes Architects
150 S. Independence Mall West, #1200, Philadelphia, PA 19106
www.blackneyhayes.com

FILE UNDER
EDUCATIONAL
Wayne, Pennsylvania

CONSTRUCTION TEAM

CONSTRUCTION MANAGER
Reynolds Construction Management, Inc.
3300 North Third Street, Harrisburg, PA 17110
www.reynoldsconstruction.com

STRUCTURAL ENGINEER:
Pennoni Associates
2041 Avenue C, #100, Bethlehem, PA 18107

MEP ENGINEER:
Concord Engineering Group, Inc.520 South Burnt Mill
Road, Voorhees, NJ 08043

CIVIL ENGINEER:
Gilmore & Associates, Inc.
350 East Butler Avenue, New Britain, PA 18901

EDUCATION PLANNER:
Ingraham Dancu Associates
1265 Lakevue Drive, Butler, PA 16002

GENERAL DESCRIPTION

SITE: 10.421 acres (9.173 less right of way).

NUMBER OF BUILDINGS: One.

BUILDING SIZE: First floor, 65,141; second floor, 52,485, third floor, 41,580; fourth floor, 29,616; total, 188,822 square feet.

BUILDING HEIGHT: First floor, 14'; second, third, fourth, 13'4"; total, 54'.

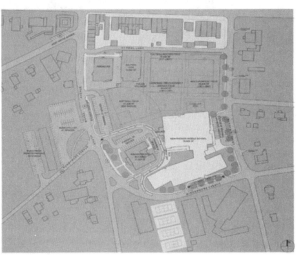

BASIC CONSTRUCTION TYPE: New/Composite structural steel.
FOUNDATION: Cast-in-place.
EXTERIOR WALLS: CMU, brick, limestone.
ROOF: Vegetated roof assembly, asphalt shingles, membrane.
FLOORS: Concrete. **INTERIOR WALLS:** CMU (Classrooms), metal stud drywall (Offices), movable partitions.

RADNOR MIDDLE SCHOOL
Construction Period: Jan 2006 to Sep 2007 • Total Square Feet: 188,822

C.S.I. Divisions (1 through 16)			COST	% OF COST	SQ.FT. COST	SPECIFICATIONS
		PROCUREMENT & CONT. REQ.	-	-	-	-
1.	1.	GENERAL REQUIREMENTS	-	-	-	Price & payment procedures, administrative requirements, quality requirements, temporary facilities & controls, product requirements, execution & closeout requirements, performance requirements (cost spread through General Trades).
3.	3.	CONCRETE	4,226,928	14.00	26.11	Cast-in-place (600 cubic yards foundation & walls, 3,700 cubic yards floors).
4.	4.	MASONRY	7,339,900	20.85	38.87	Unit, manufactured.
5.	5.	METALS	5,055,000	14.36	26.77	Structural metal framing, joists, decking, cold-formed metal framing, fabrications.
6.	6.	WOOD/PLASTICS/COMPOSITE	1,291,378	3.00	1.87	Rough carpentry, architectural woodwork.
7.	7.	THERMAL & MOIST. PROTECT	1,056,582	2.67	4.97	Dampproofing & waterproofing, thermal protection, steep sloop roofing, roofing & siding panels, membrane roofing, flashing & sheet metal, roof & wall specialties & accessories, fire & smoke protection, joint protection.
8.	8.	OPENINGS	1,526,174	5.67	10.57	Doors & frames, specialty doors & frames, entrances, storefronts, & curtain walls, windows, hardware, glazing, louvers & vents.
9.	9.	FINISHES	3,287,144	9.34	17.41	Plaster & gypsum board, tiling, ceilings, flooring, wall finishes, acoustic treatment, painting & coating.
10.	10.	SPECIALTIES	234,796	0.34	0.62	Information, interior, safety, storage, other.
11.	11.	EQUIPMENT	363,350	1.03	1.92	Foodservice, education & scientific, entertainment, athletic & recreational, other.
12.	12.	FURNISHINGS	511,823	1.45	2.71	Casework, furnishings & accessories, multiple seating.
14.	14.	CONVEYING SYSTEMS	117,398	0.33	0.62	Elevators 2 (1 passenger, 1 freight/passenger).
15.	21.	FIRE SUPRESSION	560,000	1.59	2.97	Water-based fire-suppression system, fire-extinguishing systems, fire pumps.
15.	22.	PLUMBING	1,188,400	3.38	6.29	Facility fuel systems, piping & pumps.
15.	23.	HVAC	4,924,200	13.99	26.08	Piping & pumps, equipment, fixtures, pool & fountain systems.
16.	26.	ELECTRICAL	2,857,780	8.12	15.13	Power generating & storing equipment, lighting.
16.	27.	COMMUNICATIONS	660,487	1.88	3.50	Structured cabling, data, voice.
TOTAL BUILDING COST			**35,201,340**	**100.00**	**$186.43**	
2.	31.	EARTHWORK	5,832,365			Site clearing, earth moving, earthwork methods, excavation support & protection, special foundations & load-bearing elements.
TOTAL PROJECT COST			**41,033,705**	*(Excluding architectural and engineering fees)*		

UPDATED ESTIMATE TO JAN 2011: $256.86 SF

DCD Subscribers: Access this case study and hundreds more for instant date and location calculations at www.dcd.com.

Radnor Middle School
Wayne, Pennsylvania

Architect: Blackney Hayes Architects

Radnor Township School District, located in southeastern Pennsylvania, serves the children of St. Davids and parts of Wayne, Rosemont, Bryn Mawr, Villanova, Ithan, Newtown Square, and Radnor. They have accepted the goal of inspiring and empowering students within the district to become lifelong learners. Until 2005 the District had been utilizing a middle school that was originally constructed in 1923 and the Administration and School Board recognized the need for a new facility. The District hired Reynolds Construction Management, Inc. to provide pre-construction and construction management services on a new state-of-the-art middle school.

Radnor Township School District asked Reynolds to provide assistance in developing their preliminary budget. Reynolds' team of electrical, mechanical, plumbing, architectural and structural estimators worked with the District to create a budget that could be used by the District as they started the selection of a design team and made decisions regarding the project's financing.

The design team, led by Blackney Hayes Architects of Philadelphia, began the design for the new middle school and Reynolds provided assistance to the District by ensuring that the design matched the original intent, program and budget. Reynolds also worked closely with the District and the design team to develop the correct bidding strategy for the project. The key consideration was to determine the appropriate number of bid packages for the project that would maximize participation and competition among bidders. Once the correct number of bid packages had been agreed to, Reynolds held a number of pre-bid meetings where they provided more in-depth information to interested contractors and ensured that the prospective bidders understood the bid documents and the construction schedule. This process led to more precise and lower bids from the contractors and provided Radnor Township School District the best possible price on their project.

Upon completion of the pre-construction and bidding phases, Reynolds provided the District with on-site management and supervision of the contractors. Through the direction and expertise of Reynolds, the Radnor Township School District received the

new middle school on schedule and within budget.

The middle school is designed and utilizes materials that are in-tune with the center of downtown. Wayne and the nearby residential neighborhoods. The new facility uses many of the latest technologies for lowering energy and operating costs, while improving indoor air quality and creating an optimal learning environment for the District's students. Even though the new building is 4-stories high, its scale was broken down through geometric shifts and use of the site's topography to decrease the apparent height at street level.

Environmentally sustainable design and construction have been a hallmark of recent projects at Radnor Township School District and the new Radnor Middle School

is no exception. The building includes many many and installed by Magco Inc., a Tecta America Company, recycled materials, heat, motion and light sensors and other features that will help the District achieve their goal of a gold level certification from the US Green Building Council's LEED® program.

Photos Courtesy of Reynolds Construction Management, Inc.

LEED® GOLD		
MANUFACTURERS/SUPPLIERS		
DIV 04:	*Brick* McAvoy Brick Company; *Architectural Block:* Beavertown Block Company: *Gray Block:* Fizzano Brothers.	
DIV 07:	*Membrane & Vegetated Roof:* Magco Inc., a Tecta America Company.	
DIV 08:	*Windows:* Eagle; *Curtainwall, Entrances:* Kawneer.	
DIV 09:	*Carpet:* Collins & Aikman; *VCT:* Armstrong; Expanko Cork.	
DIV 14:	*Elevators:* Thyssen Krupp.	

COST PER SQUARE FOOT $256.86

ARCHITECT
RENAISSANCE ARCHITECTS, P.C.
48 South 14th Street
Pittsburg, PA 15203
www.r3a.com

FILE UNDER
INDUSTRIAL
Munhall, Pennsylvania

CONSTRUCTION TEAM

DEVELOPER & GENERAL CONTRACTOR:
Continental Real Estate/Continental Building Systems
285 East Waterfront Dr., #150, Homestead, PA 15120
STRUCTURAL ENGINEER: The Kachele Group
1014 Perry Highway, #100, Pittsburgh, PA 15237
ELECTRICAL/MECHANICAL/PLUMBING ENGINEER:
H.F. Lenz Company
1407 Scalp Avenue, Johnstown, PA 15904
SITE SURVEYOR: Senate Engineering Company
U-PARC, 420 William Pitt Way, Pittsburgh, PA 15238
SITE/CIVIL ENGINEER:
Civil & Environmental Consultants, Inc.
333 Baldwin Road, Pittsburgh, PA 15205
LANDSCAPE ARCHITECT: Hanson Design Group, Ltd.
2333 East Carson Street, Pittsburgh, PA 15203

GENERAL DESCRIPTION

SITE: 19.54 acres.
NUMBER OF BUILDINGS: One.
BUILDING SIZES: First floor, 169,420; second floor, 21,670; total, 191,090 square feet.
BUILDING HEIGHT: First floor office, 11'4"; second floor, 11'2"; floor to floor, 15'; first floor process area, 40'; total height, 45'.
BASIC CONSTRUCTION TYPE: New/steel frame with precast panels.

FOUNDATION: Concrete.
EXTERIOR WALLS: Precast.
ROOF: Membrane.
FLOORS: Concrete.
INTERIOR WALLS: Gypsum.

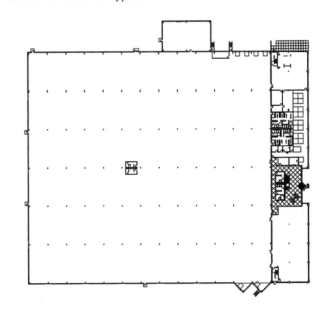

SIEMENS WESTINHOUSE FUEL CELL FACILITY
Construction Period: Oct 2001 to June 2002 • Total Square Feet: 191,090

C.S.I. Divisions (1 through 16)	COST	% OF COST	SQ.FT. COST		SPECIFICATIONS
BIDDING REQUIREMENTS	233,419	2.42	1.22		General conditions.
1. GENERAL REQUIREMENTS	940,631	9.75	4.92	1	Builder's fee, construction staking, overhead, winter protection, permits.
3. CONCRETE	1,665,345	17.26	8.72	3	Precast, sealer, shell, testing, concrete.
4. MASONRY	57,172	0.59	0.30	4	-
5. METALS	1,583,821	16.41	8.29	5	Metal wall panels, structural steel, miscellaneous metals.
6. WOOD & PLASTICS	121,550	1.26	0.64	6	Rough carpentry, finish carpentry.
7. THERMAL & MOIST. PROTECTION	325,056	3.37	1.70	7	Insulation, roofing, caulking.
8. DOORS AND WINDOWS	446,804	4.63	2.34	8	Entrances & storefronts, metal doors, overhead doors, glass &glazing.
9. FINISHES	692,333	7.17	3.62	9	Metal support systems, acoustical ceilings, drywall, painting.
10. SPECIALTIES	41,189	0.43	0.22	10	Toilet partitions & accessories.
11. EQUIPMENT	2,322	0.02	0.01	11	-
12. FURNISHINGS	-	-	-	12	-
13. SPECIAL CONSTRUCTIONS	-	-	-	13	-
14. CONVEYING SYSTEMS	31,375	0.33	0.16	14	Elevators (1).
15. MECHANICAL	2,419,751	25.08	12.66	15	HVAC, plumbing, fire protection.
16. ELECTRICAL	1,089,207	11.28	5.70	16	Basic materials & methods, electrical.
TOTAL BUILDING COST	**9,649,975**	**100%**	**$50.50**		
2. SITE WORK LANDSCAPING & OFFSITE WORK	1,400,000			2	Landscaping, improvements, foundations. Included in Site Work.
TOTAL PROJECT COST	**11,049,975**	*(Excluding architectural and engineering fees)*			

UPDATED ESTIMATE TO JANUARY 2011: $86.10 SF

Siemens Westinghouse Fuel Cell Facility
Munhall, Pennsylvania

Architect: Renaissance 3 Architects, P.C

The Siemens Westinghouse Fuel Cell Facility exemplifies Pittsburgh's efforts at remediation of land where steel mills once stood. Siemens' decision to manufacture solid oxide fuel cells as a clean, sustainable energy source of the future, at a location so intrinsically tied to Pittsburgh's industrial past validates its commitment to progress.

From the early stages of the project, Siemens embraced the goals of achieving LEED® 2.0/2.1 Certification for the entire building, which includes a two-story office and one-story high-bay manufacturing area for a total of 190,000 square feet. Conference rooms, laboratories, cafeteria, restrooms and other support spaces are included in the office wing.

The combination of demand reduction, system efficiencies, and careful material selection, produce the total efficient life cycle and energy cost reductions in this building.

The hourly energy consumption for the building was modeled under two scenarios; a base case that indicated the energy performance to be in compliance with ASHRAE Standard 90.1-1999, and a low energy design to evaluate various green building strategies. The facility's efficient low energy design resulted in a total building energy consumption which is 52.7% lower than what an "energy efficient" building consumes in a year.

The perimeter walls are 8-inch precast concrete with insulation integral to each panel, and are considered as mass walls. The insulated precast wall panels achieve an effective u-value which is 5% more efficient than what is required. The largest energy saving feature of the building's skin is the vertical glazing system. The total glazing area is 19% of the gross wall area, and conducts 16% less heat than the required glazing.

The heating, ventilating, and air conditioning (HVAC) systems serving the building consist of a chilled water plant, hydronic hot water boiler plant, ten constant volume air handling units for air distribution and ventilation in the manufacturing area, and a variable air volume, roof top air handling unit serving the office and laboratory areas.

The rooftop unit uses heat recovery to enable high indoor air quality while minimizing energy usage. CO_2 sensors monitor levels and adjust the ventilation air quantities accordingly. Air distribution

in the manufacturing area delivers cooling directly to the occupied areas, increasing ventilation effectiveness while allowing hot air to stratify, dramatically reducing the cooling load.

Water conservation measures are in place with low flow and dry type plumbing fixtures, and efficient HVAC and cooling tower equipment. The overall consumption of potable water on this project is 31% less than a facility that is compliant with the Energy Policy Act of 1992, and overall wastewater generation is reduced by 39%.

Material resources have been conserved in several ways. Nearly all products and materials were manufactured within 500 miles of the project site. Half of the products and interior finishes were selected for their recycled material content. The cafeteria and several coffee areas have built-in recycling stations to support the recycling efforts of the employees. Construction waste was separated into various dumpsters and recycled by the contractor.

A project goal was to increase indoor environmental quality through careful selection of materials. Limiting or eliminating VOC emitting materials in adhesives, sealants, paints, carpet, and wood was specified and implemented. Windows on the south and east sides enable employees to have views to the outside and access to daylight, and optimizes worker productivity. Siemens has declared the building a non-smoking facility enabling cleaner air quality and decreasing potential health risks.

LEED® 2.0/2.1 CERTIFICATION MANUFACTURERS/SUPPLIERS	
DIV 03:	*Concrete: Floor Hardener:* Ashford Formula by Curecrete; *Precaster:* Fabcon.
DIV 07:	*Membrane:* GenFlex.
DIV 08:	*Special Doors:* Overhead Door.
DIV 14:	*Elevator:* Schindler.

COST PER SQUARE FOOT $86.10

ARCHITECT
Integrated Architecture
4090 Lake Drive S.E., Grand Rapids, MI 49546
www.intarch.com

CONSTRUCTION TEAM

STRUCTURAL ENGINEER:
JDH Engineering
3000 Ivanrest S.W., #B, Grandville, MI 49418

MECHANICAL & ELECTRICAL ENGINEER:
Integrated Architecture
4090 Lake Drive S.E., Grand Rapids, MI 49546

LANDSCAPE ARCHITECT:
Michael J. Dul & Associates, Inc.
212 Daines Street, Birmingham, MI 48009

GENERAL CONTRACTOR:
Pioneer Construction
550 Kirtland Street S.W., Grand Rapids, MI 49507

GENERAL DESCRIPTION

SITE: 20.48 Acres.
NUMBER OF BUILDINGS: One.
BUILDING SIZE: First floor, 117,415; second floor, 71,560; total, 188,975 square feet.
BUILDING HEIGHT: First floor, 14'; second floor, 18'; total, 32'.
BASIC CONSTRUCTION TYPE: Addition/2-B Business, F-1 Factory/Structural Steel.
FOUNDATION: Cast-in-place, slab-on-grade.
EXTERIOR WALLS: CMU, brick, precast, curtainwall.
ROOF: Membrane, green roof over membrane.
FLOORS: Concrete.
INTERIOR WALLS: Metal stud drywall, CMU, glass in aluminum .

PECKHAM INDUSTRIES
Construction Period: Nov 2007 to Mar 2009 • Total Square Feet: 188,975

C.S.I. Divisions (1 through 16)			COST	% OF COST	SQ.FT. COST	SPECIFICATIONS
		PROCUREMENT & CONT. REQ.	85,514	0.61	0.45	Permit.
1.	1.	GENERAL REQUIREMENTS	1,083,400	7.78	5.73	General conditions, supervision, testing, LEED documentation, overhead & fee.
3.	3.	CONCRETE	1,030,589	7.40	5.45	Forming & accessories, reinforcing, cast-in-place, precast.
4.	4.	MASONRY	653,160	4.69	3.46	Unit.
5.	5.	METALS	1,905,536	13.68	10.08	Structural metal framing, joists, decking, fabrications, decorative metal.
6.	6.	WOOD/PLASTICS/COMPOSITE	262,729	1.89	1.39	Rough carpentry, finish carpentry.
7.	7.	THERMAL & MOIST. PROTECT	1,289,870	9.33	6.87	Dampproofing & waterproofing, roofing & siding panels, membrane roofing, flashing & sheet metal.
8.	8.	OPENINGS	1,143,704	8.21	6.05	Doors & frames, entrances, storefronts & curtainwalls, windows, hardware, glazing.
9.	9.	FINISHES	1,654,665	11.88	8.76	Drywall, tiling, ceilings, flooring, wall finishes, acoustic treatment, painting & coating.
10.	10.	SPECIALTIES	59,362	0.43	0.31	Signage.
11.	11.	EQUIPMENT	263,370	1.89	1.39	Kitchen
14.	14.	CONVEYING SYSTEMS	101,517	0.73	0.54	Elevators (2).
15.	21.	FIRE SUPRESSION	191,380	1.37	1.01	—
15.	22.	PLUMBING	573,692	4.12	3.04	—
15.	23.	HVAC	1,518,371	10.90	8.03	—
16.	26.	ELECTRICAL	2,102,853	15.10	11.13	—
TOTAL BUILDING COST			**13,928,712**	**100%**	**$73.71**	
2.	1.	SITEWORK	842,090			Demolition & structure moving, site remediation, facility remediation, shoring & underpinning, special foundations & load-bearing elements.
2.	32.	EXTERIOR IMPROVEMENTS	843,320			—
2.	33.	UTILITIES	6,000			—
TOTAL PROJECT COST			**15,620,122**	*(Excluding architectural and engineering fees)*		

UPDATED ESTIMATE TO JAN 2011: $90.02 SF

DCD Subscribers: Access this case study and hundreds more for instant date and location calculations at www.dcd.com.

Peckham Industries
Lansing, Michigan

Architect: Integrated Architecture

Photos Courtesy of Justin Maconochie

I n the middle of the worst economy in decades, in the middle of a state that is suffering a $600 million dollar budgetary shortfall, double digit unemployment and record home foreclosures - there is a bright spot. Actually, it is a bright green spot: Peckham, Inc. Peckham is a unique, Lansing, Michigan non-profit, community rehabilitation organization with a mission of "providing a wide range of opportunities to maximize human potential for persons striving for independence and self-sufficiency." Essentially, they offer vocational training to persons with barriers to employment.

Peckham celebrated the grand opening of their manufacturing plant and corporate headquarters on Earth Day 2009. Utilizing universal and sustainable design standards, the new facility balances the triple bottom line of people, profits and planet, supporting all Peckham employees by offering world-class workspace for disabled and able-bodied, no matter if they work on the sewing line, the cafeteria line or on the bottom line in the administrative offices.

Located on the main bus line in an industrial park adjacent to Lansing's Capital City Airport, the project is a 189,000 square foot expansion of an existing 9-year-old 140,000 square foot warehouse, adding 117,560 square feet of manufacturing space and a 70,000 square foot office, training, cafe, and open green space. The resulting facility consolidates Peckham's manufacturing and administrative functions in one location, while literally expressing their commitment to all employees, clients and staff.

The manufacturing addition on the south side of the existing building provides ample daylight through the use of curtain wall windows, sky lights and an interior light well. The light well also acts as a central connection, bridging the gap between white and blue collar workers. It also provides an identifiable icon in the large manufacturing facility.

Following the tenants of universal design the facility is designed to provide a supportive, safe environment. Examples include a covered entrance that protects clients from the elements between the bus and the door, the heated sidewalk, automatic door sensors and zero-step design. Brightly colored walls support way finding and compliment directional signage. All vertical circulation is highlighted in yellow. In the manufacturing area, purple walls indicate the restroom locations.

LEED® SILVER CERTIFICATION

COST PER SQUARE FOOT $90.02

ARCHITECT
Browne Penland McGregor Architects
520 Post Oak Boulevard, #880, Houston, TX 77027
www.bpmarch.com

CONSTRUCTION TEAM

ENERGY MODELING COMMISSIONING:
DBR Engineering Consultants, Inc.
9990 Richmond Avenue, #300, Houston, Texas 77042
LEED® CONSULTANT:
Browne Penland McGregor Architects
520 Post Oak Boulevard, #880, Houston, TX 77027
GENERAL CONTRACTOR:
Metzger Construction Company
2055 Silber, #100, Houston, TX 77055

GENERAL DESCRIPTION

SITE: 4.7617 Acres.
NUMBER OF BUILDINGS: One.
BUILDING SIZE: First floor, 27,190; second floor, 27,190;
total, 54,380 square feet.
BUILDING HEIGHT: First floor, 14'6"; second floor, 16'; total, 32'6".
BASIC CONSTRUCTION TYPE: New/Tilt-Up
FOUNDATION: Pier & grade, reinforced concrete, slab-on-grade.
EXTERIOR WALLS: Tilt-up, glass.
ROOF: Membrane.
FLOORS: Concrete.
INTERIOR WALLS: Metal stud drywall.

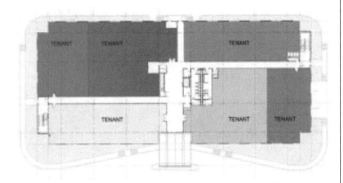

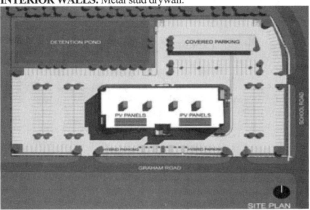

C.S.I. Divisions (1 through 16)			COST	% OF COST	SQ.FT. COST	SPECIFICATIONS
TOMBALL MEDICAL PLAZA						
Construction Period: Apr 2008 to Dec 2008 • Total Square Feet: 54,380						
1.	1.	GENERAL REQUIREMENTS	295,500	8.57	5.43	Summary, permitting.
3.	3.	CONCRETE	581,000	16.86	10.68	Cast-in-place, precast.
5.	5.	METALS	610,000	17.70	11.22	Structural steel.
6.	6.	WOOD/PLASTICS/COMPOSITE	28,200	0.82	0.52	Rough carpentry, architectural woodwork.
7.	7.	THERMAL & MOIST. PROTECTION	307,000	8.91	5.65	Dampproofing & waterproofing, roofing & siding panels, roof & wall specialties & accessories.
8.	8.	OPENINGS	381,000	11.06	7.01	Specialty doors & frames, hardware, entrances, storefronts, & curtainwalls.
9.	9.	FINISHES	305,000	8.85	5.61	Plaster & gypsum board, ceilings, flooring, wall finishes, painting & coating.
10.	10.	SPECIALTIES	33,500	0.97	0.62	Other.
14.	14.	CONVEYING SYSTEMS	43,000	1.25	0.79	Elevators (1).
15.	21.	FIRE SUPRESSION	65,000	1.89	1.20	Water-based fire-suppression systems.
15.	22.	PLUMBING	110,000	3.19	2.02	Equipment.
15.	23.	HVAC	425,000	12.33	7.82	Air distribution.
16.	26.	ELECTRICAL	240,000	6.96	4.40	Medium-voltage distribution.
16.	27.	ELECTRONIC SAFETY & SECURITY	22,000	0.64	0.40	Detection & alarm.
TOTAL BUILDING COST			**3,446,200**	**100%**	**$63.37**	
2.	31.	EARTHWORK	154,4000			Earth moving.
2.	32.	EXTERIOR IMPROVEMENTS	616,610			Bases, bollards, & paving, planting.
2.	33.	UTILITIES	239,500			Sanitary sewer, storm, fire spinkler lines, electrical.
TOTAL PROJECT COST			**4,456,310**	*(Excluding architectural and engineering fees)*		

UPDATED ESTIMATE TO JAN 2011: $88.04 SF

Tomball Medical Plaza
Tomball, Texas

Architect: Browne Penland McGregor Architects

The Tomball Medical Plaza, developed by Tomball Physicians Investments, LLC, is the first project in Tomball, Texas, registered for LEED® CS Silver Certification. Strategies to obtain this certification included an emphasis on reduced energy consumption, community connectivity, use of building materials with recycled content and education regarding sustainable features for building tenants and visitors.

The building is oriented with its longest facades facing to the north and south, respectively, to minimize heat gain from the western sun. A roof-top array of solar panels generates a portion of the building's electricity on-site. The balance of the electrical demand is provided through a long-term renewable energy contract. The glazing system is composed of three colors of insulated glass with coatings to reject heat gain in summer and limit heat loss during winter. The three colors are arranged in a Mondrian like pattern to accent the corner stair towers, the punched window openings and the freestanding, glass curtain-wall at the covered entrance. Building commissioning upon project completion confirmed success in meeting the design intent for reduced energy consumption.

The two story building contains 54,380 square feet and occupies a 4.8 acre, Greenfield site located one block north of the Tomball Regional Medical Center and just south of the city's main street. Residential neighborhoods are located within a half mile radius, and bicycle racks and showers promote non-automobile transportation. Reserved spaces are provided for patients and staff driving hybrid vehicles.

The use of materials with recycled content includes the building's interior steel frame, as well as, the concrete reinforcing steel of which 100% comes from recycled sources.

A flat screen panel in the ground floor lobby provides a real-time display of the electrical generation of the roof-top solar array for edification of building tenants and visitors. A written manual informs tenants of the building's sustainable features and directs them in the use of the recycling center on the ground floor. Finally the maintenance staff has been instructed in the use of non-toxic cleaning materials.

LEED® CERTIFIED

COST PER SQUARE FOOT $88.04

Photos Courtesy of Judd Haggard Photography

ARCHITECT
STANLEY BEAMAN & SEARS, INC.
135 Walton Street, N.W.
Atlanta, GA 30303
www.stanleybeamansears.com

GENERAL CONTRACTOR
TURNER CONSTRUCTION, INC.
3424 Peachtree Rd., N.E., #1900
Atlanta, GA 30326
www.turnerconstruction.com

FILE UNDER
MEDICAL
Atlanta, Georgia

CONSTRUCTION TEAM

STRUCTURAL ENGINEER:
Stanley D. Lindsey & Associates
2300 Windy Ridge Pkwy., #200S, Atlanta, GA 30339

ELECTRICAL & MECHANICAL ENGINEER:
Newcomb & Boyd Consulting Engineering Group
303 Peachtree Center Ave., #525, Atlanta, GA 30303

CIVIL ENGINEER:
Jordan Jones & Goulding, Inc.
6801 Governors Lake Pkwy., Bldg., #200, Norcross, GA 30071

GENERAL DESCRIPTION

SITE: 2 acres. **NUMBER OF BUILDINGS:** One
BUILDING SIZE: Basement, 45,000; first floor, 35,000; second floor, 30,000; each additional floor, 30,000; total, 260,000 square feet.
BUILDING HEIGHT: Basement, 16'; first floor, 14'; second floor, 14'; each additional floor, 14'; penthouse, 26'; total, 96'.
BASIC CONSTRUCTION TYPE: Concrete Frame Type II.

Winship Cancer Institute
EMORY UNIVERSITY

FOUNDATION: Concrete. **EXTERIOR WALLS:** Precast.
ROOF: Membrane, tile. **INTERIOR WALLS:** Gypsum.
FLOORS: Carpet, resilient, VCT, terrazzo, tile.

WINSHIP CANCER INSTITUTE, EMORY UNIVERSITY
Construction Period: Mar 2002 to July 2003 • Total Square Feet: 260,000

C.S.I. Divisions (1 through 16)	COST	% OF COST	SQ.FT. COST		SPECIFICATIONS
BIDDING REQUIREMENTS	-	-	-		Included in Div. 1.
1. GENERAL REQUIREMENTS	8,194,786	14.92	31.52	1	-
3. CONCRETE	7,651,088	13.93	29.43	3	Formwork, reinforcement, accessories, cast-in-place, curing, cementitious decks & toppings, grout, mass.
4. MASONRY	756,687	1.38	2.91	4	Masonry & grout, accessories, unit, stone.
5. METALS	1,700,272	3.09	6.54	5	Materials, coatings, fastening, cold formed framing, fabrications, sheet metal, ornamental, expansion control.
6. WOOD & PLASTICS	1,607,378	2.93	6.18	6	Fasteners & adhesives, rough & finish carpentry, wood & metal systems, wood treatment, woodwork, solid polymer fabrications.
7. THERMAL & MOIST. PROTECTION	1,633,812	2.97	6.28	7	Waterproofing, dampproofing, water repellents, vapor retarders, air barriers, insulation, EIFS, fireproofing, firestopping, shingles & roof tiles, membrane roof, traffic coatings, flashing & sheet metal, roof specialties, skylights, joint.
8. DOORS AND WINDOWS	2,160,860	3.93	8.31	8	Metal doors & frames, wood & plastic doors, door opening assemblies, special doors, entrances & storefronts, metal windows, hardware, glazing, glazed curtainwalls.
9. FINISHES	6,539,025	11.90	25.15	9	Metal support systems, lath & plaster, gypsum, tile, terrazzo, acoustical, special wall & ceiling surfaces, resilient, carpet, special flooring & coatings, painting, wall coverings.
10. SPECIALTIES	484,460	0.88	1.86	10	-
11. EQUIPMENT	620,506	1.13	2.39	11	-
12. FURNISHINGS	2,130,870	3.88	8.20	12	-
13. SPECIAL CONSTRUCTIONS	574,456	1.05	2.21	13	Special purpose rooms, sound, vibration, & seismic, radiation, pre-engineered structure, liquid & gas storage tanks, utility control, building automation, fire suppression, special security.
14. CONVEYING SYSTEMS	1,157,642	2.11	4.45	14	Elevators, lifts, material handling systems.
15. MECHANICAL	11,977,487	21.80	46.07	15	Basic materials & methods, insulation, specialties, plumbing, HVAC, heat generation, refrigeration, heat transfer, air distribution, controls, testing, adjusting & balancing.
16. ELECTRICAL	7,749,354	14.11	29.81	16	Basic materials & methods, medium voltage distribution, service & distribution, lighting, special systems, communications, electric resistance heating, controls, testing, adjusting & balancing.
TOTAL BUILDING COST	**54,938,683**	**100%**	**$211.30**		
2. SITE WORK	4,361,317			2	-
LANDSCAPING & OFFSITE WORK	-				-
TOTAL PROJECT COST	**59,300,000**	*(Excluding architectural and engineering fees)*			

UPDATED ESTIMATE TO JANUARY 2011: $428.64 SF

DCD Subscribers: Access this case study and hundreds more for instant date and location calculations at www.dcd.com.

Winship Cancer Institute, Emory University
Atlanta, Georgia

Architect: Stanley Beaman & Sears, Inc.

Emory University has had a long history of providing cancer care, research and medical training since its clinics first opened in 1937. As the program grew, Emory realized that a new clinical and research facility was required to help meet its goal to become a designated Comprehensive Cancer Center. The Senior Leadership envisioned a facility that would not only meet the University's research mission, but also provide an exceptional environment for the care of patients and their families.

Situated on a dense site, with restrictions, the site posed enormous challenges, both technically and functionally. Seven stories high, with subterranian spaces and tunnel connections, this building was partially constructed on top of active linear accelerator vaults, and surrounded on nearly all sides by other structures. For success, the project required a collaborative approach between the architect and contractor to meet the client's schedule and budget.

Photos Courtesy © Gary Knight: Gary Knight + Associates, Inc.

Although healthcare and research facilities are historically large energy consumers, the architecture and engineering team was challenged with designing the facility to qualify for LEED® certification. In January of 2005, the Cancer Institute formally received certification in the US Green Building Council's LEED® program ("Leadership in Energy and Environmental Design); the first building of its type to achieve this recognition.

The Cancer Center Director wanted a design that could speak to both researchers and patients in tangible ways sending a message of optimism to patients, and a reminder to researchers to "accelerate discovery."

Inspired by these ideas, the design team worked to embed a language of hope, caring and imagination into the details of the building both literally and symbolically. The building's exterior design echoes the University's more traditional architectural style, but the main entry invites patients into a crisp, modern interior, reinforcing the medical sophistication and the sense of confidence a patient seeks for cancer treatment.

The artistic centerpiece of the building is clearly the illuminated entry tower, which houses the monumental stair linking all clinical and research floors. With inspirational phrases embedded into

the landings, and compelling reminders to researchers to tap into their imagination, the stair prompts a daily dialogue. This design element keeps both patients and researchers in mind at every turn, reminding one of the other.

Input from patients influenced the design of the Infusion Center, which, at 80 stations, was daunting in size. The design was developed in more intimate clusters of 4 patients each, with half-walls that personalize the space for family members, while maintaining visibility for good nursing care. The clusters allow patients the opportunity to converse with their newly-found "support group", or pull a curtain for privacy.

In short, the building design seeks to engage in a conversation with its occupants about life, health, and the ultimate hope for a cure.

LEED® CERTIFIED MANUFACTURERS/SUPPLIERS	
DIV 07:	*Roofing Insulation:* Atlas; *Exterior Architectural Coating:* **ParexLahabra.**
DIV 08:	*Curtainwall:* Kawneer; *Windows:* EFCO Corp; *Glazing:* Viracon; *Skylights:* Gammon Architectural.

COST PER SQUARE FOOT $428.64

ARCHITECT
Vision 3 Architects
225 Chapman Street, Providence, RI 02905
www.vision3architects.com

FILE UNDER
MEDICAL
Providence, Rhode Island

CONSTRUCTION TEAM

STRUCTURAL ENGINEER:
Odeh Engineers, Inc.
1223 Mineral Spring Avenue, North Providence, RI 02904
ELECTRICAL & MECHANICAL ENGINEER:
Creative Environment Corp.
50 Office Parkway, East Providence, RI 02914
GENERAL CONTRACTOR & COST ESTIMATOR:
New England Construction
293 Bourne Avenue, Rumford, RI 02916

GENERAL DESCRIPTION

SITE: 1.41 acres.
NUMBER OF BUILDINGS: One.
BUILDING SIZE: First floor, 12,400; second floor, 12,924; third floor, 11,205; fourth floor, 11,205; total, 47,734 square feet.
BUILDING HEIGHT: First floor, 11'4"; second floor, 10'8"; each additional floor, 10'8"; floor to floor, 10'8"; total, 43'11".
BASIC CONSTRUCTION TYPE: Renovation/Steel Frame.
FOUNDATION: Cast-in-place.
EXTERIOR WALLS: Brick.
ROOF: TPO/Ballasted.
FLOORS: Concrete.
INTERIOR WALLS: Metal stud drywall.

HOME & HOSPICE CARE CENTER
Construction Period: Aug 2008 to May 2009 • Total Square Feet: 47,734

C.S.I. Divisions (1 through 16)		COST	% OF COST	SQ.FT. COST	SPECIFICATIONS
1.	1.	GENERAL REQUIREMENTS	1,230,143	16.25	25.77	Mobilization, change orders, temporary facilities, permits, insurance, fees, superintendent.
3.	3.	CONCRETE	88,230	1.17	1.85	Forming & accessories, reinforcing, cast-in-place.
4.	4.	MASONRY	49,490	0.65	1.04	—
5.	5.	METALS	459,528	6.07	9.63	Structural steel.
6.	6.	WOOD/PLASTICS/COMPOSITE	607,719	8.03	12.73	Rough carpentry, finish carpentry, architectural woodwork.
7.	7.	THERMAL & MOIST. PROTECTION	188,098	2.49	3.94	Waterproofing & dampproofing, roofing.
8.	8.	OPENINGS	484,745	6.40	10.16	Glass work, storefront & windows, doors, frames & hardware.
9.	9.	FINISHES	878,018	11.60	18.39	Plaster & gypsum board, ceilings, flooring, wall finishes, painting & coating.
10.	10.	SPECIALTIES	61,311	0.81	1.28	Fire place, toilet accessories, toilet partitions, signage, fire extinguishers.
11.	11.	EQUIPMENT	5,133	0.07	0.11	Kitchen.
14.	14.	CONVEYING SYSTEMS	189,715	2.51	3.97	Elevators (2 existing refurbished).
15.	21.	FIRE SUPRESSION	196,608	2.60	4.12	—
15.	22.	PLUMBING	518,630	6.85	10.87	—
15.	23.	HVAC	1,839,798	24.30	38.54	—
16.	26.	ELECTRICAL	771,995	10.20	16.71	—
TOTAL BUILDING COST			**7,569,161**	**100%**	**$158.57**	
2.	2.	EXISTING CONDITIONS	154,000			Demolition, abatement demolition.
2.	31.	EARTHWORK	336,992			Excavation and backfill.
2.	32.	EXTERIOR IMPROVEMENTS	48,650			Landscaping.
TOTAL PROJECT COST			**8,108,803**	*(Excluding architectural and engineering fees)*		

UPDATED ESTIMATE TO JAN 2011: $177.72 SF

DCD Subscribers: Access this case study and hundreds more for instant date and location calculations at www.dcd.com.

Home & Hospice Care of Rhode Island
Providence, Rhode Island

Architect: Vision 3 Architects

Photos Courtesy of Aaron Usher

Home & Hospice Care of Rhode Island is the state's largest and most comprehensive provider of hospice and palliative care, and is the third oldest hospice in the country. In 2006, Home & Hospice purchased 1085 North Main Street in Providence, with plans to consolidate their hospice facility, administrative offices, and education and bereavement center into one building. On May 31, 2009, a crowd of nearly 500 people celebrated the grand opening of Home & Hospice's new headquarters.

The renovation of the four-story building is currently pursuing LEED®Gold certification and is expected to be the first fully operational LEED® certified health care facility in Rhode Island. Sustainable design aligns with Home & Hospice's philosophy on the cycles of life and the cycles of nature. During design, Home & Hospice consulted with a cultural anthropologist on critical design issues.

The first, major sustainable design commitment Home & Hospice made was to convert an abandoned building, instead of building new. "Not only does reusing an existing facility significantly divert demolition and construction waste from landfills," states David Sluter, CEO of New England Construction, contractor for the renovation, "it enhances the neighborhood by converting a vacant building into a thriving healthcare facility that is open to community use." Throughout construction, 92.6% of all construction waste was recycled. In addition, 95% of the existing wall, floor, and roof construction was reused. "When walking through the new Home & Hospice," says Diana Franchitto, President and CEO of Home & Hospice Care of Rhode Island, "you would never believe that 95% of what you see existed here before. Everything looks brand new."

Other sustainable design features include a reflective roof to prevent heat absorption; low-flow water fixtures with motion sensors; high-performing and energy-efficient building mechanical and electrical systems; and low or no VOC-emitting carpets, paints, adhesives, and wood products. Home & Hospice has also committed to obtaining at least 35% of their electricity from renewable sources, and using only green cleaning methods and products to reduce chemicals in the environment.

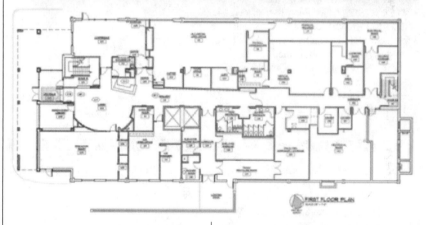

FIRST FLOOR PLAN

Besides the project's sustainable design features, the goal of the project was to provide a facility in which Home & Hospice Care could fulfill their mission to provide compassionate, professional, state of the art physical, emotional and spiritual care for all people facing life-threatening illness. "Vision 3 Architects wrapped the entire design of the facility around this mission," affirms Keith Davignon, Principal of Vision 3 Architects. "We listened closely to Home & Hospice's staff, and provided them with a comfortable and dignified environment for patients and their families."

"Our new home reflects thoughtful planning geared toward the needs of our patients, families and staff. Our goals included creating a sustainable hospice environment that offers patients and family members comfort, peace and plenty of space for reflection and quiet time," states Franchitto. "With the creativity and guidance of Vision 3 and New England Construction, we've achieved these goals and look forward to continuing our important role in the state"s health care scheme."

LEED® GOLD PENDING

COST PER SQUARE FOOT $177.72

ARCHITECT
CDG Architects, Ltd.
2102 N. Country Club Road, #9, Tucson, AZ 85716
www.cdg-architects.com

FILE UNDER
MEDICAL
Tucson, Arizona

CONSTRUCTION TEAM

STRUCTURAL ENGINEER:
Turner Structural Engineering
3026 North Country Club Road, Tucson, AZ 85716

MECHANICAL ENGINEER:
PH Mechanical Engineering
1660 N. Alvernon Way, Tucson, AZ 85712

ELECTRICAL ENGINEER:
Bowden Engineering, Inc.
7337 E. Tanque Verde Road, Tucson, AZ 85715

GENERAL CONTRACTOR:
Chestnut Construction Corporation
2127 E. Speedway, #101, Tucson, AZ 85719

COMMISSIONING AGENT:
Adams and Associates Engineers PLLC
4067 East Grant Road, #200, Tucson, AZ 85712

GENERAL DESCRIPTION

SITE: 1.32 Acres.
NUMBER OF BUILDINGS: One.
BUILDING SIZE: First floor, 14,000; total, 14,000 square feet.
BUILDING HEIGHT: First floor, 18'; total, 18'.
BASIC CONSTRUCTION TYPE: New/Type V-B/Structural Steel.
FOUNDATION: Slab-on-grade.
EXTERIOR WALLS: EIFS: Outsulation system.
ROOF: Built-up.
FLOORS: Concrete.
INTERIOR WALLS: Metal stud drywall.

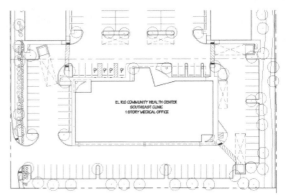

SITE PLAN

EL RIO COMMUNITY HEALTH CARE CENTER
Construction Period: Jan 2009 to August 2009 • Total Square Feet: 14,000

C.S.I. Divisions (1 through 16)			COST	% OF COST	SQ.FT. COST	SPECIFICATIONS
1.	1.	GENERAL REQUIREMENTS	382,390	20.81	27.31	Available information, conditions of the contract, summary, price & payment procedures, administrative requirements, quality requirements, temporary facilities & controls, execution & closeout.
3.	3.	CONCRETE	107,017	5.82	7.64	Forming & accessories, reinforcing, cast-in-place.
4.	4.	MASONRY	34,000	1.85	2.43	Unit.
5.	5.	METALS	46,600	2.54	3.33	Structural metal framing, decorative.
6.	6.	WOOD/PLASTICS/COMPOSITE	195,828	10.66	13.99	Architectural woodwork, structural composites, composite fabrications.
7.	7.	THERMAL & MOIST. PROTECTION	52,108	2.84	3.72	Thermal protection, weather barriers, flashing & sheet metal.
8.	8.	OPENINGS	111,150	6.05	7.94	Doors & frames, entrances, storefronts, & curtainwalls, windows, roof windows & skylights, hardware, glazing.
9.	9.	FINISHES	332,503	18.10	23.75	Plaster & gypsum board, tiling, ceilings, flooring, painting & coating.
10.	10.	SPECIALTIES	26,155	1.42	1.87	—
15.	21.	FIRE SUPRESSION	35,600	1.94	2.54	Water-based fire-suppression systems.
15.	22.	PLUMBING	87,030	4.74	6.22	Equipment, fixtures.
15.	23.	HVAC	167,691	9.13	11.98	Piping & pumps, central HVAC equipment.
16.	26.	ELECTRICAL	234,900	12.77	16.78	Medium-voltage distribution, lighting.
16.	27.	ELECTRICAL POWER GENERATION	24,456	1.33	1.75	Solar panel installation (54 photovohaic panels awarded to El Rio by Tucson Electric Power)
TOTAL BUILDING COST			**1,837,488**	**100%**	**$131.25**	
2.	2.	EXISTING CONDITIONS	155,964			—
2.	31.	EARTHWORK	378,387			Site clearing, earth moving.
2.	32.	EXTERIOR IMPROVEMENTS	61,112			Bases, bollards, & paving, improvements, irrigation, planting.
2.	33.	UTILITIES	217,905			Water, sanitary sewer, storm drainage, electrical, communications.
TOTAL PROJECT COST			**2,650,856**	*(Excluding architectural and engineering fees)*		

UPDATED ESTIMATE TO JAN 2011: $ 158.27 SF

DCD Subscribers: Access this case study and hundreds more for instant date and location calculations at www.dcd.com.

El Rio Community Health Center
Southeast Clinic
Tucson, Arizona

Architect: CDG Architects, Ltd.

Photos Courtesy of Cindy Hogan

E l Rio Community Health Center is at the forefront of healthcare trends, offering quality care for underserved, low-income patients. Since 1970, many have depended on this agency as their primary source of medical care and have come to rely on the distinctively "pro-patient" character of their facilities. El Rio continuously works to exceed industry standards for preventive and chronic care.

In 2007, an El Rio Board decision expanded the organization's definition of "community wellness" to include sustainable and environmentally responsible clinic facilities and work environments. The Board is carrying out this decision by building sustainable clinics that consume fewer local resources such as electric power and potable water, utilize local and recycled building materials, use energy-efficient building design and heating/cooling technologies and plant water-sensitive landscapes. The Southeast Clinic is their first LEED®Silver certified clinic, also the first of its kind in Tucson, Arizona.

El Rio expresses their concern for staff and client welfare within the "fabric" of their buildings. Abundant natural light was introduced through careful placement of windows with respect to the Arizona sun and roof overhangs. "Solatubes" were used to bring natural light into hallways; clear transom windows at the exam rooms allow this light into the rooms for the benefit of both clients and staff. Arizona facilities managers continuously evaluate their approach to cooling systems for the comfort of building users. The balance between the up-front cost of high-efficiency cooling units and the desire to lower

utility fees is ongoing. El Rio selects its equipment based on experience with unit longevity and ease of maintenance. Mechanical system decisions were carefully coordinated between the design team consultants to ensure that rating of the units, the zoning design and the distribution of heat producing light fixtures result in maximum efficiency. Tucson Electric Power awarded El Rio a grant of 54 photovoltaic panels for the building that will provide nearly 5% of the building's annual energy cost. The inverter is now on display in the entrance court so that patrons can see that renewable energy is part of El Rio's commitment to the health of the community.

The success of El Rio is due in large part to cost-aware decisions made by facilities managers and maintenance staff. El Rio facilities are used intensively and finish materials have been carefully selected to be easy for maintenance staff to clean and sturdy enough to withstand substantial wear and tear. Colored concrete floors were used to bear up under heavy lobby traffic. Graffiti-proof finishes were provided at public restrooms. For the Southeast Clinic, staff elected to use high quality, LEED®-responsive systems furniture for workstations; these types of furnishings are sturdy, long lasting and flexible enough for today's changing work environment.

El Rio is eager to continue to develop new facilities that reinforce their positive presence in the community. LEED® Silver certification has proved to be an exciting avenue available to El Rio, serving as a tangible demonstration of their commitment to the overall health of the community.

LEED® Points Achieved	35 Total
Sustainable Sites	9
Water Efficiency	2
Energy & Athmosphere	8
Materials & Resources	6
Indoor Environmental Quality	9
Innovation Design Process	1

LEED® SILVER MANUFACTURERS/SUPPLIERS	
DIV 07:	*Roofing:* **GAFGLAS®** by **GAF**; *Built-Up Roofing:* **Elastek Solar Tek Extreme.**
DIV 08:	*Automatic Entrances:* **NABCO**; *Storefront:* **Arcadia**.
DIV 09:	*Flooring:* **Armstrong**; *Carpet Tile:* **Miliken**
DIV 16:	*Lighting:* **Cooper, IOTA** *Daylighting:* **Solatube**

COST PER SQUARE FOOT $158.27

ARCHITECT

Baer Architectural Group, Inc.
265 Main Street, Northborough, MA 01532
www.baerarchitecture.org

FILE UNDER

EDUCATIONAL
Marlborough, Massachusetts

CONSTRUCTION TEAM

STRUCTURAL ENGINEER:
 Johnson Structural Engineering, Inc.
 30 Faith Avenue, Auburn MA 01501
GENERAL CONTRACTOR:
 Erland Construction Company, Inc.
 83 Second Avenue, Burlington, MA 01803
ELECTRICAL ENGINEER:
 Shepherd Engineering, Inc.
 1308 Grafton Street, Worcester, MA 01604
MECHANICAL ENGINEER:
 Seaman Engineering Corporation
 30 Faith Avenue, Auburn, MA 01501

GENERAL DESCRIPTION

SITE: 3-acres project site on a 150-acre campus.
NUMBER OF BUILDINGS: One.
BUILDING SIZE: First floor new, 16,000; first floor renovation, 18,000; total, 34,000 square feet.
BUILDING HEIGHT: First floor, 16'3"; total, 16'3".
BASIC CONSTRUCTION TYPE: Addition/Renovation.
FOUNDATION: Concrete, structural steel.
EXTERIOR WALLS: Brick.
ROOF: EPDM.
FLOORS: Rubber flooring, carpet, ceramic tile, VCT.
INTERIOR WALLS: Gypsum.

ACADEMIC & HEALTH CENTER, HILLSIDE SCHOOL

Construction Period: Jun 2007 to Mar 2008 • Total Square Feet: 34,000

C.S.I. Divisions (1 through 16)			COST	% OF COST	SQ.FT. COST	SPECIFICATIONS
		PROCUREMENT & CONT. REQ.	446,043	9.67	13.12	-
1.	1.	GENERAL REQUIREMENTS	372,852	8.08	10.97	Summary, price & payment procedures, administrative requirements, quality requirements, temporary facilities & controls, product requirements, execution & closeout requirements, performance requirements, life cycle activities.
3.	3.	CONCRETE	159,789	3.46	4.70	Forming & accessories, reinforcing, cast-in-place, precast.
4.	4.	MASONRY	109,478	2.37	3.22	Unit.
5.	5.	METALS	404,084	8.76	11.88	Structural metal framing, joists, decking, cold-formed metal framing, fabrications.
6.	6.	WOOD/PLASTICS/COMPOSITE	202,249	4.38	5.95	Rough carpentry, finish carpentry, architectural woodwork, plastic fabrications.
7.	7.	THERMAL & MOIST. PROTECT	392,885	8.52	11.56	Dampproofing & waterproofing, thermal protection, weather barriers, roofing & siding panels, membrane roofing, flashing & sheet metal, joint protection.
8.	8.	OPENINGS	524,413	11.37	15.42	Doors & frames, specialty doors & frames, entrances, storefronts & curtainwalls, windows, hardware, glazing.
9.	9.	FINISHES	521,426	11.30	15.34	Plaster & gypsum board, tiling, ceilings, flooring, wall finishes, acoustic treatment, painting & coating.
10.	10.	SPECIALTIES	86,159	1.87	2.53	Interior, safety, other.
11.	11.	EQUIPMENT	10,448	0.23	0.31	Residential, educational & scientific, athletic & recreational, healthcare.
12.	12.	FURNISHINGS	19,019	0.41	0.56	Casework, multiple seating.
15.	21.	FIRE SUPRESSION	-	-	-	Included in HVAC: Water-based fire-suppression systems.
15.	23.	HVAC	831,055	18.01	24.44	Water-based fire-suppression, plumbing piping & pumps, plumbing equipment, plumbing fixtures, HVAC piping & pumps, HVAC air distribution, HVAC air cleaning devices.
16.	26.	ELECTRICAL	533,694	11.57	15.69	Medium-voltage distribution, low-voltage transmission, lighting.
TOTAL BUILDING COST			**4,613,594**	**100.00**	**$315.69**	
2.	2.	EXISTING CONDITIONS	16,615			Assessment, demolition & structure moving.
2.	31.	EARTHWORK	185,046			Site clearing, earth moving, utilities.
TOTAL PROJECT COST			**4,815,255**	*(Excluding architectural and engineering fees)*		

UPDATED ESTIMATE TO JAN 2011: $161.54 SF

Academic & Health Center, Hillside School
Marlborough, Massachusetts

Architect:Baer Architecture Group, Inc.

Photos Courtesy of Baer Architecture Group, Inc.

The Hillside Academic and Health Center (AHC) is the largest capital improvement stemming from a comprehensive campus master plan. It serves to grow and enhance the academic spaces for the growing Hillside community and provide an improved home for academics, arts, and athletics while incorporating and fostering the school's mission to help students develop academically, physically, spiritually, morally, and socially while building their confidence and maturity.

The goal of creating a "green" academic building was a clear imperative from the start of the project. Hillside curriculum teaches students to be stewards of the environment; the sustainable design of the AHC supports this educational principle while benefiting from the cost savings associated with sustainable energy usage recovery and management systems.

Baer Architecture Group's role was to create a facility which, through addition to and renovation of the existing space (Stevens Hall), doubled the number of academic classrooms in the building, created an enhanced infirmary, a small studio apartment for the on-call nurse, a fitness center/weight room, and a multi-purpose/wrestling room along with giving the students a protected outdoor space adjacent to the classrooms.

Stevens Hall, originally built in the early 1950's, was in dire need of life safety and finish updates. Part of the scope of work for this project was to introduce a sprinkler system throughout the new and existing buildings and update the fire alarm throughout the existing building. In order to better the energy efficiency in Stevens Hall, the original steel-framed, single-glazed storefront was removed and the same system used in the new addition was installed.

The exterior design solutions needed to be aesthetically consistent with the existing building, yet meet the goal of improved energy efficiency. Because the existing academic building had a brick pier exterior with an inefficient single-pane steel curtain-wall, the challenge was to mimic the detailing and aesthetics while improving energy-efficiency and functionality. To achieve this, a similar water-struck face brick was used, zinc metal panels, a highly efficient aluminum storefront system, and low-e glass in five different color types to help reduce the amount of solar heat gain and glare in the classroom spaces. Roofing was a white EPDM membrane.

The interior of the building was organized around two main circulation spines. Connecting the existing building to the Academic and Health Center is a space entitled the Student Commons. This space organizes the main public spaces within the AHC, those being the multi-purpose/wrestling room and the flex room. The Student Common is used by the community for program functions, fundraising events, and student gatherings. To the east of the commons are the self-contained classrooms, serving the fifth and sixth grades. To the west of the commons are three general purpose classrooms, and the three state-of-the art science classrooms and laboratories. Located at the western edge of the center are a series of offices, and the health center. The center houses six general beds, and four quarantine rooms.

The interior design used recycled rubber flooring products (100% post-consumer, 50% post-industrial waste) in the community spaces, corridors, wrestling room, and fitness center, and low-VOC paints throughout. All of the finished wood products were rift-sawn Hickory; supplied by a friend of the school. The HVAC system is an energy-efficient system of water-source heat pumps coupled with a standing column geothermal system. There is an estimated 18-month payback period for the client on the installation of the system.

The project was completed on schedule and within budget, and provides the Hillside Community with an expanded facility to foster the academic, social, physical, and environmental initiatives to enrich the lives of Hillside students.

MANUFACTURERS/SUPPLIERS	
DIV 04:	*Brick:* **The Belden Brick Company.**
DIV 07:	*EPDM:* Carlisle; *Metal Panels:* Rheinzink
DIV 08:	*Aluminum Storefront, Curtainwall & Windows:* EFCO Corporation.
DIV 09:	*Rubber Flooring:* EcoSurfaces; *Carpet:* Blue Ridge; *Tile:* American Olean; *VCT:* Stonescape by Estrie, Inc.

EXTENDED PRODUCT INFORMATION
Brick:
The Belden Brick Company.

COST PER SQUARE FOOT $161.54

ARCHITECT
BROWNIE PENLAND MCGREGOR STEPHENS ARHITECTS, INC.
520 Post Oak Boulevard, #880
Houston, TX 77027
www.bpmsa.com

FILE UNDER
MEDICAL
Pearland, Texas

CONSTRUCTION TEAM

STRUCTUREL ENGINEER:
ASA Consulting Engineers, Inc.
1155 Dairy Ashford, #111, Houston, TX 77079
GENERAL CONTRACTOR:
Farnsworth & Lott Construction Services
3103 Nottingham, Pearland, TX 77581
MECHANICAL & ELECTRICAL ENGINEER:
DBR Engineering Consultants, Inc.
9990 Richmond Avenue, South Bldg., #300,
Houston, TX 77042
COST ESTIMATOR:
Farnsworth & Lott Construction Services
3103 Nottingham, Pearland, TX 77581

GENERAL DESCRIPTION

SITE: 1.9 acres. **NUMBER OF BUILDINGS:** One.
BUILDING SIZE: First floor, 10,388; total, 10,388 square feet.
BUILDING HEIGHT: First floor,18'7"; total, 18'7".
BASIC CONSTRUCTION TYPE: New/Type II-B.
FOUNDATION: Cast-in-place.
EXTERIOR WALLS: Limestone CMU, tilt-up.
ROOF: Membrane. **FLOORS:** Concrete, linoleum.
INTERIOR WALLS: Metal stud drywall.

PEARLAND PEDIATRICS

Construction Period: Jan 2005 to Oct 2005 Total Square Feet 10,388

C.S.I. Divisions (1 through 16)			COST	% OF COST	SQ.FT. COST	SPECIFICATIONS
		PROCUREMENT & CONT. REQ.	-	-	-	-
1.	1.	GENERAL REQUIREMENTS	124,423	11.05	11.98	-
3.	3.	CONCRETE	148,643	13.20	14.31	Forming & accessories, reinforcing, cast-in-place, precast.
4.	4.	MASONRY	14,600	1.30	1.41	Unit.
5.	5.	METALS	104,811	9.31	10.09	Structural metal framing, joists, decking, cold formed metal framing, fabrications.
6.	6.	WOOD/PLASTICS/COMPOSITE	70,879	6.30	6.82	Finish carpentry, composite fabrications.
7.	7.	THERMAL & MOIST. PROTECT	63,335	5.63	6.10	Dampproofing & waterproofing, thermal protection, membrane roofing.
8.	8.	OPENINGS	85,505	7.59	8.23	Doors & frames, entrances, storefronts, & curtainwalls, windows.
9.	9.	FINISHES	176,800	15.70	17.02	Plaster & gypsum, ceilings, flooring, wall finishes, painting & coating.
10.	10.	SPECIALTIES	64,553	5.73	6.21	Information, safety, exterior, other.
11.	11.	EQUIPMENT	2,257	0.20	0.22	Residential.
12.	12.	FURNISHINGS	2,386	0.21	0.23	-
14.	14.	CONVEYING SYSTEMS	-	-	-	-
15.	21.	FIRE SUPRESSION	-	-	-	-
15.	22.	PLUMBING	66,056	5.87	6.36	Piping & pumps, equipment, fixtures.
15.	23.	HVAC	71,640	6.36	6.89	Air distribution, central HVAC.
16.	26.	ELECTRICAL	130,000	11.55	12.51	Medium-voltage distribution, lighting.
TOTAL BUILDING COST			**1,125,888**	**100%**	**$108.38**	
2.	31.	EARTHWORK	74,474			Site clearing, earth moving.
2.	32.	EXTERIOR IMROVEMENTS	135,469			Bases, ballasts, & paving, planting.
2.	33.	UTILITIES	47,096			Water, sanitary sewerage, storm drainage.
TOTAL PROJECT COST			**1,386,927**	*(Excluding architectural and engineering fees)*		

UPDATED ESTIMATE TO JAN 2011: $194.31 SF

DCD Subscribers: Access this case study and hundreds more for instant date and location calculations at www.dcd.com.

Pearland Pediatrics
Pearland, Texas

Architect: Browne Penland McGregor Stephens Architects, Inc.

The owner and lead physician, Dr. Deborah Gant, of Pearland Pediatrics wanted to provide an environmentally friendly workplace for her staff and a healing environment for her young patients, and to make a statement about the kind of future her patients will inherit.

To meet the owner's goal of an environmentally responsible building, Browne Penland McGregor Stephens Architects and the design team used the United States Green Building Council's (USGBC) sustainable design criteria to create a holistic environment. The result is the first privately owned LEED® certified healthcare facility in Texas.

The single-story, limestone clad, tilt wall, concrete building is 10,388 square feet with a 6,000 square foot pediatric clinic space. With the original plan of utilizing natural daylight and preserving the oak trees on the property, the design team incorporated energy efficiency and indoor air quality to their design. The design team chose a white reflective roof, daylight monitors and an exterior canopy/light shelf to illuminate the interior, while glazing transoms and high performance insulated glazing were used to reject heat, yet let in daylight. Indoor air quality was maintained by using low-emitting materials such as adhesives, paints, sealers, and flooring. A mechanical ventilation system and CO_2 monitoring system were installed to minimize the exposure of young patients and staff to hazardous indoor particulates and chemical pollutants. Also, to help reduce waste, water saving fixtures and a dual flush plumbing system was installed.

The owners operate the building on wind generated power and have a storm water retention pond behind the building.

Photos Courtesy of Jud Haggard Photography

Rapidly renewable resources were used such as linoleum flooring. The split faced limestone, masonry perimeter wall, fly ash concrete tilt wall structure, and ceiling tile are all local Texas building materials found within 500 miles of the site. With the first energy bills the owners noticed half the energy consumption and cost in the new clinic, which is double the size of the previous building.

The design team utilized a coastal theme, with two aquarium eco-systems and a special educational area that includes nature-oriented books and games that makes the waiting area special for the patients. The waiting areas, exam rooms, corridors, and work stations are illuminated by natural daylight and have views to the outdoors.

Jim McGregor, AIA, LEED® AP of Browne Penland McGregor Stephens Architects was project manager for Pearland Pediatrics.

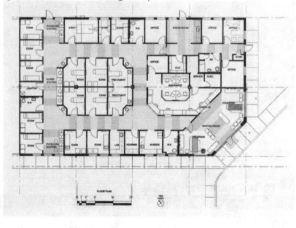

LEED® CERTIFIED	
MANUFACTURERS/SUPPLIERS	
DIV 07:	*White Reflective Membrane:* Firestone.
DIV 08:	*Entrances & Storefronts, Windows:* Vistawall; *High Performance Insulated Glazing:* **Arch Aluminum & Glass Co.;** *Doors, Frames & Interior Window Frames: Versatrac.*
DIV 09:	*Linoleum:* Forbo; *Low Emitting Paint:* Harmony™ by Sherwin Williams.
EXTENDED PRODUCT INFORMATION	
High Performance Insulated Glazing: **Arch Aluminum & Glass Co.;**	

COST PER SQUARE FOOT $ 194.31

ARCHITECT
SYNTHESIS LLP
162 Jay Street
Schenectady, NY 12305
www.synthesisllp.com

GENERAL CONTRACTOR
TURNER CONSTRUCTION COMPANY
54 State Street
Albany, NY 12207
www.turnerconstruction.com

FILE UNDER
OFFICE
Schenectady, New York

CONSTRUCTION TEAM

STRUCTURAL ENGINEER:
Almy & Associates Consulting Engineers
238 Genesee Street, Utica, NY 13502
ELECTRICAL & MECHANICAL ENGINEER:
Plumb Engineering PC Consulting Engineers
4 Wolfert Avenue, Albany, NY 12204
LANDSCAPE ARCHITECT: Synthesis LLP
162 Jay Street, Schenectady, NY 12305
COST ESTIMATOR: Turner Construction Company
54 State Street, Albany, NY 12207

GENERAL DESCRIPTION

SITE: 0.33 acres.
NUMBER OF BUILDINGS: One.
BUILDING SIZE: First floor, 31,250; second floor, 31,250;
each additional floor, 31,250; total, 125,000 square feet.
BUILDING HEIGHT: First floor, 14'6"; second floor, 14'6";
each additional floor, 14'6": floor to floor, 14'6"; penthouse 15',
total, 73'.
BASIC CONSTRUCTION TYPE: New/Steel frame.
FOUNDATION: Mat slab on grade.
EXTERIOR WALLS: Masonry.
ROOF: Membrane.

First Floor Plan

FLOORS: Carpet, VCT, tile.
INTERIOR WALLS: Drywall.

NEW YORK STATE D.O.T. REGION ONE HEADQUARTERS
Construction Period: Sep 2001 to Feb 2003 • Total Square Feet: 125,000

C.S.I. Divisions (1 through 16)	COST	% OF COST	SQ.FT. COST		SPECIFICATIONS
BIDDING REQUIREMENTS	1,378,530	7.50	11.03		Bonds & certificates, LEED® designer, performance bond, permits & fees, builder's risk insurance.
1. GENERAL REQUIREMENTS	1,795,126	9.77	14.36	1	General conditions, testing, documents, administration.
3. CONCRETE	1,120,300	6.10	8.96	3	Formwork, lightweight concrete, miscellaneous, stair pans.
4. MASONRY	1,472,100	8.01	11.78	4	Masonry & grout, stone, washdowns.
5. METALS	1,899,100	10.34	15.19	5	Structural steel, lintels, door frames, miscellaneous. framing,stairs.
6. WOOD & PLASTICS	72,700	0.40	0.58	6	Counter top, cabinets, architectural woodwork, moistureresistant window sills.
7. THERMAL & MOIST. PROTECTION	343,150	1.87	2.75	7	EPDM, manufactured roofing & siding, hatches, control joints.
8. DOORS AND WINDOWS	416,600	2.27	3.33	8	Doors, exterior doors, hardware, miscellaneous glazing,overhead doors, entrances & storefronts, windows.
9. FINISHES	1,801,946	9.81	14.42	9	Gypsum, interior partitions, VCT, wall coverings, tile, carpet,painting.
10. SPECIALTIES	676,200	3.68	5.41	10	Identifying devices, fire protection, bicycle racks, flagpoles,laminates, toilet compartments, operable partitions, toiletaccessories, window blinds.
11. EQUIPMENT	-	-	-	11	-
12. FURNISHINGS	446,684	2.43	3.57	12	Furnishings & equipment.
13. SPECIAL CONSTRUCTION	-	-	-	13	-
14. CONVEYING SYSTEMS	190,300	1.04	1.52	14	Elevators 2 (1 freight, 1 passanger).
15. MECHANICAL	2,698,530	14.68	21.59	15	Basic materials & methods, fire protection, plumbing, HVAC,vibration & seismic control, insulation, air curtains, diffusers, registers, & grilles, louvers, testing, adjusting & balancing.
16. ELECTRICAL	4,060,290	22.10	$32.48	16	Basic materials & methods, service & distribution, lighting,communications, controls, testing, special systems, fire alarm.
TOTAL BUILDING COST	**18,371,556**	**100%**	**$146.97**		
2. SITE WORK	542,500			2	Preparation, earthwork, paving & surfacing, connecting walkway, landscaping, stormwater & connection.
LANDSCAPING & OFFSITE WORK					
TOTAL PROJECT COST	**18,914,056**	*(Excluding architectural and engineering fees)*			

UPDATED ESTIMATE TO JANUARY 2011: $243.48 SF

DCD Subscribers: Access this case study and hundreds more for instant date and location calculations at www.dcd.com.

New York State D.O.T. Region One Headquarters
Schenectady, New York

Architect: Synthesis LLP
General Contractor: Turner Construction Company

Synthesis LLP was the lead design consultant for the construction of the new $20 million regional headquarters for the New York State Department of Transportation in downtown Schenectady. Synthesis worked with NYS Office of General Services (OGS) team leaders, Turner Construction Company, and AJS Masonry, Inc. to develop a state-of-the-art design solution that is integrated with the traditional city fabric. The building replaces historic buildings that were unsuitable for reuse and there was significant interest in creating a building that would be sensitive to its central downtown core location. The 125,000-square-foot, four-story office building is a U.S. Green Building Council LEED® Certified Silver level "green building." The design uses locally produced construction materials and a high degree of detail and craft.

The building respects traditional building structure by using base, middle and top proportions. The cast stone base at the ground floor enframes glass bays that provide a high level of transparency, create pedestrian comfort, and allow the entrance of natural light. The brick upper floors are modulated to create character at each level and are interlaced with cast stone to cap the building. The four-inch deep full-height pilasters, at 30 feet, express the steel structure of the building and

Photos Courtesy of Paul Newman, Synthesis LLP

traditional building widths. The building maintains its depth and detailing even at night with carefully designed lighting.

The building exhibits a high degree of quality materials and craftsmanship. Brick and cast stone are woven together to create a richly textured structure. The building uses a red wire struck "FBS" clay brick that showcases diverse bricklaying

techniques. Included are true brick arches, saw-tooth panels, rowlock coursing, soldier courses, header courses, stack bonds and striations. The cast stone uses 10-inch returns at the ground floor window bays to create building depth. Offset bands of cast stone create horizontal modulation and deep striations with beveled edges to create significant building shadows. Cast stone inset panels, sills and accents highlight the intricate brick detailing. The cast stone is engraved at the main building corner with primary street names and an engraved cornerstone at the entry marks the completion of the building and pays tribute to the design professionals and construction team.

LEED® SILVER		
MANUFACTURERS/SUPPLIERS		
DIV 07:	*Membrane:* **Firestone.**	
DIV 09:	*Acoustical Treatment:* **Armstrong;** *VCT:* **Armstrong;** *Carpet:* **Lees.**	
DIV 14:	*Elevators:* **Otis.**	

COST PER SQUARE FOOT $ 243.48

ARCHITECT
CLARK NEXSEN ARCHITECTURE & ENGINEERING
6160 Kempsville Circle, #200A
Norfolk, VA 23502
www.clarknexsen.com

FILE UNDER
OFFICE
Norfolk, Virginia

CONSTRUCTION TEAM

GENERAL CONTRACTOR:
Hourigan Construction Corp.
5544 Greenwich Road, #100, Virginia Beach, VA 23462
STRUCTURAL ENGINEER:
Clark Nexsen Architecture & Engineering
6160 Kempsville Circle, #200A, Norfolk, VA 23502
ELECTRICAL & MECHANICAL ENGINEER:
Clark Nexsen Architecture & Engineering
6160 Kempsville Circle, #200A, Norfolk, VA 23502
LANDSCAPE ARCHITECT:
Siska Aurand
523 West 24th Street, Norfolk, VA 23517

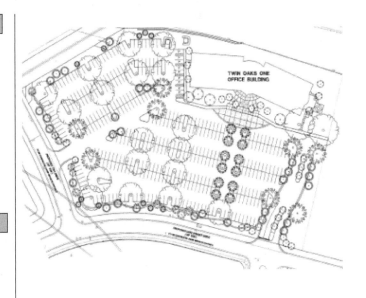

GENERAL DESCRIPTION

SITE: 6 Acres.
NUMBER OF BUILDINGS: One; 4-story, Class "A" office building.
BUILDING SIZE: First floor, 22,030; second floor, 22,030; each of two floors, 22,900; total 89,860 square feet.
BUILDING HEIGHT: First floor, 13'8"; second floor, 13'8"; third floor, 13'8"; fourth floor, 14'8"; total 58'2".
BASIC CONSTRUCTION TYPE: New/2C.
FOUNDATION: Cast-in-place.
EXTERIOR WALLS: Brick, precast concrete.
ROOF: EPDM single-ply membrane.
FLOORS: Carpet, tile.
INTERIOR WALLS: Gypsum board.

TWIN OAKS I OFFICE TOWER
Construction Period: Aug 1998 to May 1999 • Total Square Feet: 89,860

C.S.I. Divisions (1 through 16)			COST	% OF COST	SQ.FT. COST	SPECIFICATIONS
1.	1.	GENERAL REQUIREMENTS	482,057	9.54	5.37	Summary of work, measurement & payment, coordination, field engineering, regulatory requirements, project meetings, submittals, quality control, construction facilities & temporary controls, facility startup/commissioning, contract closeout, overhead & profit, modification procedures.
3.	3.	CONCRETE	736,400	14.57	8.20	Formwork, reinforcement, accessories, cast-in-place, precast, grout. 2.84 4 Masonry & grout, accessories, unit.
4.	4.	MASONRY	255,600	5.06	2.84	Masonry & grout, accessories, unit.
5.	5.	METALS	879,800	17.41	9.79	Structural framing, joists, decking, cold formed framing, fabrications, sheet metal fabrications.
6.	6.	WOOD/PLASTICS/COMPOSITE	42,660	0.84	0.47	Rough carpentry, finish carpentry, wood treatment, architectural woodwork.
7.	7.	THERMAL & MOIST. PROTECTION	140,751	2.78	1.57	Waterproofing, vapor retarders, insulation, fireproofing, firestopping, manufactured roofing & siding, membrane roofing, flashing & sheet metal, roof specialties & accessories, joint sealers.
8.	8.	OPENINGS	481,586	9.53	5.36	Metal doors & frames, wood & plastic doors, special doors, entrances & storefronts, metal windows, glazing, hardware.
9.	9.	FINISHES	640,982	12.68	7.13	Metal support systems, gypsum board, tile, acoustical treatment, resilient flooring, carpet, painting, wall coverings.
10.	10.	SPECIALTIES	38,180	0.76	0.42	Compartments & cubicles, pedestrian control devices, toilet & bath accessories.
12.	12.	FURNISHINGS	23,080	0.46	0.26	Furniture & accessories, rugs & mats.
14.	14.	CONVEYING SYSTEMS	98,500	1.95	1.10	Elevators (2).
15.	15.	MECHANICAL	861,165	17.04	9.58	Basic materials & methods, insulation, fire protection, plumbing, HVAC, air distribution, controls, testing, adjusting & balancing.
16.	16.	ELECTRICAL	372,900	7.38	4.15	Basic materials & methods, medium voltage distribution, service & distribution, lighting, communications, controls, testing.
TOTAL BUILDING COST			**5,053,661**	**100%**	**$56.24**	
2.	2.	SITE WORK	695,345			Included in Site Work.
TOTAL PROJECT COST			**5,749,006**	*(Excluding architectural and engineering fees)*		

UPDATED ESTIMATE TO JAN 2011: $120.37 SF

DCD Subscribers: Access this case study and hundreds more for instant date and location calculations at www.dcd.com.

Twin Oaks I Office Tower
Norfolk, Virginia

Architect: Clark Nexsen Architecture & Engineering

Photos Courtesy of David J. Moniot, AIA

One of the more notable energy-saving features of the project was that this existing developed site, which had been abandoned for years, was recycled into a thriving office community by actually incorporating existing items into the new design: old asphalt paving was recycled and used in the site fill under roads and parking areas; existing light poles were refurbished and used throughout the site and along the streets.

Twin Oaks I received a "2000 Merit Award" for Best Office Building in Hampton Roads over 75,000 square feet. Awarded by The Hampton Roads Association of Commercial Real Estate, "HR ACRE."

Together, Clark Nexsen Architecture and Engineering, the City of Norfolk Economic Development Department and the private developer The GEES Group, transformed Norfolk's old Met Baseball Park into a 40-acre upscale executive center. The existing site was an abandoned baseball stadium and parking area that had served for many years as home to the Tidewater Tides (a minor-league, farm-club for the New York Mets.)

After arrangements and designs were completed, the entire construction process took only six months to complete both the infrastructure and the first building. Twin Oaks I is the first of two buildings to be constructed in the office park. It was necessary to complete the construction for occupancy within a 6-month time frame in order to meet the Spring leasing season and to accommodate previously-leased tenants. The building was strategically sited immediately adjacent to I-64 and is in the most discernible location in the Executive Center in order to provide excellent visibility and also establish market awareness. The building includes a 2-story entrance lobby with Italian marble floors and a grand circular staircase. It is constructed with brick veneer, accented with precast concrete elements and aluminum windows over a steel substructure.

Sustainable design features played a key part of the project's design. Because the site was adjacent to the City of Norfolk's supply for water intake, water quality was a primary concern. Stormwater

runoff was an absolute design requirement. Retention ponds were included in the executive center's design to manage runoff and to add a pleasant landscape amenity feature.

Energy efficiency was also a major factor in selecting the heating and air conditioning systems for Twin Oaks I. A complete cost analysis was performed and several systems evaluated. The siting of the building and natural lighting from window units were also considered along with landscaping around the building providing shade for the first floor.

MANUFACTURERS/SUPPLIERS	
DIV 07:	*Membrane:* **Firestone.**
DIV 08:	*Doors:* **Marshfield DoorSystems**
	Hardware: **Corbin Russwin, Hager Dorma, Pempko.**
DIV 09:	*Toilet & Bath Accessories:* **Bobrick**
	Ceiling Tile: **Armstrong;**
	Carpet: **SHAW**
DIV 14:	*Elevators:* **Schindler**

COST PER SQUARE FOOT $ 120.37

ARCHITECT
Cooke Douglass Farr Lemons Architects & Engineers,
PA 3100 North State Street, #200, Jackson, MS 39216
www.cdfl.com

CONSTRUCTION TEAM

STRUCTURAL ENGINEER:
Spencer-Engineers, Inc.
P. O. Box 4328, Jackson, MS 39296
MECHANICAL ENGINEER & COMMISSIONING AGENT:
Atherton Consulting Engineers, Inc.
1900 Lakeland Drive, Jackson, MS 39216
ELECTRICAL ENGINEER:
Cooke Douglass Farr Lemons Architects & Engineers,
PA 3100 North State Street, #200, Jackson, MS 39216
GENERAL CONTRACTOR:
Evan Johnson & Sons Construction
1165 Weems Street, Pearl, MS 39208

GENERAL DESCRIPTION

SITE: 1.59 Acres.
NUMBER OF BUILDINGS: One.
BUILDING SIZE: Garage, 7,212; first floor, 11,978; second thru fifth floor, 20,685; sixth floor, 19240; total, 121,170 square feet.
BUILDING HEIGHT: First floor, 14'; second thru fifth floor, 11'1"; floor to floor, 11'1"; penthouse, 17'; total, 86'7".
BASIC CONSTRUCTION TYPE: Renovation/Concrete frame.
FOUNDATION: Pier & grade beam, reinforced concrete.
EXTERIOR WALLS: CMU, curtainwall, insulated metal wall panels, porcelain tile.
ROOF: Built-up. **FLOORS:** Concrete.
INTERIOR WALLS: Metal stud drywall.

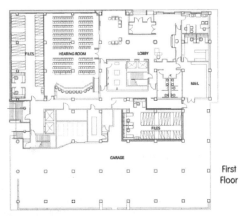

First Floor

Second Floor

MISSISSIPPI DEPARTMENT OF ENVIRONMENTAL QUALITY
Construction Period: Aug 2007 to Sep 2008 • Total Square Feet: 121,170

C.S.I. Divisions (1 through 16)			COST	% OF COST	SQ.FT. COST	SPECIFICATIONS
		PROCUREMENT & CONT. REQ.	316,640	5.05	2.61	Bonds, insurance, tax.
1.	1.	GENERAL REQUIREMENTS	280,975	4.48	2.32	Temporary facilities & controls.
3.	3.	CONCRETE	67,145	1.07	0.55	Forming & accessories, reinforcing, cast-in-place, precast, cast decks & underlayment, grouting, mass.
4.	4.	MASONRY	20,545	0.33	0.17	Unit.
5.	5.	METALS	97,180	1.55	0.80	Structural metal framing, joists, decking, cold-formed metal framing, fabrications.
6.	6.	WOOD/PLASTICS/COMPOSITE	210,930	3.36	1.74	Rough carpentry, finish carpentry, architectural woodwork.
7.	7.	THERMAL & MOIST. PROTECT	2,694,350	42.98	22.24	Dampproofing & waterproofing, thermal protection, weather barriers, roofing & siding panels, membrane roofing, flashing & sheet metal, roof & wall specialties & accessories, fire & smoke protection, joint protection.
8.	8.	OPENINGS	650,650	10.38	5.37	Doors & frames, speciality doors & frames, entrances, storefronts, & curtainwalls, windows, hardware, glazing, louvers & vents.
9.	9.	FINISHES	1,281,075	20.44	10.57	Plaster & gypsum board, tiling, ceilings, flooring, wall finishes, painting & coating, acoustic treatment.
10.	10.	SPECIALTIES	78,530	1.25	0.65	Information, interior, safety.
11.	11.	EQUIPMENT	72,390	1.15	0.60	—
14.	14.	CONVEYING SYSTEMS	39,375	0.63	0.32	Elevators.
15.	21.	FIRE SUPRESSION	44,200	0.71	0.37	Water-based fire-suppression systems, fire-extinguishing systems.
15.	22.	PLUMBING	5,460	0.09	0.05	—
15.	23.	HVAC	172,200	2.75	1.42	Piping & pumps, air distribution, air cleaning devices.
16.	26.	ELECTRICAL	237,350	3.78	1.96	Medium-voltage distribution, lighting.
TOTAL BUILDING COST			**6,268,995**	**100%**	**$51.74**	
2.	2.	EARTHWORK	41,415			Earth moving, shoring & underpinning, excavation support & protection.
2.	32.	EXTERIOR IMPROVEMENTS	21,590			Bases, bollards, & paving, improvements, irrigation, planting.
TOTAL PROJECT COST			**6,332,000**	*(Excluding architectural and engineering fees)*		

UPDATED ESTIMATE TO JAN 2011: $72.17 SF

Mississippi Department of Environmental Quality

Jackson, Mississippi

Architect: Cooke Douglass Farr Lemons Architects & Engineers, PA

Photos Courtesy of De Shields-Marley

Cooke Douglass Farr Lemons Architects and Engineers, PA was awarded a large renovation in late 2007 by the Mississippi Bureau of Building, Grounds and Real Property Management that presented the firm with many opportunities to employ environmentally sustainable techniques and materials. The CDFL team was charged with maximizing the indoor environmental quality of two buildings with failed exterior surfaces and ineffective mechanical systems. The Mississippi Department of Environmental Quality currently occupies both buildings.

The first structure that was renovated was a 6-story, 110,000 square-foot facility located at 515 East Amite Street which previously housed the LDDS WorldCom headquarters. Exterior renovations included removal of the ineffective Exterior Insulated Fenestration System (EIFS) building envelope, or "skin," that was replaced with a 2-inch insulated metal panel system with integrated windows providing a water-tight facade. All of the windows were replaced with insulated "Low-E" glass to keep out harmful ultra-violet light rays and solar heat gain further enhancing the new HVAC system and occupant comfort. The primary sub-contractor, F. L. Crane and Sons, recycled all existing aluminum window frames removed during the renovation.

Interior upgrades involved removing 8,500 yards of old carpet that was recycled through Mohawk Group's Recover Program. This was the largest quantity of carpet ever recycled not only in Mississippi but also in the surrounding states. CDFL's Interior Architecture department chose one of Mohawk's newest modular carpets for the renovation. Mohawk's products are created from recycled materials, such as plastic soda and water bottles, and specifically designed to meet the United States Green Building Council's (USGBC) stringent low-chemical emissivity standards to improve indoor air quality. Furthermore, all interior walls received low-VOC paints to minimize odor

and the harmful Volatile Organic Compounds found in most paints. Evan Johnson & Sons Construction served as the general contractor for the project that was budgeted at $6,000,000.

The second phase of the project will focus on the 5-story, 62,670-square-foot building at 700 State Street where similar interior and mechanical renovations are necessary.

MANUFACTURERS/SUPPLIERS	
DIV 07:	*Insulated Metal Panels::* **Formawall®** **Dimension Series** *Built-Up Roof::* **Johns Manville.**
DIV 08:	*Curtainwall, Entrances & Storefronts:* **Kawneer.**
DIV 09:	*Tile:* **American Olean, Crossville.** *Carpet:* **Mohawk, Invision, Shaw;**
DIV 26:	*Light Fixtures:* **Visa, Omega, Daybrite, Kirlin.**

COST PER SQUARE FOOT $ 72.17

ARCHITECT
Durrant Group, Inc.
1066 Executive Parkway, #100, St. Louis, MO 63141
www.durrant.com

FILE UNDER
OFFICE
Highland, Illinois

CONSTRUCTION TEAM

STRUCTURAL ENGINEER:
Netemeyer & Associates
3300 Highline Road, Aviston, IL 62216
MECHANICAL ENGINEER:
Toennies Service
219 E. Main Street, Damiensville, IL 62215
ELECTRICAL ENGINEER:
DRDA Electric Company
24 Kettle River Drive, Glen Carbon, IL 62034
GENERAL CONTRACTOR:
Korte & Luitjohan Contractors, Inc.
12052 Highland Road, Highland, IL 62249

GENERAL DESCRIPTION

SITE: 10 Acres.
NUMBER OF BUILDINGS: Two.
BUILDING SIZE: Warehouse, 16,000; basement, 1,000; first floor, 9,000; total, 26,000 square feet.
BUILDING HEIGHT: Warehouse, 22'; basement, 10'; first floor, 20'; total, 20'.
BASIC CONSTRUCTION TYPE: New.
FOUNDATION: Cast-in-place, pier & grade beam, reinforced concrete, slab-on-grade, insulated concrete forms (ICF).
EXTERIOR WALLS: Brick, ICF.
ROOF: Membrane, EPDM. **FLOORS:** Concrete.
INTERIOR WALLS: Metal stud drywall.

KORTE & LUITJOHAN OFFICE & SHOP
Construction Period: June 2006 to Sep 2007 • Total Square Feet: 26,000

C.S.I. Divisions (1 through 16)			COST	% OF COST	SQ.FT. COST	SPECIFICATIONS
1.	1.	GENERAL REQUIREMENTS	103,495	5.49	3.98	Summary, administrative requirements, quality requirements, temporary facilities & controls, product requirements, execution & closeout requirements, performance requirements, life cycle activities.
3.	3.	CONCRETE	409,712	21.72	15.76	Includes concrete for floors and walls forming & accessories, reinforcing, mass, (Concrete breakdown: cubic yards foundation, 202; cubic yards walls, 350; cubic yards floors, 598).
4.	4.	MASONRY	107,344	5.69	4.13	Unit.
5.	5.	METALS	235,239	12.47	9.05	Structural metal framing, joists, decking.
6.	6.	WOOD/PLASTICS/COMPOSITE	110,362	5.85	4.24	Rough carpentry, finish carpentry.
7.	7.	THERMAL & MOIST. PROTECT	131,190	6.95	5.05	Dampproofing & waterproofing, thermal protection, weather barriers, roofing & siding panels, membrane roofing.
8.	8.	OPENINGS	134,541	7.13	5.17	Doors & frames, specialty doors & frames, entrances, storefronts, & curtainwalls, windows, roof windows & skylights.
9.	9.	FINISHES	294,773	15.62	11.34	Plaster & gypsum board, tiling, flooring, painting & coating.
10.	10.	SPECIALTIES	5,952	0.32	0.23	Interior, exterior, other.
15.	22.	PLUMBING	29,659	1.57	1.14	Piping & pumps, fixtures.
15.	23.	HVAC	132,805	7.04	5.11	Central heating, central cooling.
16.	26.	ELECTRICAL	191,318	10.14	7.36	Medium-voltage distribution, low-voltage transmission, lighting.
16.	27.	COMMUNICATIONS	8,753	0.01	0.33	Data.
TOTAL BUILDING COST			**1,895,143**	**100%**	**$72.89**	
2.	2.	EARTHWORK	124,067			Assessment, demolition & structure moving, water remediation, facility remediation.
2.	3.	SITEWORK	62,400			Site clearing.
2.	32.	SITE CONCRETE	264,919			Site clearing.
TOTAL PROJECT COST			**2,346,529**	*(Excluding architectural and engineering fees)*		

UPDATED ESTIMATE TO JAN 2011: $93.49 SF

DCD Subscribers: Access this case study and hundreds more for instant date and location calculations at www.dcd.com.

Korte & Luitjohan Office & Shop
Highland, Illinois

Architect: Durrant Group, Inc.

The new headquarters for Korte & Luitjohan Contractors, Inc. includes 9,000 square feet of office space and a separate 16,000-square-foot warehouse maintenance building that serves a multitude of uses. The entire complex is located on 10 acres in Highland, Illinois.

The office facility has a total of 13 private offices for the President, General Manager, Project Managers and Estimators, with adjoining open layout for assistant project management and accounting departments. The office facility also includes a conference room which comfortably seats 12 people, a plan room for subcontractors and suppliers to review plans to prepare pricing, a fully functioning kitchen and break room including Energy Star rated appliances, and a training room that accommodates approximately 50 people. This space also includes a 1,000-square-foot basement which houses the facility's geo-thermal pump units, air handlers, and miscellaneous storage. Airlocks control exterior air flow. Heating and cooling for the office building is divided into seven zones and controlled by programmable thermostats.

The warehouse maintenance building includes an office for the shop foreman, a break room for the service personnel, storage of equipment parts, small construction equipment and tools, service garage for heavy equipment and company vehicles, a wash bay, and storage area for materials for construction projects.

Both buildings were built with "Green" building practices and materials in mind. The exterior of both buildings, including the perimeter basement walls were constructed utilizing Insulated Concrete Forms (ICF). The ICF walls of the buildings have an R-value over 25. The roof decks are constructed of Oriented Strand Board (OSB) instead of plywood, covered by a white, Thermoplastic Polyolefin (TPO) single-ply membrane to reflect solar heat, and have an R-value of 30. The buildings contain recycled components including carpeting, shelving,

Photos Courtesy of Teri Allison

ceiling tile, steel joists and studs, roof decking,doors and door frames, OSB wall sheathing, ICF wall forms, ISO roof insulation, TPO roof membrane. Recycled concrete backfill and subgrade were used as site materials.

Energy efficient "T5" fluorescent lighting illuminates 99.9% of the buildings, and while the office utilizes motion detectors to switch off many of the lights when not needed, the shop utilizes 21 skylights to reduce the need for light during the day.

All windows have Low "E" glass with encapsulated blinds. Plus most feature an operable lower unit to allow fresh air and reduce HVAC demand during the fall and spring seasons.

The warehouse is heated with radiant heat in the floors produced by a 99% clean burn waste oil boiler. The company is able to reuse the waste oil generated by its heavy equipment and trucks over the course of a year to heat the warehouse. The remainder of the company's waste oil is sold to a recycler.

Storm water is collected in an excess capacity detention pond for use in washing equipment coming in for maintenance or repair, and for filling water trucks and trailers for construction site needs

COST PER SQUARE FOOT $ 93.49

ARCHITECT
Architect Lott + Barber
110 East President Street, Savannah, GA 31410
www.lottbarber.com

CONSTRUCTION TEAM

STRUCTURAL ENGINEER:
W. Hunter Saussy, P.E.
400E Johnny Mercer Boulevard, Savannah, GA 31410
MECHANICAL & ELECTRICAL ENGINEER:
Rosser International, Inc.
109 Park of Commerce Drive, #6, Savannah, GA 31405
GENERAL CONTRACTOR & COST ESTIMATOR:
J.T. Turner Construction
2250 East Victory Drive, #104, Savannah, GA 31404

GENERAL DESCRIPTION

SITE: 1.4 Acres.
NUMBER OF BUILDINGS: One.
BUILDING SIZE: First floor, 7,195; second floor, 965; total, 8,160 square feet.
BUILDING HEIGHT: First floor, 12'9"; second floor, 15'1"; total, 27'10".
BASIC CONSTRUCTION TYPE: New/Wood Frame.
FOUNDATION: Slab-on-grade.
EXTERIOR WALLS: CMU, brick, curtain-wall.
ROOF: Metal. **FLOORS:** Concrete.
INTERIOR WALLS: Wood stud drywall, CMU.

LOTT ✦ BARBER

INTERNATIONAL LONGSHOREMAN'S ASSOCIATION BUILDING
Construction Period: Dec 2007 to Jan 2009 • Total Square Feet: 8,160

C.S.I. Divisions (1 through 16)			COST	% OF COST	SQ.FT. COST	SPECIFICATIONS
		PROCUREMENT & CONTRACTING	155,410	8.35	19.05	—
1.	1.	GENERAL REQUIREMENTS	110,489	5.94	13.54	—
3.	3.	CONCRETE	63,005	3.39	7.72	Forming & accessories, reinforcing, cast-in-place (Concrete breakdown: 88 cubic yards foundation, 88 cubic yards floors).
4.	4.	MASONRY	106,376	5.72	13.04	Stone assemblies.
5.	5.	METALS	147,700	7.94	18.10	Structural metal framing, joists, decking, cold-formed metal framing, fabrications.
6.	6.	WOOD/PLASTICS/COMPOSITE	321,407	17.27	39.39	Rough carpentry, finish carpentry, architectural woodwork.
7.	7.	THERMAL & MOIST. PROTECT	210,378	11.31	25.78	Weather barriers, roofing & siding panels, joint protection.
8.	8.	OPENINGS	150,944	8.11	18.50	Doors & frames, entrances, storefronts, & curtainwalls, hardware, glazing, louvers & vents.
9.	9.	FINISHES	131,820	7.08	16.15	Plaster & gypsum board, ceilings, flooring, wall finishes, painting & coating.
10.	10.	SPECIALTIES	20,343	1.09	2.49	Fountain, other.
11.	11.	EQUIPMENT	2,500	0.13	0.31	Security, detention & banking, foodservice.
15.	22.	PLUMBING	63,096	3.39	7.73	Piping & pumps, equipment, fixtures, water-based fire-suppression systems.
15.	23.	HVAC	108,900	5.85	13.35	Air distribution, central heating, central cooling, central HVAC equipment.
16.	26.	ELECTRICAL	183,946	9.88	22.54	Medium-voltage distribution, lighting.
16.	27.	COMMUNICATIONS	66,375	3.57	8.13	—
16.	28.	ELECTRONIC SAFETY & SECURITY	18,203	0.98	2.23	—
TOTAL BUILDING COST			**1,860,892**	**100%**	**$228.05**	
2.	2.	EXISTING CONDITIONS	14,084			Demolition & structure moving.
2.	31.	EARTHWORK	21,642			Site clearing, earth moving.
2.	32.	EXTERIOR IMPROVEMENTS	80,028			Bases, bollards, & paving.
2.	33.	UTILITIES	17,241			Water, sanitary sewer, storm.
TOTAL PROJECT COST			**1,993,887**	*(Excluding architectural and engineering fees)*		

UPDATED ESTIMATE TO JAN 2011: $321.58 SF

International Longshoreman's Association, Employee's Welfare Fund Building
Savannah, Georgia

Architect: Lott + Barber

A ward-winning commercial architecture firm Lott + Barber recently completed the new state-of-the-art International Longshoremen's Association Pension and Welfare Building at 10 Mersey Way in Savannah, Georgia. The office building, which features a dramatic covered arch over the front entrance, is currently pending LEED®-NC certification from the U.S. Green Building Council.

This energy-efficient structure houses the benefits office for the local International Longshoremen's Association chapters and serves as a model of sustainable commercial architecture in coastal Georgia. In addition to developing the architectural design, Lott + Barber also assisted in site selection and the LEED® certification application process for the handsome brick structure. J.T. Turner Construction of Savannah, Georgia served as the general contractor for the $2 million project.

"This new building incorporates a wide range of energy-efficient features that will reduce operating costs while minimizing the impact on the environment," said Lott + Barber principal Forrest R. Lott, FAIA, LEED®AP. "We were delighted to work with the International Longshoremen's Association to design an office building that would meet the needs of the workers of today and tomorrow, while also incorporating key elements of sustainable design."

Lott + Barber used state-of-the-art Building Information Modeling (BIM) technology, which allows the project team to "construct" a virtual building before the actual construction process begins, to reduce overall construction costs and maximize the efficiency of the building design. The construction project features water- saving plumbing fixtures designed to reduce water usage by 55%, ultra-efficient spray insulation and 100% recycled drywall. The building, which was occupied by the International Longshoremen's Association in November of 2008, also incorporates low-VOC carpet. A waste management plan helped divert 65% of construction waste from the local landfill, resulting in the recycling of more than 50 tons of material during the construction process.

Photos Courtesy of Lott + Barber

In another nod to sustainability, all the major building materials came from within a 500-mile radius. Lott + Barber elected to use Georgia pine for the framing, recycled drywall from Florida, recycled steel from around the Southeast and window frames manufactured in Dublin, Georgia in an effort to minimize transportation costs and resulting carbon emissions that can negatively impact the environment.

The structure has a reduced environmental footprint, which extends only 10 feet out from the building on all sides. All landscaping includes drought tolerant native plants eliminating the need for an irrigation system, saving over 600,000 gallons of drinking water each year.

Together, the building's sustainable features have reduced total energy use byapproximately 18%. "This building serves as an ideal example of how smart planning can help reduce costs at every level, from the design through the construction phase and beyond," said Steven G. Stowers, AIA, LEED®AP, of Lott + Barber. "The International Longshoremen"s Association will continue to reap the benefits of this building"s lower utility bills for many years to come."

**LEED® NC
CERTIFICATION PENDING**

COST PER SQUARE FOOT $ 321.58

ARCHITECT
Designstream LLC
30 Isabella Street, #101, Pittsburgh, PA 15212
www.dsstudio.biz

FILE UNDER
OFFICE
Southpointe, Canonsburg, Pennsylvania

CONSTRUCTION TEAM

OWNER/DEVELOPER:
Horizon Properties
375 Southpointe Boulevard, #410, Canonsburg, PA 15317
STRUCTURAL ENGINEER:
Churches Consulting Engineers
347 Locust Avenue, Washington, PA 15301
ELECTRICAL & MECHANICAL ENGINEER:
Claitman Engineering Associates, Inc.
1340 Old Freeport Road, Pittsburgh, PA 15238
GENERAL CONTRACTOR:
Continental Building Systems
285 E. Waterfront Drive, #150, Homestead, PA 15120

GENERAL DESCRIPTION

SITE: 24.8 Acres.
NUMBER OF BUILDINGS: One.
BUILDING SIZE: Lower level (mechanical/support/parking),
71,500; first floor, 77,000; second floor, 76,600; third floor, 46,000;
fourth floor, 46,400; exterior roof area (roof garden, 18,500); total,
365,000* square feet.
BUILDING HEIGHT: Basement, 14' 8"; first floor, 16'; second
floor, 14'; third floor, 14'; fourth floor, 16'; penthouse, 14'; total,
74'8" above grade, 88' plus basement.

BASIC CONSTRUCTION TYPE: New/Structural Steel.
FOUNDATION: Cast-in-place, pier & grade beam.
EXTERIOR WALLS: CMU, curtainwall, granite.
ROOF: Built-up, metal, membrane, green roof with pavers.
FLOORS: Concrete.
INTERIOR WALLS: CMU, metal stud drywall.

CONSOL ENERGY CORP. HEADQUARTERS
Construction Period: Nov 2006 to Aug 2008 • Total Square Feet: 365,000

	C.S.I. Divisions (1 through 16)	COST	% OF COST	SQ.FT. COST	SPECIFICATIONS
1.	1.	GENERAL REQUIREMENTS	5,010,606	9.79	15.78	Price & payment procedures, administrative requirements, quality requirements, temporary facilities & controls.
3.	3.	CONCRETE	3,691,012	7.21	11.63	Cast-in-place, cast decks & underlayment, grouting, mass (Concrete Breakdown: 1,200 cubic yards foundation, 5,630 cubic yards floor).
4.	4.	MASONRY	2,840,000	5.55	8.94	Unit, stone assemblies, corrosion-resistant masonry.
5.	5.	METALS	6,585,762	12.86	20.74	Structural metal framing, joists, decking, fabrications, decorative metal.
6.	6.	WOOD/PLASTICS/COMPOSITE	1,741,458	3.40	5.48	Rough carpentry, finish carpentry.
7.	7.	THERMAL & MOIST. PROTECT	3,134,260	6.12	9.87	Thermal protection, weather barriers, roofing & siding panels, membrane roofing, flashing & sheet metal, roof & wall specialties & accessories, fire & smoke protection, joint protection.
8.	8.	OPENINGS	4,674,709	9.13	14.73	Doors & frames, specialty doors & frames, entrances, storefronts, curtainwalls, windows, roof windows & skylights, hardware, glazing, louvers & vents.
9.	9.	FINISHES	6,528,168	12.75	20.56	Plaster & gypsum board, tiling, ceilings, flooring, wall finishes, acoustic treatment, painting & coating.
10.	10.	SPECIALTIES	924,603	1.81	2.91	Interior, exterior, other.
11.	11.	EQUIPMENT	40,000	0.08	0.13	Food service.
		FURNISHINGS	1,292,831	2.53	4.07	Casework, other.
		CONVEYING SYSTEMS	658,154	1.29	2.07	Elevators (4 Passenger, 1 Freight, 1 lift.)
		FIRE SUPRESSION	707,600	1.38	2.23	Water-based fire-suppression system.
15.	22.	PLUMBING	990,627	1.93	3.12	Piping & pumps, equipment.
15.	23.	HVAC	4,279,000	8.36	13.48	Piping & pumps, air distribution, air cleaning devices, central heating equipment, central cooling equipment, central HVAC equipment.
		INTEGRATED AUTOMATION	710,778	1.39	2.24	Network, instrumentation and terminal devices, facility controls, control sequences.
16.	26.	ELECTRICAL	6,497,000	12.68	20.46	Medium-voltage distribution, electrical & cathodic protection, lighting.
16.	27.	COMMUNICATIONS	576,159	1.12	1.82	Data.
16.	28.	ELECTRONIC SAFETY & SECURITY	318,226	0.62	1.00	Monitoring and control.
		TOTAL BUILDING COST	**51,200,953**	**100%**	**$161.26**	
2.	2.	EXISTING CONDITIONS	112,000			Assessment.
2.	31.	EARTHWORK	3,041,420			Site clearing, earth moving, earthwork methods, special foundations and load-bearing elements.
2.	32.	EXTERIOR IMPROVEMENTS	2,317,358			Bases, ballasts, & paving, improvements, planting.
		TOTAL PROJECT COST	**56,671,731**	*(Excluding architectural and engineering fees)*		

UPDATED ESTIMATE TO JAN 2011: $208.12 SF

DCD Subscribers: Access this case study and hundreds more for instant date and location calculations at www.dcd.com.

CONSOL Energy Corporation Headquarters

Southpointe, Canonsburg, Pennsylvania

Architect: Designstream LLC

Photos Courtesy of Ed Massery, Massery Photography

With gestures towards elements of the earth and the energy industry, CONSOL Energy's new CNX Center, headquarters incorporates architecture as a reflection of mining history. Tinted sloping glass and polished granite panels are used as abstractions of coal seams. Sweeping metal roofs reference the conveyor belts essential to product delivery, while other elements symbolize the ventilation towers and lanterns. Main entrances are clearly defined with the "erosion of coal" granite cutaways. Strategic night lighting uses blue low voltage flame to represent the CNX Gas Corporation, a CONSOL Energy company, as well as energy conservation in general. In an effort to generate a unique architectural signature; the building shape and design is crafted into the CONSOL logo when viewed from above.

While railings and other interior elements reflect the pilings and entry portals indicative to this business, the two-story atrium is flooded with natural light and generous seating. The two translucent skylights are each 220 feet long, and eliminate the need for artificial light in almost 10,000 square feet of the building during day time operation. This open-air, naturally lit environment promotes co-mingling of staff in "no-wall" meetings. The free exchange of ideas strengthens communications between departments and the company.

From the project onset the client requested the architect explore options to break down the "departmental silos" and find ways for departmental staff to interact more. Subsequently, all common conference rooms were located on the first floor and lower level to promote physical travel and interaction. Some were left "open air" to invite participation and the sharing of information, while most use sophisticated room scheduling software integrated with all desktop personal calendars. Kitchenettes and coffee stations

are located between departments rather than within, promoting the same cross-fertilization of ideas.

The main lobby and second floor atrium offer multiple seating venues for work, discussions, research, or lunch breaks during inclement weather. Gathering spaces outside of group training rooms double as welcome centers and a museum which displays the history of the mining industry and CONSOL Energy. Both Interior and exterior signage are integrated, reflecting the shape of the building down to the last detail. The "green" earthen roof offers a respite from the work day with café-style tables, landscaping and plaza seating for all employees.

Implementing LEED® practices into 365,000 square feet of useable real estate, the building is full of highly efficient technology and energy monitoring equipment to improve communications and reduce consumption. CONSOL "leads by example." As one of the nation's largest producers of energy; the new much larger headquarters building will consume significantly less energy than their previous location.

A quote from Continental Building Systems regarding project safety: "The CONSOL Energy Headquarters construction project earned the most prestigious recognition offered by the Occupational Safety and Health Administration (OSHA) when it was named a Star Site in the agency's Voluntary Protection Program (VPP). The VPP certification process entailed an exhaustive review of the project's safety and health programs, including management commitment, employee involvement, worksite analyses such as inspections and Job Hazard Analyses, extensive training including the OSHA 30-Hour Course, and subcontractor involvement. In addition, a team of four OSHA representatives verified the site's safety culture with a week-long onsite assessment and interviews with over 120 contractor employees working on the project."

LEED® SILVER

COST PER SQUARE FOOT $ 208.12

ARCHITECT
Webb Architects
3701 Kirby Drive, Houston, TX 77098

CONSTRUCTION TEAM

OWNER:
Jacob White Development LLC
2000 West Parkwood, #200, Friendswood, TX 77546
MECHANICAL & ELECTRICAL ENGINEER:
C.F. McDonald Electric, Inc.
5044 Timber Creek, Houston, TX 77017
GENERAL CONTRACTOR & COST ESTIMATOR:
Jacob White Construction Company
2000 West Parkwood, #100, Friendswood, TX 77546

GENERAL DESCRIPTION

SITE: 1.659 Acres.
NUMBER OF BUILDINGS: One.
BUILDING SIZE: First floor, 12,042; second floor, 12,042; total, 24,084 square feet.
BUILDING HEIGHT: First floor, 14'3"; second floor, 14'3"; total, 32'6".
BASIC CONSTRUCTION TYPE: New/Shell Only.
FOUNDATION: Pier & grade beam, slab-on-grade.
EXTERIOR WALLS: CMU, curtainwall.
ROOF: Membrane, living.
FLOORS: Concrete.
INTERIOR WALLS: Metal stud drywall.

First Floor Plan

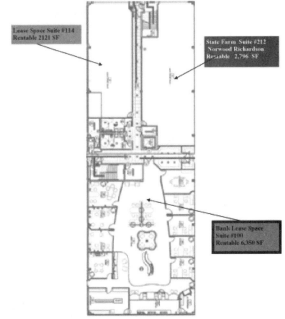

FC GULF FREEWAY					
Construction Period: Aug 2009 to April 2010 • Total Square Feet: 24,084					
C.S.I. Divisions (1 through 16)			COST	% OF COST	SQ.FT. COST	SPECIFICATIONS
1.	1.	GENERAL REQUIREMENTS	733,801	21.89	30.47	—
3.	3.	CONCRETE	235,930	7.04	9.80	Forming & accessories, reinforcing, cast-in-place.
5.	5.	METALS	522,560	15.59	21.70	Structural steel, fabrications.
6.	6.	WOOD/PLASTICS/COMPOSITE	13,000	0.39	0.54	Rough carpentry, finish carpentry.
7.	7.	THERMAL & MOIST. PROTECT	222,280	6.63	9.23	Dampproofing & waterproofing, thermal protection, metal roofing, joint protection.
8.	8.	OPENINGS	403,660	12.04	16.76	Doors & frames, entrances, windows, hardware, glazing.
9.	9.	FINISHES	287,960	8.59	11.96	Plaster & gypsum board, ceilings, flooring, wall finishes, acoustic treatment, painting & coating.
10.	10.	SPECIALTIES	7,412	0.22	0.31	Exterior.
14.	14.	CONVEYING SYSTEMS	59,600	1.78	2.47	Elevators (1).
15.	15.	MECHANICAL	579,184	17.28	24.05	Fixtures, plumbing, HVAC, rough plumbing.
16.	26.	ELECTRICAL	286,716	8.55	11.89	Rough in, service, fixtures, fire alarm.
TOTAL BUILDING COST			**3,352,103**	**100%**	**$139.18**	
2.	2.	SITE WORK	907,677			Demolition, sidewalks, paving, excavation, fencing, irrigation, landscaping.
TOTAL PROJECT COST			**4,259,780**	*(Excluding architectural and engineering fees)*		
UPDATED ESTIMATE TO JAN 2011: $165.21 SF						

FC Gulf Freeway
Houston, Texas

Architect: Webb Architects

Photos Courtesy of Jacob White Construction Company

Jacob White Construction Company strives to exceed client expectations in all areas and did so with the FC Gulf Freeway, a state-of-the-art commercial building completed in April 2010. Strategically located at the entrance of the Metro Park and Ride near the corner of IH45 and Beltway 8, the FC Gulf Freeway building is easily accessible from the entire south and southeast regions of Houston, Texas.

Jacob White focused on Green technology in the design and completion of the 24,000-square-foot LEED® Platinum registered building. Green technology can reduce water consumption, electricity and waste that will improve efficiency, air quality, and increase the overall value of the building. Studies show that improved air quality and the use of materials that do not emit volatile organic compounds can increase productivity from five to 28 percent based on the type of business. Indoor Air Quality Standards (IAQ) were met using filtration, air stream purification, tobacco smoke control and CO_2 monitoring with enhanced ventilation effectiveness.

Indoor climate controls were designed and distributed to allow both perimeter and non-perimeter spaces to be independently controlled with thermal comfort meeting the applicable ASHRAE standards; monitored on a real-time basis by the building's energy management system. Ninety percent of the spaces within the building have access to daylight and a view of the exterior. The building also includes the latest in "touch technologies," a developed anti-microbial surface shown to eliminate the threat of common colds and flu.

The building is home to Houston's largest living (green) roof, as well as many green features such as storing over 465,000 gallons of storm water runoff for irrigation and water closet use, illuminating the exterior with highly efficient LED lighting, and power supplied by 100% green power.

High efficiency, CFC free cooling equipment with electric heat was used in the building. It included MERV 13 filtration, CO2 monitoring and UV-C light air stream purification treatment. Equipment is monitored and controlled by an extensive energy management system.

The building has a calculated roof R-value of 61 (excluding the evaporative cooling effect of the green roof); wall R-values of 24+ and 1-inch insulated glazing with minimal UV transmittance and low emissivity coatings.

The building is projected to use an average of 30,000 kWh per month of electrical power. Overall energy reduction techniques (equipment, construction and operation) resulted in an energy usage level of 55% less than a baseline building.

Upon completion of 12941 Gulf Freeway, Jacob White Construction now officially has four " LEED® buildings" under its belt and is working on its own new 10,000 square foot headquarters being built to LEED® Platinum certified standards.

LEED® PLATINUM PENDING

COST PER SQUARE FOOT $ 165.21

ARCHITECT
BNIM ARCHITECTS
106 W. 14th Street, #200
Kansas City, MO 64105
www.bnim.com

CONSTRUCTION TEAM
STRUCTURAL ENGINEER:
Structural Engineering Associates
1000 Walnut, #1570, Kansas City, MO 64106
GENERAL CONTRACTOR:
Professional Contractors and Engineers, Inc.
5900-C North Tower Drive, Columbia, MO 65202
ELECTRICAL ENGINEER: FSC, Inc.
3100 South 24th Street, Kansas City, KS 66106
MECHANICAL ENGINEER: Smith Boucher
25501 W. Valley Parkway, #200, Olathe, KS 66061

GENERAL DESCRIPTION
SITE: 5.6 acres. **NUMBER OF BUILDINGS:** One.
BUILDING SIZE: First floor, 30,000; second - fourth floor, 30,000; total, 120,000 square feet.
BUILDING HEIGHT: First floor, 14'; second and third floor, 14'; fourth floor, 23'; total, 65'.
BASIC CONSTRUCTION TYPE: New.
FOUNDATION: Pier & grade beam, reinforced concrete, slab-on-grade.

EXTERIOR WALLS: CMU, curtainwall, fiber cement siding.
ROOF: Metal. **FLOORS:** Concrete.
INTERIOR WALLS: Metal stud drywall.

LEWIS AND CLARK STATE OFFICE BUILDING

Construction Period: Jun 2003 to Mar 2005 Total Square Feet 120,000

C.S.I. Divisions (1 through 16)			COST	% OF COST	SQ.FT. COST	SPECIFICATIONS
		PROCUREMENT & CONT. REQ.	824,000	4.80	6.87	Project forms, conditions of contract, revisions, clarifications, & modifications.
1.	1.	GENERAL REQUIREMENTS	442,000	2.57	3.68	Quality requirements, temporary facilities & controls, product requirements.
3.	3.	CONCRETE	3,390,000	19.74	28.25	Forming & accessories, reinforcing, cast-in-place, precast.
4.	4.	MASONRY	157,000	0.91	1.31	Unit.
5.	5.	METALS	855,000	4.98	7.13	Structural metal framing, cold-formed framing, fabrications, decorative.
6.	6.	WOOD/PLASTICS/COMPOSITE	1,344,000	7.83	11.20	Rough & finish carpentry, architectural woodwork, structural composites.
7.	7.	THERMAL & MOIST. PROTECT	1,076,000	6.27	8.97	Dampproofing & waterproofing, thermal protection, weather barriers, roofing & siding panels, membrane roofing, flashing & sheet metal, fire & smoke protection, joint protection.
8.	8.	OPENINGS	943,000	5.49	7.86	Doors & frames, specialty doors & frames, entrances, storefronts, & curtainwalls, windows, roof windows & skylights, hardware, glazing, louvers & vents.
9.	9.	FINISHES	2,047,000	11.92	17.06	Plaster & gypsum, tiling, ceilings, flooring, acoustic treatment, painting & coating.
10.	10.	SPECIALTIES	292,500	1.70	2.44	Information, interior, safety, exterior, other.
11.	11.	EQUIPMENT	19,000	0.11	0.16	Food service.
12.	12.	FURNISHINGS	52,000	0.30	0.43	Furniture & accessories.
13.	13.	SPECIAL CONSTRUCTIONS	140,000	0.83	1.17	Special facility components.
14.	14.	CONVEYING SYSTEMS	208,500	1.21	1.74	Elevators (3), other.
15.	21.	FIRE SUPRESSION	250,000	1.46	2.08	Water-based fire-suppression.
15.	22.	PLUMBING	988,000	5.75	8.23	Piping & pumps, equipment, fixtures.
15.	23.	HVAC	2,452,000	14.29	20.43	Piping & pumps, air distribution, central heating, central cooling, central HVAC.
16.	26.	ELECTRICAL	1,514,000	8.82	12.62	Medium-voltage distribution, facility power generating & shoring, lighting.
16.	27.	COMMUNICATIONS	175,000	1.02	1.45	-
TOTAL BUILDING COST			**17,169,000**	**100%**	**$143.08**	
2.	2.	EXISTING CONDITIONS	94,000			Assessment, demolition & structure moving.
2.	31.	EARTHWORK	815,000			Site clearing, earthwork.
2.	32.	EXTERIOR IMROVEMENTS	495,000			Bases, ballasts, & paving, planting.
2.	33.	UTILITIES	-	-	-	-
TOTAL PROJECT COST			**18,573,000**	*(Excluding architectural and engineering fees)*		

UPDATED ESTIMATE TO JAN 2011: $247.21 SF

DCD Subscribers: Access this case study and hundreds more for instant date and location calculations at www.dcd.com.

Lewis and Clark State Office Building
Jefferson City, Missouri

Architect: BNIM Architects

Photos Courtesy of Farshid Assassi

As part of its mission to protect and restore Missouri's natural resources, the Missouri Department of Natural Resources commissioned a new 120,000-square-foot office building to consolidate all of their employees under one roof and to showcase affordable, cost-effective and replicable "green" strategies.

The site, selected after an extensive evaluation process, originally housed part of the Jefferson City Correctional Facilities, which was carefully deconstructed prior to construction. One goal of the restorative site plan was to completely eliminate stormwater runoff from the site. To accomplish this, the site was carefully designed with level spreaders, bioswales, native prairie plants and ecosystems. Rainwater from the roof is collected in a 50,000 gallon storage tank, which provides flushing water to the building's toilets.

The building is organized around a four-story atrium space that opens to a grand view of the Missouri River Valley. Oriented along an east-west axis, the building was developed with a narrow footprint to maximize daylighting strategies. Each elevation responds to its respective climatic condition. Precast concrete sunshades protrude from the south facade and allow low sun angles to provide supplemental heat during the winter months yet block heat gain during the intense summer months. The north facade was developed without shading devices, a design strategy that maximizes daylight and views toward the river. Different glazing was specifically selected for each elevation. Operable windows are distributed throughout the building to allow for natural ventilation.

Maximizing daylight and minimizing energy costs were primary goals of the owner. By locating open office space around the perimeter of the building, occupants are exposed to natural light and views. Exterior sunshades and interior fabric light shelves "harvest" sunlight, directing it into the core of the building. Several strategies maximized the efficiency of the HVAC system: a dedicated outside air system, hybrid ventilation air handlers, natural free cooling through the use of a chilled water thermal storage tank charged by the tower during cool night hours. Daylighting controls combined with a thermal storage system significantly reduce peak power use and sizing. Photovoltaics provide 2.5% of the building's energy needs, and a solar hot water system is used to help supply domestic hot water.

The building has received a LEED® Platinum rating from the U.S. Green Building Council.

LEED® PLATINUM

MANUFACTURERS/SUPPLIERS

DIV 07:	*Roofing & Siding Panels :* Hardiplank® Lap Siding; *Metal Roofing:* Berridge Manufacturing.
DIV 08:	*Entrances & Storefronts, Windows:* Kawneer; *Glazing:* **Viracon.**
DIV 16:	*Lighting:* Architectural Area Lighting, KIM, Hydrel, Peerless, Niedhardt,, Lithonia, Williams, LSI Industries, Cole, Edison Price, Columbia, Focal Point, Prudential, Elliptiar, Dialight, Gradco, Karlin, Alkco.

EXTENDED PRODUCT INFORMATION
Glazing: **Viracon.**

COST PER SQUARE FOOT $ 247.21

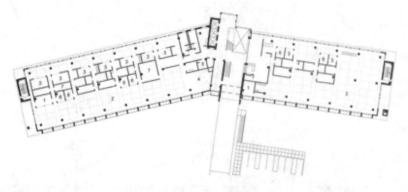

ARCHITECT
CALLOWAY JONSON MOORE & WEST, P.A.
119 Brookstown Avenue, #100,
Winston-Salem, NC 27101

CONSTRUCTION TEAM

GENERAL CONTRACTOR:
WEAVER COOKE CONSTRUCTION, LLC
8401 Key Boulevard, Greensboro, NC 27409
www.weavercooke.com
STRUCTURAL ENGINEER:
Calloway Johnson Moore & West, P.A.
119 Brookstown Avenue, #100, Winston-Salem, NC 27101
ELECTRICAL ENGINEER:
Johnson's Modern Electric Company, Inc.
6629 Old U.S. 421 Highway, East Bend, NC 27018
MECHANICAL ENGINEER: Superior Mechanical
P.O. Box 877, Randleman, NC 27317
INTERIOR DESIGN: Small Kane Architects, P.A.
3105 Glenwood Avenue, #100, Raleigh, NC 27612

GENERAL DESCRIPTION

SITE: 3 acres. **NUMBER OF BUILDINGS:** One.
BUILDING SIZE: First floor, 15,464; total, 15,464.
BUILDING HEIGHT: First floor, 26'; total, 26'.
BASIC CONSTRUCTION TYPE: Tilt-Up/New.
FOUNDATION: Cast-in-place, slab-on-grade.
EXTERIOR WALLS: Tilt-up, brick, curtainwall.
ROOF: Metal, membrane. **FLOORS:** Concrete.
INTERIOR WALLS: Metal stud drywall.

WEAVER COOKE CONSTRUCTION HEADQUARTERS
Construction Period: May 2006 to Jan 2007 Total Square Feet 15,464

C.S.I. Divisions (1 through 16)			COST	% OF COST	SQ.FT. COST	SPECIFICATIONS
		PROCUREMENT & CONT. REQ.	-	-	-	-
1.	1.	GENERAL REQUIREMENTS	127,000	6.00	8.21	Administrative requirements, quality requirements, temporary facilities & controls, execution & closeout requirements, life cycle activities.
3.	3.	CONCRETE	292,000	13.80	18.88	Forming & accessories, reinforcing, cast-in-place, grouting.
4.	4.	MASONRY	58,000	2.74	3.75	Unit.
5.	5.	METALS	167,000	7.89	10.80	Structural metal framing, joists, decking, fabrications.
6.	6.	WOOD/PLASTICS/COMPOSITE	122,000	5.76	7.89	Rough carpentry, architectural woodwork.
7.	7.	THERMAL & MOIST. PROTECT	135,000	6.38	8.73	Roofing & siding panels, membrane roofing, flashing & sheet metal, roof & wall specialties & accessories, joint protection.
8.	8.	OPENINGS	268,000	12.66	17.33	Doors & frames, entrances, storefronts, & curtain walls, windows.
9.	9.	FINISHES	363,000	17.15	23.47	Plaster & gypsum board, tiling, ceilings, flooring, painting & coating.
10.	10.	SPECIALTIES	86,500	4.09	5.60	Interior, safety, exterior.
11.	11.	EQUIPMENT	27,000	1.28	1.75	Food service, other.
15.	21.	PLUMBING	56,000	2.65	3.62	Piping & pumps, equipment, fixtures.
15.	22.	HVAC	223,000	10.54	14.42	Air distribution, air cleaning devices, central HVAC equipment.
16.	26.	ELECTRICAL	162,000	7.65	10.48	Medium-voltage distribution, lighting.
16.	27.	COMMUNICATIONS	30,000	1.41	1.94	Data.
TOTAL BUILDING COST			**2,116,500**	**100.00**	**$136.87**	
2.	2.	EXISTING CONDITIONS	-	-	-	-
2.	31.	EARTHWORK	165,000			Site clearing, earth moving.
2.	32.	EXTERIOR IMPROVEMENTS	140,000			Bases, bollards, & paving, planting.
2.	33.	UTILITIES	43,000			Water, sanitary sewerage, storm.
TOTAL PROJECT COST			**2,464,500**	*(Excluding architectural and engineering fees)*		

UPDATED ESTIMATE TO JAN 2010: $215.41 SF

DCD Subscribers: Access this case study and hundreds more for instant date and location calculations at www.dcd.com.

Weaver Cooke Construction Headquarters
Greensboro, North Carolina

Architect: Calloway Johnson Moore & West, P.A.

Weaver Cooke Construction's corporate headquarters building is the embodiment of the company's progressive environmental philosophy and its passion for art. The company, which recently dedicated itself to becoming a leader in green construction, wanted a headquarters that not only expressed this new mission, but that also highlights the aesthetic qualities inherent in different construction techniques.

A significant segment of Weaver Cooke's work consists of the renovation and adaptive re-use of old mill buildings throughout the southeast. The co-mingling of past and present that is a part of much of Weaver Cooke's work is echoed in the design of the new headquarters at all scales from the form of the building itself down to the detailing of individual parts.

The building is organized around a light-filled, wedge-shaped central gallery in which the company's extensive art collection is displayed. The roof line combines the sawtooth skylighting of old mill buildings with a forceful sloping plane. The materials used exposed wood, steel, and concrete bring a strong sense of materiality to the gallery; texture, surface, and detailing are as much on display as the art. Two office wings flank the gallery; corridors flooded with natural light create a vibrant working environment.

Weaver Cooke Construction's progressive environmental philosophy ensured that sustainability was a key factor in the design of every space within the building. The structure's transparency, high floor-to-ceiling relationship, interior corridor glazing, and skylights allow everyone in the building, regardless of location, to feel connected to the outside. The building is sited to take maximum advantage of natural daylighting; exterior sunshading devices control glare. Landscaping and high reflectance surfaces reduce heat gain. Individual lighting and zoned temperature controls conserve energy and allow employees control over their environment. Low flow fixtures and the use of indigenous vegetation that requires no irrigation resulted in a substantial reduction in water use.

Much of the construction debris was recycled rather than sent to the landfill and many of the building's materials have a high-recycled content. Regionally manufactured and extracted materials were used and the building itself has an area dedicated for the storage and collection of recyclables.

The Weaver Cooke Construction Company Corporate Headquarters is a physical statement that reinforces the ideals and goals of the company. The building embodies Weaver Cooke's dedication to sustainability while expressing the company's passion for both artistic and tectonic beauty.

Photos Courtesy of Bill Kund Photography

LEED® GOLD

MANUFACTURERS/SUPPLIERS	
DIV 03:	*Building Insulation System:* **THERMOMASS®.**
DIV 07:	*Membrane Roof:* Firestone.
DIV 08:	*Curtainwall, Entrances & Storefronts:* YKK AP America; *Windows:* Lincoln; *Sunscreens:* Peachtree Protective Covers.
DIV 09:	*Carpet:* Shaw, Interface.

EXTENDED PRODUCT INFORMATION
Building Insulation System:
THERMOMASS®.

COST PER SQUARE FOOT $215.41

ARCHITECT
PERKINS + WILL
655 Winding Brook Drive, Glastonbury, CT 06033
www.perkinswill.com

FILE UNDER
COMMERCIAL
Hartford, Connecticut

CONSTRUCTION TEAM

STRUCTURAL ENGINEER:
Szewczak Associates Consulting Engineers
200 Fisher Drive, Avon, CT 06001

GENERAL CONTRACTOR:
Barlett Brainard Eacott
70 Griffin Road South, Bloomfield, CT 06002
www.bbeinc.com

MECHANICAL & ELECTRICAL ENGINEER:
Kohler Ronan, LLC
301 Main Street, Danbury, CT 06810

SOUND CONSULTANT:
Acentech
33 Moulton Street, Cambridge, MA 02138

LIGHTING CONSULTANT:
HLB Lighting Design
200 Park Avenue South, #1401, New York, NY 10003

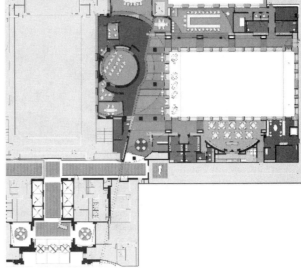

GENERAL DESCRIPTION

SITE: 37.75 acres. **NUMBER OF BUILDINGS:** One.

BUILDING SIZE: First floor, 12,000; mezzanine 2,600;

total 14,600 square feet. **BUILDING HEIGHT:** First floor, 19'; mezzanine, 8'5" total 22' from grade to roof.

BASIC CONSTRUCTION: Renovation/Tenant Build Out
FOUNDATION: Concrete, **EXTERIOR WALLS:** Brick
ROOF: Membrane **Floors:** Concrete.
INTERIOR WALLS: Metal stud, masonry.

AETNA CUSTOMER CENTER
Construction Period: Sep 2007 to March 2008 • Total Square Feet: 14,600

C.S.I. Divisions (1 through 16)			COST	% OF COST	SQ.FT. COST	SPECIFICATIONS
		PROCUREMENT & CONT. REQ.	-	-	-	-
1.	1.	GENERAL REQUIREMENTS	390,492	7.72	26.75	Summary, price & payment procedures, administrative requirements, quality requirements, temporary facilities & controls, product requirements, execution & closeout requirements, performance requirements, life cycle activities.
3.	3.	CONCRETE	120,650	2.39	8.26	Forming & accessories, reinforcing, cast-in-place, cast decks & underlayment, grouting.
4.	4.	MASONRY	6,000	0.12	0.41	Stone assemblies.
5.	5.	METALS	39,000	0.77	2.67	Structural metal framing, joists, decking, cold-formed metal framing, fabrications, decorative.
6.	6.	WOOD/PLASTICS/COMPOSITE	526,405	10.41	36.06	Rough carpentry, finish carpentry, architectural woodwork, plastic fabrications, composite fabrications.
7.	7.	THERMAL & MOIST. PROTECT	19,400	0.38	1.33	Membrane roofing, flashing & sheet metal, fire & smoke protection.
8.	8.	OPENINGS	291,378	5.76	19.96	Doors & frames, specialty doors & frames, entrances, storefronts & curtain walls, hardware, glazing.
9.	9.	FINISHES	905,539	17.91	62.02	Plaster & gypsum board, tiling, ceilings, flooring, wall finishes, acoustic treatment, painting & coating.
10.	10.	SPECIALTIES	63,875	1.26	4.38	Interior, safety.
11.	11.	EQUIPMENT	80,920	1.60	5.54	Security, detention, & banking, foodservice.
12.	12.	FURNISHINGS	96,800	1.91	6.63	Multiple seating, other.
13.	13.	SPECIAL CONSTRUCTIONS	276,420	5.47	18.93	Special facility components.
14.	14.	CONVEYING SYSTEMS	39,000	0.77	2.67	Lift to mezzanine.
15.	21.	FIRE SUPRESSION	121,000	2.40	8.29	Water-based fire suppression.
15.	22.	PLUMBING	67,000	1.33	4.59	Piping & pumps, equipment, fixtures.
15.	23.	HVAC	392,000	7.75	26.85	Piping & pumps, air distribution, air cleaning devices, central heating, central cooling.
16.	26.	ELECTRICAL	1,177,645	23.29	80.66	Medium-voltage distribution, low-voltage transmission, electrical & cathodic protection, lighting.
16.	27.	COMMUNICATIONS	290,362	5.74	19.89	Structured cabling, data, voice, audio-visual, distributed communications & monitoring systems.
16.	28.	ELECTRONIC SAFETY & SECURITY	152,867	3.02	10.46	Electronic access control & intrusion detection, surveillance, detection & alarm, monitoring & control.
TOTAL BUILDING COST			**5,056,753**	**100.00**	**$346.35**	
2.	2.	EXISTING CONDITIONS	187,850			Demolition & Structure moving, facility remediation.
TOTAL PROJECT COST			**5,244,603**	*(Excluding architectural and engineering fees)*		

UPDATED ESTIMATE TO JAN 2011: $387.55 SF

DCD Subscribers: Access this case study and hundreds more for instant date and location calculations at www.dcd.com.

Aetna Customer Center
Hartford, Connecticut

Architect: Perkins + Will

The new Aetna Customer Center represents the future in executive briefing centers and business relationship-enhancing venues in many ways. The new Center incorporates several forms of presentation technology to effectively communicate and reinforce the values of Aetna to new and existing clients.

The new Customer Center is located just yards away from the geographic center of Hartford, the "insurance capital of the world." This historic building is the largest colonial revival-style in the country and was originally built in 1931. Here cutting-edge technology plays well against the backdrop of a traditional landmark building and generates a sense of progressive service balanced with the permanence of the institution.

After completely demolishing all of the existing finishes, mechanical, electrical, and topping slabs the new facility began in-filling the space. Due to its proximity to large meeting and office areas, special concern was paid to isolating sound and vibration so that disturbances would be kept to a minimum. Several special features were incorporated to enhance the design of the Center and again reinforce the Aetna image. The project was designed and built under the guidelines of the LEED® CI program and qualifies as a registered project. Certain elements of the original building were re-incorporated into the project conserving resources and preserving the character of the original construction.

The extensive use of technology to present Aetna's message provided the impetus to upgrade the mechanical and electrical systems of the original base building to provide service while improving energy efficiency.

The presentation and meeting areas are complemented by several small break-out and conference areas. Additionally the Center attends to all the needs of today's business travelers so that the comfort and focus of visitors is maintained.

The space makes excellent use of the large south-facing windows by channeling the light deep into the space using the original double height ceilings. Program area was maximized by building a mezzanine space to house certain back-of-house functions and mechanical/electrical support equipment. The mezzanine space is strategically located to

Photos Courtesy of Gerry Holland ©

minimize impact on the original high ceilings and isolate mechanical noise from the presentation areas.

A food service preparation area was placed in a way to serve the Center via discrete routes and allow ready access to the building's central service corridor thus allowing for full dining support without interruption of business presentations.

LEED® CI	
MANUFACTURERS/SUPPLIERS	
DIV 03:	*Gypsum Floor:* Level Rock® 3500 by USG.
DIV 09:	*Floating Concrete Floor:* Kinetics Floor Noise Control System; *Carpet:* Shaw, Bentley, *Wall Panels:* 3 form.

COST PER SQUARE FOOT $387.55

ARCHITECT
Cornerstone Design-Architects
48-50 West Chestnut Street, #400, Lancaster, PA 17603
www.cornerstonedesign.com

FILE UNDER
RECREATIONAL
Lancaster, Pennsylvania

CONSTRUCTION TEAM

STRUCTURAL ENGINEER:
  C. S. Davidson, Inc.
  315 West James Street, #102, Lancaster, PA 17603
GENERAL CONTRACTOR:
  Wohlsen Construction Company
  548 Steel Way, Lancaster, PA 17601
ELECTRICAL & MECHANICAL ENGINEER:
  Z & F Consulting, Inc.
  57 West Avenue, Wayne, PA 19087
INTERIOR DESIGNER:
  Design for Functional Interiors
  245 East King Street, Lancaster, PA 17602

GENERAL DESCRIPTION

SITE: 1.79 Acres.
NUMBER OF BUILDINGS: One.
BUILDING SIZE: First floor, 21,105; second floor, 14,614; third floor, 6,783; total, 42,502 square feet.
BUILDING HEIGHT: First floor, 16'; second floor, 12'8"; third floor, 14'8"; total, 43'4".
BASIC CONSTRUCTION TYPE: New/IIB.
FOUNDATION: Reinforced concrete, slab-on-grade.
EXTERIOR WALLS: CMU, brick, EIFS, curtainwall.
ROOF: Membrane, metal.
FLOORS: Concrete.
INTERIOR WALLS: CMU.

LANCASTER YMCA HARRISBURG AVENUE
Construction Period: Sep 2008 to Sep 2009 • Total Square Feet: 42,502

C.S.I. Divisions (1 through 16)			COST	% OF COST	SQ.FT. COST	SPECIFICATIONS
1.	1.	GENERAL REQUIREMENTS	2,207,587	21.83	51.94	-
3.	3.	CONCRETE	365,407	3.61	8.60	Forming & accessories, cast-in-place.
4.	4.	MASONRY	909,325	8.99	21.39	Unit.
5.	5.	METALS	1,071,550	10.59	25.21	Structural metal framing, fabrications.
6.	6.	WOOD/PLASTICS/COMPOSITE	134,677	1.33	3.17	Rough carpentry, finish carpentry, architectural woodwork, composite fabrications.
7.	7.	THERMAL & MOIST. PROTECTION	315,628	3.12	7.43	Dampproofing & waterproofing, thermal protection, membrane roofing, flashing & sheet metal, fire & smoke protection, joint protection.
8.	8.	OPENINGS	473,175	4.68	11.13	Doors & frames, entrances, storefronts & curtainwalls, windows, hardware.
9.	9.	FINISHES	725,057	7.17	17.06	Plaster & gypsum board, tiling, ceilings, flooring, acoustic treatment, painting & coating.
10.	10.	SPECIALTIES	102,220	1.01	2.41	Other.
11.	11.	EQUIPMENT	18,920	0.19	0.45	Athletic & recreational.
12.	12.	FURNISHINGS	7,110	0.07	0.17	Other.
13.	13.	SPECIAL CONSTRUCTION	684,349	6.77	16.10	Special purpose rooms, special structures.
14.	14.	CONVETING SYSTEMS	64,900	0.64	1.53	Elevators (1).
15.	21.	FIRE SUPPRESSION	171,531	1.70	4.04	Water-based fire-suppression systems.
15.	22.	PLUMBING	471,548	4.66	11.09	Piping & pumps.
15.	23.	HVAC	1,554,461	15.37	36.57	Air distribution.
16.	26.	ELECTRICAL	836,440	8.27	19.67	Medium-voltage distribution.
TOTAL BUILDING COST			**10,113,885**	**100%**	**$237.96**	
2.	2.	EXISTING CONDITIONS	83,000			Demolition & structure moving.
2.	32.	EXTERIOR IMPROVEMENTS	292,166			Bases, bollards, & paving, improvements, planting.
2.	33.	UTILITIES	8,000			Electrical.
TOTAL PROJECT COST			**10,497,051**	*(Excluding architectural and engineering fees)*		

UPDATED ESTIMATE TO JAN 2011: $282.59 SF

DCD Subscribers: Access this case study and hundreds more for instant date and location calculations at www.dcd.com.

Lancaster YMCA Harrisburg Avenue
Lancaster, Pennsylvania

Architect: Cornerstone Design-Architects

Photos Courtesy of Nathan Cox

Before moving into their new facility in September of 2009, the Lancaster Family YMCA had felt the need for the consolidation of spaces and a facility that would consume less energy and resources. In May of 2007, the Lancaster Family YMCA held a green design charrette with the intention of generating design concepts and ideas for the new facility in conjunction with the adoption of certain green building principles. Representatives from the YMCA, surrounding community, major stakeholders, design professionals, contractors, and YMCA members attended the charrette and had opportunity to provide input into the design of the facility. Ultimately, the Owner decided to pursue a LEED® certification level of silver.

The building is located on a 1.78 acre downtown Lancaster, brownfield site adjacent to the newly constructed Clipper Stadium. Hazardous materials remediation efforts took place on the site to ensure that all contaminated substances were removed. The building is comprised of three floor levels due to the constraints of the site and the need to locate a certain quantity of parking spaces on the site as well. Major spaces in the facility include a high-school regulation size gymnasium with an elevated running track, wellness area with weightlifting and cardio equipment, and a six-lane competition pool and spa. Other spaces include child watch, men's, women's, and special needs/family locker rooms, youth areas, aerobics, yoga/pilates, office area, and miscellaneous support areas.

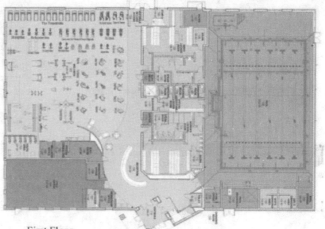

First Floor

During the course of the green design charrette, sustainable building principles were incorporated into the design of the project. The location of the site was selected primarily for its easy access to public transportation. Preference in parking spaces was given to fuel-efficient vehicles and the parking lot area was sized to the minimum allowable zoning standards. Light pollution was reduced so that the trespass of building and site lighting was minimal on adjoining properties. Native landscaping was selected to avoid the need for potable water irrigation and water efficient plumbing fixtures were selected for the interior of the building. The mechanical and electrical systems were designed to result in an energy savings of twenty to twenty-five percent. Materials with recycled content and available locally were used in the project and over 82% of the construction waste was diverted from landfills. Natural light was incorporated into the design of the facility to ensure that occupants had a connection with the exterior. Finally, materials with low volatile organic compounds (VOC) were specified in the project and a significant amount of occupants were given direct control over their lighting and temperature controls in the facility.

Another unique characteristic of this new facility is the incorporation of an older remnant of the original YMCA in downtown Lancaster. The original nine foot high and twelve foot wide stained glass window was restored and placed in the lobby of the new facility.

The project has been considered a success by the Owner, the YMCA members, and the surrounding community. Cornerstone Design has been proud to assist in the YMCA's mission - building strong kids, strong families, and strong communities.

**LEED®
SILVER PENDING**

COST PER SQUARE FOOT $ 282.59

ARCHITECT

Integrated Architecture
4090 Lake Drive, S.E., Grand Rapids, MI 49546
www.intarch.com

FILE UNDER

RECREATIONAL
Allendale, Michigan

CONSTRUCTION TEAM

STRUCTURAL ENGINEER:
  JDH Engineering, Inc.
    3000 Ivanrest, S.W., #B, Grandville, MI 49418
GENERAL CONTRACTOR:
  Erhardt Construction Company
    6060 East Fulton, Ada, MI 49301
ELECTRICAL & MECHANICAL ENGINEER:
  Integrated Architecture
    4090 Lake Drive, S.E., Grand Rapids, MI 49546

GENERAL DESCRIPTION

SITE: 8 Acres.
NUMBER OF BUILDINGS: One.
BUILDING SIZE: First floor, 123,053; second floor, 14,609; total, 137,662 square feet.
BUILDING HEIGHT: First floor, 75' to peak; second floor, 61' to peak; floor to floor, 14'; total, 75'
BASIC CONSTRUCTION TYPE: New/Pre-Engineered.
FOUNDATION: Slab-on-grade.
EXTERIOR WALLS: CMU, brick, curtainwall.
ROOF: Metal, membrane.
FLOORS: Concrete, precast.
INTERIOR WALLS: Metal stud drywall, CMU.

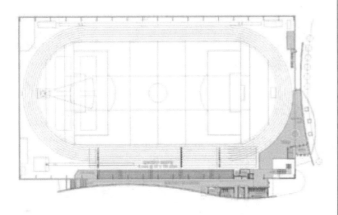

GRAND VALLEY STATE UNIVERSITY BUILDING
Construction Period: May 2007 to Aug 2008 • Total Square Feet: 137,662

C.S.I. Divisions (1 through 16)			COST	% OF COST	SQ.FT. COST	SPECIFICATIONS
		PROCUREMENT & CONT. REQ.	85,500	0.65	0.62	Bonds.
1.	1.	GENERAL REQUIREMENTS	870,850	6.61	6.33	General requirements, testing.
3.	3.	CONCRETE	1,720,560	13.05	12.50	Cast-in-place, precast.
4.	4.	MASONRY	608,670	4.62	4.42	Unit.
5.	5.	METALS	869,880	6.60	6.32	Structural metal framing, joists, decking, fabrications, decorative.
6.	6.	WOOD/PLASTICS/COMPOSITE	316,660	2.40	2.30	Rough carpentry, finish carpentry, architectural woodwork, composite fabrications.
7.	7.	THERMAL & MOIST. PROTECTION	374,170	2.84	2.72	Dampproofing & waterproofing, roofing & siding panels, membrane roofing, roof & wall specialties (vegetated roof), joint protection.
8.	8.	OPENINGS	219,010	1.66	1.59	Doors & frames, entrances, storefronts, & curtainwalls, special doors.
9.	9.	FINISHES	1,349,500	10.24	9.80	Plaster & gypsum board, flooring, artificial turf, track surface, painting & coating.
10.	10.	SPECIALTIES	15,425	0.12	0.11	Interior.
11.	11.	EQUIPMENT	316,100	2.40	2.30	Foodservice, athletic, recreational.
13.	13.	SPECIAL CONSTRUCTION	3,300,340	25.04	23.97	Pre-engineered building.
14.	14.	CONVEYING SYSTEMS	46,075	0.35	0.33	Elevators (1).
15.	21.	FIRE SUPPRESSION	59,525	0.45	0.43	Water-based fire-suppression systems.
15.	22.	PLUMBING	14,880	0.11	0.11	Gas service.
15.	23.	HVAC	1,926,790	14.62	14.00	HVAC & plumbing, geothermal well system.
16.	26.	ELECTRICAL	1,085,590	8.24	7.89	Medium-voltage distribution, low-voltage transmission, facility power generating & storing equipment.
TOTAL BUILDING COST			**13,179,525**	**100%**	**$95.74**	
2.	2.	EARTHWORK	1,759,364			Site clearing, earth moving, earthwork methods, excavation.
2.	32.	EXTERIOR IMPROVEMENTS	626,655			
2.	33.	UTILITIES	430,080			
TOTAL PROJECT COST			**15,995,624**	*(Excluding architectural and engineering fees)*		

UPDATED ESTIMATE TO JAN 2011: $121.83 SF

Grand Valley State University
Laker Turf Building
Allendale, Michigan

Architect: Integrated Architecture

Photos Courtesy of Nathan Cox

The term "Fast Track" has an entirely different meaning at the Grand Valley State University Laker Turf Building. Offering a six-lane, 300-meter indoor track with nine sprint lanes, the building literally houses a fast track that is destined to be among the Midwest's premier competition indoor venues. Since its opening last August, NCAA track and field records have been falling like rain in April with seven events posting top speed and distance numbers in everything from the hurdles to the pole vault.

The track surrounds the 100-yard indoor football / lacrosse / soccer field, which serves the University's intramural and club sports while providing indoor practice space for GVSU Division II championship teams. A true multi-use facility, even football has to take its turn. When scheduling the initial spring practice, the coach discovered the field was booked for a 2-day soccer tournament. As a result, the inaugural practice was scheduled for midnight!

Functional, sustainable, cost-effective and student focused, the facility's design was shaped by the program, specifically the 100-yard field and indoor track. While the Laker Turf Building has an uncomplicated feel, it is a structurally complex long span building. It is intentionally positioned on the site to maximize available sunlight, solar heating and natural ventilation. Designed to earn a Gold LEED® certification, its sustainable attributes include a geothermal heat pump that heats and cools the floors, an array of photovoltaic solar panels providing 20 KW of electricity, and 3,000 square feet of SolarWall paneling on the building's South wall that warms exterior air which is then added to the building's HVAC system, lowering its winter work load. The system also provides cooling in the summer by shading the south wall and assisting in the ventilation of hot interior air. Storm water is managed with a 9,000 square foot rain garden/retention pond system which collects the runoff from the massive roof to irrigate the adjacent athletic fields.

The clean, simple silver metal wall and roof panels along with oversized block walls help reduce the massive feel of the facility which earns much of its WOW factor from the sheer size and open plan.

With a check in desk at the main entrance and movement science multipurpose rooms, lockers, showers, and a training room on the lower level, the building offers a strength of purpose that comes from power of the open design. Spectator seating and circulation on the main level along with a Laker Wall of Fame, support GVSU's winning tradition. Durable, practical interior block masonry is appropriately finished in GVSU's Laker team colors: Blue and White.

Filled with natural light, the facility provides the GVSU winning teams a venue that reflects their championship status and gives students, faculty, and alumni an indoor place to play throughout Michigan's long winter.

LEED® GOLD
MANUFACTURERS/SUPPLIERS
DIV 07: *Roofing:* **Carlisle Syntech**

COST PER SQUARE FOOT $121.83

ARCHITECT OF RECORD
Brown Reynolds Watford Architects
3535 Travis Street, #250, Dallas, TX 75204
www.brwarch.com

DESIGN ARCHITECT
Antoine Predock Architect PC
300 12th Street NW, Albuquerque, NM 87102

FILE UNDER
RECREATIONAL
Dallas, Texas

CONSTRUCTION TEAM

STRUCTURAL ENGINEER:
Jaster-Quintanilla/Dallas
2105 Commerce St. #300, Dallas, TX 75201
COST ESTIMATOR:
Hill International, Inc.
5307 East Mockingbird Lane, #700, Dallas, TX 75206
ELECTRICAL & MECHANICAL ENGINEER:
Lopez Garcia Group
1950 North Stemmons Freeway, #6000, Dallas, TX 75207
LEED® CONSULTANT:
Rocky Mountain Institute
2317 Snowmass Creek Road, Snowmass, CO 81654

GENERAL DESCRIPTION

SITE: 120 Acres.
NUMBER OF BUILDINGS: One.
BUILDING SIZE: First floor, 20,336; mezzanine, 455;
total, 20,791 square feet.
BUILDING HEIGHT: Varies.
BASIC CONSTRUCTION TYPE: New/Steel Frame.
FOUNDATION: Cast-in-place, pier & grade beam, slab-on-grade,
elevated deck over crawl space.
EXTERIOR WALLS: Curtainwall, wood siding, metal panel.
ROOF: Membrane, garden.
FLOORS: Concrete.
INTERIOR WALLS: Metal stud drywall.

TRINITY RIVER AUDUBON CENTER
Construction Period: Apr 2007 to Sep 2008 • Total Square Feet: 20,791

C.S.I. Divisions (1 through 16)			COST	% OF COST	SQ.FT. COST	SPECIFICATIONS
		PROCUREMENT & CONT. REQ.	1,316,048	14.45	63.30	General conditions, bond, permit.
1.	1.	GENERAL REQUIREMENTS	298,854	3.28	14.37	Mobilization, change orders, temporary facilities.
3.	3.	CONCRETE	1,129,000	12.40	54.30	Forming & accessories, reinforcing, cast-in-place (Concrete Breakdown: cubic yards foundation including pier concrete, 512; cubic yards walls including exterior wall and integral color wall 288; cubic yards floors, 460; cubic yards roof structural slab, 197; cubic yards other, 259).
5.	5.	METALS	1,037,357	11.39	49.89	Steel erection, miscellaneous.
6.	6.	WOOD/PLASTICS/COMPOSITE	256,620	2.82	12.34	Rough carpentry, finish carpentry, architectural woodwork, wood siding.
7.	7.	THERMAL & MOIST. PROTECTION	651,041	7.15	31.31	Waterproofing & dampproofing, insulation, roof & wall panels, membrane roofing, modified bitumen plaza deck roofing.
8.	8.	OPENINGS	565,583	6.21	27.20	Doors & frames, entrances, storefronts, hardware, glazing.
9.	9.	FINISHES	677,412	7.44	32.58	Plaster & gypsum board, ceilings, flooring, wall finishes, painting & coating.
10.	10.	SPECIALTIES	48,260	0.53	2.32	Fountain, other.
11.	11.	EQUIPMENT	472,934	5.19	22.75	Exhibits.
12.	12.	FURNISHINGS	40,000	0.44	1.93	Window treatment, mats & frames.
13.	13.	SPECIAL CONSTRUCTION	404,484	4.44	19.46	—
15.	21.	FIRE SUPPRESSION	42,900	0.47	2.06	Water-based fire-suppression systems.
15.	22.	PLUMBING	611,369	6.71	29.41	Piping & pumps, equipment, fixtures.
15.	23.	HVAC	743,939	8.17	35.78	Submittals, piping, ductwork, equipment, instrumentation & controls, testing, adjusting & balancing.
16.	26.	ELECTRICAL	785,056	8.62	37.76	—
16.	27.	COMMUNICATIONS	26,700	0.29	1.29	—
TOTAL BUILDING COST			**9,107,557**	**100%**	**$438.05**	
2.	2.	EARTHWORK	342,763			—
2.	32.	EXTERIOR IMPROVEMENTS	2,044,858			—
2.	33.	UTILITIES	369,691			Water, sanitary sewer, storm.
TOTAL PROJECT COST			**11,864,869**	*(Excluding architectural and engineering fees)*		

UPDATED ESTIMATE TO JAN 2011: $602.75 SF

Trinity River Audubon Center
Dallas, Texas

Architect: Brown Reynolds Watford Architects

Photos Courtesy of Michael Lyon Photography

The Trinity River Audubon Center resides within the largest urban hardwood forest in the United States within a few miles of the nearby urban downtown environment. At this intersection of city and nature, the Center is sculpted by elements of the natural world - earth, water, and air. This site, previously overlooked for decades, was improperly used as a waste landfill while coexisting with century old native trees and serpentine, deeply cut riverbanks. The closed landfill is a visionary restoration project, reclaiming 120 acres of the Great Trinity River Corridor for the use of future generations. The Center was conceived as a teaching tool, promoting both a greater appreciation of surrounding ecosystems and an understanding of man"s impact on the environment. Additionally, the Center allows visitors to appreciate the uniqueness of this forgotten place, while understanding the evolution of the land caused by man's impact through domestic development. The facility distills the energy of this restored place through exhibits, classrooms, and research laboratories that celebrate the land.

The Center is not only part of the landscape, carefully placed within, but also a reflection of its surroundings. Each wing reflects a different biome of the Great Trinity Forest. The dichotomy of concrete feather walls softens the envelope while reminding visitors of Audubon"s

focus. The spaces are flexible and energy efficient. An outdoor corridor to classrooms reduces the need for conditioned space. Partitioned rooms are used as small classrooms or opened for large meeting areas. The great hall and cafe are also open to house various types of events and provide visitors a comfortable space to appreciate nature from the indoors. Perhaps more significant however, is the dedication shown in restoring this expansive site. Following the state"s criteria and with the goal of returning the land to its natural state for use with future generations, the plan consolidated the waste into capped hills replanted

with prairie grass and native trees. At the base of the hills, a series of cascading wetland marshes and ponds captures and filters runoff from adjoining neighborhoods and prairies before returning the cleansed water to the river.

ARCHITECT
Vision 3 Architects
225 Chapman Street, Providence, RI 02905-4592
www.vision3architects.com

CONSTRUCTION TEAM

STRUCTURAL ENGINEER:
Odeh Engineers, Inc.
1223 Mineral Spring Avenue, North Providence, RI 02904
GENERAL CONTRACTOR:
E. Turgeon Construction Corp.
1 Harry Street, Cranston, RI 02907
ELECTRICAL ENGINEER:
Gaskell Associates, a Div. of Thielsch Engineering, Inc.
1341 Elmwood Avenue, Cranston, RI 02910
MECHANICAL ENGINEER:
Creative Environment Corp.
450 Warren Avenue, East Providence, RI 02914
CIVIL ENGINEER:
Casali & D'Amico Engineering, Inc.
300 Post Road, Warwick, RI 02888

GENERAL DESCRIPTION

SITE: 1.272 Acres.
NUMBER OF BUILDINGS: One.
BUILDING SIZE: Basement, 8,425; first floor, 21,843; total, 30,268 square feet.
BUILDING HEIGHT: Basement, 12'; floor to floor, 13'4"; gym, 31'6"; main building, 14'.
BASIC CONSTRUCTION TYPE: New/ 2B.
FOUNDATION: Cast-inplace, slab-on-grade.
EXTERIOR WALLS: CMU, brick, cement siding boards.
ROOF: Membrane, asphalt shingles.
FLOORS: Concrete, wood.
INTERIOR WALLS: CMU, metal stud drywall.

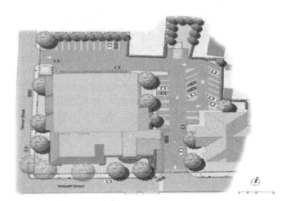

ST. RAPHAEL ACADEMY ALUMNI HALL ATHLETIC & WELLNESS CENTER
Construction Period: May 2006 to Sep 2007 • Total Square Feet: 30,268

C.S.I. Divisions (1 through 16)			COST	% OF COST	SQ.FT. COST	SPECIFICATIONS
		PROCUREMENT & CONT. REQ.	139,588	2.47	4.61	Permits, insurance, bond.
1.	1.	GENERAL REQUIREMENTS	596,615	10.56	19.71	General requirements, fees.
3.	3.	CONCRETE	450,333	7.97	14.88	Forming & accessories, cast-in-place.
4.	4.	MASONRY	694,135	12.28	22.93	CMU, brick, grout, precast, accessories, reinforcing steel.
5.	5.	METALS	517,428	9.15	17.09	Structural metal framing, joists, decking, trusses.
6.	6.	WOOD/PLASTICS/COMPOSITE	254,603	4.50	8.41	Rough carpentry, finish carpentry, millwork.
7.	7.	THERMAL & MOIST. PROTECTION	404,795	7.16	13.37	Dampproofing & waterproofing, membrane, asphalt shingles, fire & smoke protection, joint protection.
8.	8.	OPENINGS	292,665	5.18	9.67	Doors & frames, specialty doors & frames, entrances, storefronts, & curtainwalls, windows, hardware, glazing, skylight.
9.	9.	FINISHES	641,099	11.34	21.18	Plaster & gypsum board, tiling, ceilings, flooring, wall finishes, acoustic treatment, painting & coating.
10.	10.	SPECIALTIES	107,287	1.90	3.54	Information, toilet partitions & accessories, signage.
11.	11.	EQUIPMENT	50,724	0.90	1.68	Athletic.
12.	12.	FURNISHINGS	96,500	1.71	3.19	Floor mats, window treatments, bleacher system.
14.	14.	CONVEYING SYSTEMS	46,000	0.81	1.52	Elevators (1).
15.	21.	FIRE SUPPRESSION	66,000	1.17	2.18	—
15.	22.	PLUMBING	199,220	3.52	6.58	Piping & pumps, equipment, fixtures.
15.	23.	HVAC	610,000	10.79	20.16	Piping & pumps, air distribution, air cleaning devices, installation.
16.	26.	ELECTRICAL	485,000	8.59	16.03	—
TOTAL BUILDING COST			**5,651,992**	**100%**	**$186.73**	
2.	2.	EXISTING CONDITIONS	13,500			Demolition.
2.	31.	EARTHWORK	242,800			Site clearing, earth moving, earthwork methods, excavation support & protection.
2.	32.	EXTERIOR IMPROVEMENTS	201,220			Bases, bollards, & paving, improvements, irrigation, planting.
2.	33.	UTILITIES	154,000			Water, sanitary sewerage, storm drainage, electrical.
TOTAL PROJECT COST			**6,263,512**	*(Excluding architectural and engineering fees)*		

UPDATED ESTIMATE TO JAN 2011: $238.21 SF

DCD Subscribers: Access this case study and hundreds more for instant date and location calculations at www.dcd.com.

St. Raphael Academy
Alumni Hall Athletic & Wellness Center
Pawtucket, Rhode Island

Architect: Vision 3 Architects

Photos Courtesy of Michael Lyon Photography

I n September 2007, St. Raphael Academy in Pawtucket, R.I. held the grand opening and dedication service for Alumni Hall, their new athletic and wellness center. Designed by Vision 3 Architects of Providence, R.I., Alumni Hall replaces the academy's 80-year old gymnasium.

At the heart of the new facility is the gymnasium. It is the home court for the St. Raphael Saints basketball team, and has two overlapping practice courts, a full size volleyball court, and seating for 850 spectators.

The building also includes a 2,500-square-foot fitness and weight training room with over 30 pieces of equipment ranging from free weights to treadmills. Adjacent to the weight room is a "smart," multi-purpose classroom for academic classes and team presentations. The lower level contains locker rooms, showers, faculty offices, a conference room, and a new training and treatment center.

The new center enhances the academy's physical education program. Not only does Alumni Hall house competitive and intramural sports, it provides options for those who do not excel in group sports or regular gym class. The weight and fitness training area, with its combination of cardiovascular equipment, free weights and circuit machines, will be utilized as part of the academy's educational program for all students.

St. Raphael Academy's campus spans two city blocks in the historic Quality Hill neighborhood of Pawtucket. Alumni Hall is the first new, stand-alone building built on the academy's campus since 1928.

Vision 3 Architects collaborated with the academy, the community, and the City to develop a design that complements the historic architectural character of the neighborhood, while meeting stringent City zoning requirements. Traditional materials, when combined with contemporary forms and architectural detailing, blend Alumni Hall into the surrounding neighborhood. By placing all support spaces in the basement level, the building footprint was minimized to create parking and landscape areas to satisfy zoning requirements.

One of Vision 3's priorities was to incorporate energy efficient features into the design, thereby reducing operating costs, and making Alumni Hall an environmentally friendly building.

Sustainable design features include:

- Orienting the building north-south to maximize daylight and minimize heat gain.
- Reducing water usage by installing water efficient plumbing fixtures and hands-free automatic flush valves and faucets.
- Incorporating glazing and insulation that exceeds energy code requirements. A white, reflective roof protects against UV-rays, lowers surface temperatures, reduces energy costs, and reduces cooling loads.
- Using low, or no omitting, VOC (volatile omitting contaminant) floor and wall finishes. Sealed and polished concrete is the predominant floor material throughout the building. The thermal mass of the concrete reduces cooling and heating loads as well as
- the potential for indoor air contaminants.
- Using large spans of high performance glass and skylights to maximize daylight and reduce the need for artificial lighting.
- Using high performance, low energy, fluorescent lighting fixtures. Room occupancy sensors automatically turn off lighting when spaces are not occupied.

MANUFACTURERS/SUPPLIERS

DIV 04:	*Brick:* **Carolina Ceramics.**
DIV 07:	*Membrane:* **Firestone;** *Shingles:* CertainTeed; *IFS:* **Dryvit.**
DIV 08:	*Windows, Entrances & Storefron:* Glass: **Pilington Gray Eclipse Advantage.**
DIV 09:	*Acoustical Treatment:* **Armstrong;** *Gypsum:* **USG;** *Concrete Coating:* **Permashine;** Floor Mats: **Mat's Inc.;** *Wood:* **Connor Sports Flooring** *Tile:* **Dal-Tile;** *Carpet:* **Shaw Contract.**
DIV 26:	*Lighting:* **Lithonia, Lightolier, Focal Point; Gardco, Sportlite.**

COST PER SQUARE FOOT $ 238.21

ARCHITECT
INTEGRATED ARCHITECTURE
4090 Lake Drive, S.E.
Grand Rapids, MI 49543
www.intarch.com

FILE UNDER
RECREATIONAL
Grand Rapids, Michigan

CONSTRUCTION TEAM

STRUCTURAL ENGINEER: JDH Engineering, Inc.
3000 Ivanrest Avenue S.W., #B, Grandville, MI 49418
MECHANICAL & ELECTRICAL ENGINEER:
Integrated Architecture
4090 Lake Drive, S.E., Grand Rapids, MI 49543
GENERAL CONTRACTOR: Visser Brothers, Inc.
1946 Turner Avenue, N.W., Grand Rapids, MI 49504
POOL DESIGN: Bill Robertson Pool Design
913 Whitegate Drive, Northville, MI 48167

GENERAL DESCRIPTION

SITE: 3.197 acres. **NUMBER OF BUILDINGS:** One.
BUILDING SIZE: Lower level, 9,936; first floor, 64,178;
second floor, 30,415; third floor, 49,583; mezzanine, 8,854; total,
162,966 square feet. **BUILDING HEIGHT:** Lower level, 15'; first
foor, 16'; second floor, 18'; third floor, 16'; mezzanine, 16'; elevator
penthouse, 12'; total, 78'4" to top of stair tower.
BASIC CONSTRUCTION TYPE: New.
FOUNDATION: Cast-in-place, reinforced, slab-on-grade.
EXTERIOR WALLS: CMU, curtainwall, Arriscraft stone.
ROOF: Membrane. **FLOORS:** Concrete, precast.
INTERIOR WALLS: CMU, metal stud drywall.

Photos Courtesy of Justin Maconochie

DAVID D. HUNTING YMCA
Construction Period: Nov 2003 to June 2005 Total Square Feet 162,966

C.S.I. Divisions (1 through 16)			COST	% OF COST	SQ.FT. COST	SPECIFICATIONS
		PROCUREMENT & CONT. REQ.	1,292,026	6.51	7.93	-
1.	1.	GENERAL REQUIREMENTS	1,812,101	9.13	11.12	Change orders, general conditions.
3.	3.	CONCRETE	2,899,873	14.61	17.79	Concrete, precast.
4.	4.	MASONRY	1,079,985	5.44	6.63	Masonry.
5.	5.	METALS	1,291,250	6.51	7.92	Steel, structural & miscellaneous.
6.	6.	WOOD/PLASTICS/COMPOSITE	452,480	2.28	2.78	Carpentry, lumber, millwork.
7.	7.	THERMAL & MOIST. PROTECT	249,810	1.26	1.53	Caulking, dampproofing & waterproofing, fire stopping, insulation, roofing.
8.	8.	OPENINGS	1,881,988	9.48	11.55	Glass, aluminum & sunshade, hollow metal doors & frames, hardware.
9.	9.	FINISHES	2,097,945	10.57	12.87	Drywall, ceilings, flooring, painting, resinous flooring, wood flooring.
10.	10.	SPECIALTIES	452,110	2.28	2.77	Specialties, lockers, partitions, toilet partitions & bath accessories.
11.	11.	EQUIPMENT	136,055	0.69	0.84	Athletic, food service equipment.
12.	12.	FURNISHINGS	-	-	-	-
13.	13.	SPECIAL CONSTRUCTIONS	1,778,079	8.96	10.91	Pool slide, pools, racquetball court glass & strongwall plaster.
14.	14.	CONVEYING SYSTEMS	227,096	1.14	1.40	Elevators (2).
15.	15.	MECHANICAL	3,166,100	15.96	19.43	Fire Protection, HVAC, plumbing.
15.	21.	FIRE SUPRESSION	-	-	-	-
15.	22.	PLUMBING	-	-	-	-
15.	23.	HVAC	-	-	-	-
16.	26.	ELECTRICAL	1,027,095	5.18	6.30	-
16.	27.	COMMUNICATIONS	-	-	-	-
TOTAL BUILDING COST			**19,843,993**	**100%**	**$121.77**	
2.	2.	SITEWORK	556,500			Site work, landscaping.
2.	31.	EARTHWORK	-			
2.	32.	EXTERIOR IMROVEMENTS	-			
2.	33.	UTILITIES	-			
TOTAL PROJECT COST			**20,400,493**	*(Excluding architectural and engineering fees)*		

UPDATED ESTIMATE TO JAN 2011: $200.52 SF

DCD Subscribers: Access this case study and hundreds more for instant date and location calculations at www.dcd.com.

David D. Hunting YMCA
Grand Rapids, Michigan

Architect: Integrated Architecture

Photos Courtesy of Mark Thomas Productions

The David D. Hunting YMCA emerges from the landscape along US 131 to offer a tangible manifestation of the Y's emphasis on community. Designed to be a 'destination' Y, this structure radiates energy and excitement by physically and visually connecting people and spaces both within the facility and to passersby.

This building's simple, uncomplicated form responds to both function and setting, filling its location on a fringe urban site with its east, west, and south edges abutting adjacent streets. The footprint, with its arched north and south facades, is derived from the third floor indoor running circuit.

Glass curtainwalls support passive solar strategies and combine with zinc panels and renaissance stone to express an image of a solid, long-term community resident. The contrast created by mixing solidity and transparency establishes a kinetic and highly animated experience, while respecting the YMCA's traditional urban construction.

The lobby entrance immerses visitors in activity, offering views through the first floor aquatic center to the Grand Valley State University Campus and beyond. The nondenominational chapel, central to the YMCA mission, floats above the circulation desk, offering an inward focus and oasis of quiet in the midst of activity. While the pools anchor the main floor, an open staircase urges members and visitors upward to second and third floor fitness areas.

As one of the largest urban Y's in the nation, the David D. Hunting YMCA serves a diverse population, from downtown business people to inner city families. Challenged to create a building

that would be a realization of the YMCA mission to build strong kids, strong families and strong communities, the building's open, interconnected spaces work together to create a sense of community, as well as opportunities for interaction, as members naturally flow from one activity to another. The two full-court basketball floors can be sectioned off for other activities, while the 1/7th mile running track, around the perimeter of the building, offers unobstructed views of the city as well as activities taking place within the building. Numerous conversation areas present opportunities to catch your breath or catch up with a friend.

The decision to seek LEED® certification, the first Y in the nation to do so, came at the urging of the Hunting family, whose $5,000,000 gift in honor of their father, David D. Hunting, provided the

foundation for the fund-raising effort. Sustainable components range from low or no VOC off-gassing finishes to photovoltaic solar roof panels. The result is an elegant facility that reflects the vision and passion of David D.Hunting, a Steelcase founder, to unite the community of Grand Rapids. The building of community is evident as everyone, regardless of income, age, ability, race or religion, is welcome. Since it's opening in July of 2005 membership has grown from 8,945 to 18,008 an increase of over 200%.

The David D. Hunting YMCA has been awarded the Athletic Business 2006 Facility of Merit and the 2006 Honor Award from the Masonry Institute of Michigan.

LEED® CERTIFIED		
MANUFACTURERS/SUPPLIERS		
DIV 04:	*Stone:* Artiscraft Stone.	
DIV 07:	*Membrane Roof:* Firestone.	
DIV 08:	*Entrances & Storefronts, Windows, Curtainwall:* Vistawall; *Special Doors:* Won-Door.	
DIV 09:	*Carpet:* Interface; *Flooring:* Mondo, Expanko Cork; *Resinous:* Stonhard; *Wood: Connor.*	
DIV 14:	*Elevators:* ThyssenKrupp.	
DIV 16:	*Lighting:* Copper, Linear, Leviton.	

COST PER SQUARE FOOT $ 200.52

<table>
<tr><td colspan="2">

ARCHITECT
Cuningham Group Architecture, P.A.
201 Main Street S.E. #325, Minneapolis, MN 55414
www.cuningham.com

</td><td>

FILE UNDER
RELIGIOUS
Plymouth, Minnesota

</td></tr>
</table>

6

CONSTRUCTION TEAM

STRUCTURAL ENGINEER:
Reigstad & Associates, Inc.
192 West 9th Street, #200, St. Paul, MN 55113

GENERAL CONTRACTOR:
Anderson Builders
4220 Park Glen Road, St. Louis Park, MN 55426

ELECTRICAL & MECHANICAL ENGINEER:
Karges-Falconbridge
670 West County Road B, St. Paul, MN 55113

GENERAL DESCRIPTION

SITE: 36.5 Acres.
NUMBER OF BUILDINGS: Two; Shell only.
BUILDING SIZE: First floor, 67,500; second floor, 2,200; total, 69,700 square feet.
BUILDING HEIGHT: First floor, 67'; second floor, 15'; total, 67'.
BASIC CONSTRUCTION TYPE: New/2B.
FOUNDATION: Cast-in-place.
EXTERIOR WALLS: Brick, metal, curtainwall.
ROOF: Metal, membrane, asphalt shingles.
FLOORS: Concrete.
INTERIOR WALLS: Metal stud drywall.

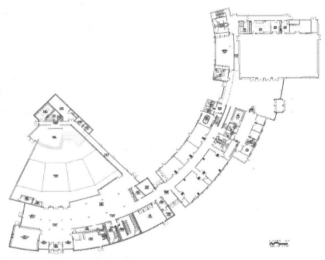

Beautiful Savior Lutheran Church
Construction Period: Sep 2005 to Feb 2007 • Total Square Feet: 69,700

C.S.I. Divisions (1 through 16)			COST	% OF COST	SQ.FT. COST	SPECIFICATIONS
		PROCUREMENT & CONTRACTING	623,178	11.95	8.94	Project forms, conditions of the contract, price & payment procedures, quality requirements, temporary facilities & controls, execution & closeout requirements, performance requirements.
1.	1.	GENERAL REQUIREMENTS	473,797	9.09	6.80	Weather allowance, contingency, change orders.
3.	3.	CONCRETE	425,079	8.15	6.10	Forming & accessories, reinforcing, cast-in-place, grouting (Concrete breakdown: cubic yards foundation, 380; cubic yards floors, 540).
5.	5.	METALS	766,385	14.70	11.00	Structural metal framing, joists, decking, cold-formed metal framing, fabrications.
6.	6.	WOOD/PLASTICS/COMPOSITE	252,693	4.85	3.63	Rough carpentry, finish carpentry, architectural woodwork.
7.	7.	THERMAL & MOIST. PROTECTION	519,810	9.97	7.46	Dampproofing & waterproofing, thermal barriers, roofing & siding panels, membrane roofing, flashing & sheet metal, roof & wall specialties & accessories, fire & smoke protection, joint protection.
8.	8.	OPENINGS	219,927	4.22	3.16	Doors & frames, entrances, storefronts, & curtainwalls, windows, hardware, glazing, louvers & vents.
9.	9.	FINISHES	718,677	13.78	10.31	Plaster & gypsum board, tiling, ceilings, wall finishes, acoustic treatment, painting & coating.
10.	10.	SPECIALTIES	26,660	0.51	0.38	—
12.	12.	FURNISHINGS	8,000	0.15	0.11	—
15.	23.	HVAC	901,230	17.27	12.92	Water-based fire-suppression systems, fire pumps, plumbing piping & pumps, equipment, fixtures, HVAC piping & pumps, air distribution.
16.	26.	ELECTRICAL	279,510	5.36	4.01	Medium-voltage distribution, low-voltage transmission.
TOTAL BUILDING COST			**5,214,946**	**100%**	**$74.82**	
2.	32.	EXTERIOR IMPROVEMENTS	166,078			Bases, bollards, & paving, improvements, irrigation, planting, site clearing, earth moving, earthwork methods.
TOTAL PROJECT COST			**5,381,024**	*(Excluding architectural and engineering fees)*		

UPDATED ESTIMATE TO JAN 2011: $103.37 SF

DCD Subscribers: Access this case study and hundreds more for instant date and location calculations at www.dcd.com.

Beautiful Savior Lutheran Church
Plymouth, Minnesota

Architect: Cuningham Group Architecture, P.A.

Photos Courtesy of Reynolds Construction Management

The floor plan is simple and direct. The progression from the front door is axial in nature, beginning in the parking lot, moving through the front door and the center of the Commons, through the Worship Center entrance and the font area, down to the main alter and ultimately to the Cross. The Commons is a place to meet and orient one to the facility. The semi-circular-shaped, sloped Sanctuary floor is designed for intimacy as well as a sense of community and togetherness. All guests in the Sanctuary can see activities as the front chancel is elevated by five steps, and a carved-out section on the right of the chancel allows a variety of music groups to perform.

The Sanctuary structural system is primarily a steel frame, incorporating large-scale trusses to allow for a column-free space. The exterior skin is composed of Belden Brick (where it joins Phase I) and two-color metal flat seam vertical paneling (denoting worship space). The space is appointed with carpeting, paint, and has many interior windows to allow visual access throughout the Church. The exterior landscaping is straight forward and mainly consists of local plantings for easy maintenance, appropriate scale and appearance for Minnesota.

At the grand opening the Building Committee, headed by the Pastor Thomas Stoebig, stated that the church had achieved all its facility goals on a functional and creative basis. They were ready to welcome the community into their space - as a Beacon of Hope for all.

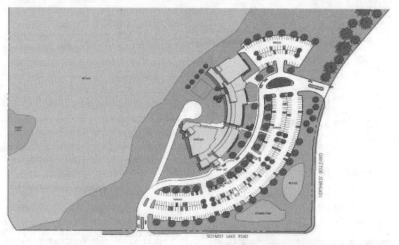

COST PER SQUARE FOOT $ 103.37

ARCHITECT
BNIM Architects
106 W. 14th Street, Kansas City, MO 64105
www.bnim.com

FILE UNDER
RELIGIOUS
Kansas City, Missouri

CONSTRUCTION TEAM

GENERAL CONTRACTOR:
McCownGordon Construction L.L.C.
422 Admiral Boulevard, Kansas City, MO 64106
STRUCTURAL ENGINEER:
Structural Engineering Associates, Inc.
1000 Walnut, #1570, Kansas City, MO 64106
ELECTRICAL & MECHANICAL ENGINEER:
BGR Consulting Engineers
908 Broadway, #200, Kansas City, MO 64105
LANDSCAPE ARCHITECT:
BNIM Architects
106 W. 14th Street, Kansas City, MO 64105

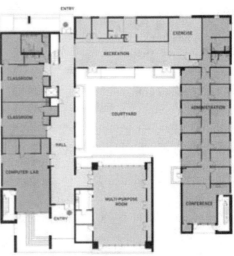

GENERAL DESCRIPTION

SITE: 0.67 Acres.
NUMBER OF BUILDINGS: One.
BUILDING SIZE: First floor, 14,320; second floor, 12,000; total, 26,320 square feet.
BUILDING HEIGHT: First floor, 14'; second floor, 26'; total, 26'.
BASIC CONSTRUCTION TYPE: New/Masonry bearing wall, precast cored slabs.

FOUNDATION: Pier & grade beam, reinforced concrete, slab-on-grade.
EXTERIOR WALLS: CMU, brick, curtainwall, reclaimed wood siding.
ROOF: Membrane.
FLOORS: Concrete, precast.
INTERIOR WALLS: CMU, metal stud drywall.

CHRISITAN LIFE CENTER
Construction Period: Feb 2007 to Mar 2008 • Total Square Feet: 26,320

C.S.I. Divisions (1 through 16)			COST	% OF COST	SQ.FT. COST	SPECIFICATIONS
		PROCUREMENT & CONTRACTING	627,335	10.04	23.83	Solicitation, instructions for procurement, contracting forms & supplements, project forms, conditions of the contract, revisions, clarifications, and modifications.
1.	1.	GENERAL REQUIREMENTS	26,584	0.43	1.01	Price & payment procedures, administrative requirements, quality requirements, temporary facilities & controls, execution & closeout requirements, performance requirements.
3.	3.	CONCRETE	1,099,202	17.59	41.76	Forming & accessories, reinforcing, cast-in-place, precast, cast decks & underlayment, grouting. (Concrete breakdown: foundation, 490 cubic yards; floors, 27,120 square feet of precast.)
4.	4.	MASONRY	652,401	10.44	24.79	Unit, stone assemblies, manufactured masonry.
5.	5.	METALS	214,420	3.43	8.15	Structural metal framing, decking, cold-formed metal framing, fabrications.
6.	6.	WOOD/PLASTICS/COMPOSITE	403,717	6.46	15.34	Rough carpentry, finish carpentry, architectural woodwork.
7.	7.	THERMAL & MOIST. PROTECTION	302,465	4.84	11.49	Dampproofing & waterproofing, thermal protection, weather barriers, membrane roofing, flashing & sheet metal, roof & wall specialties & accessories, joint protection.
8.	8.	OPENINGS	547,073	8.75	20.79	Doors & frames, specialty doors & frames, entrances, storefronts, & curtain walls, windows, hardware.
9.	9.	FINISHES	905,284	14.48	34.40	Plaster & gypsum board, tiling, ceilings, flooring, wall finishes, acoustical treatment, painting & coating.
10.	10.	SPECIALTIES	44,081	0.71	1.67	Interior, storage.
14.	14.	CONVEYING SYSTEMS	49,786	0.80	1.89	Elevators (1).
15.	21.	FIRE SUPPRESSION	53,312	0.85	2.03	Water-based fire-suppression systems.
15.	23.	HVAC	796,716	12.75	30.27	Piping & pumps, air distribution, central heating, central cooling, central HVAC.
16.	26.	ELECTRICAL	500,006	8.00	19.00	Medium-voltage distribution, low-voltage transmission, facility power generating & storing equipment, electrical & cathodic protection, lighting.
TOTAL BUILDING COST			**6,250,550**	**100%**	**$237.48**	
2.	31.	EARTHWORK	216,622			Site clearing, earth moving, earthwork methods, excavation support & protection.
2.	32.	EXTERIOR IMPROVEMENTS	22,757			Bases, bollards, & paving, improvements, planting.
2.	33.	UTILITIES	65,116			Water, wells, sanitary, storm drainage.
TOTAL PROJECT COST			**6,555,045**	*(Excluding architectural and engineering fees)*		

UPDATED ESTIMATE TO JAN 2011: $305.99 SF

DCD Subscribers: Access this case study and hundreds more for instant date and location calculations at www.dcd.com.

Christian Life Center for City Union Mission
Plymouth, Minnesota

Architect: Cuningham Group Architecture, P.A.

© Assassi - Photos Courtesy of BNIM

City Union Mission is a not-for-profit organization serving the base level needs of those whose lives have been disrupted in the Kansas City community. BNIM was hired to analyze the existing Men's Shelter Building and to address how the organization could meet an overwhelming increase in need. It quickly became clear that their existing facility was inadequate to meet rising demands, and the development of a separate structure housing the Christian Life Program was created.

This program nurtures the re-entry process for one-hundred of the most promising men who will live and attend classes in the facility during their one-year curriculum. The resulting design supports the believe that students of the program deserve experiencing the best during their course and afterward wherever their new life leads.

Located in a neglected neighborhood near the urban core, the creation of a safe and healthy environment was paramount. To address this goal, the facility was developed around a secure courtyard that connects the interior and exterior throughout. Programmatic spaces include a dormitory, living area, classrooms, recreation rooms, and administrative offices. A large-multi-purpose space is used for dining, recreation and worship. Exterior materials include a rain screen of recycled hardwood combined with brick and burnished block masonry.

The site design incorporates a variety of sustainable features and is a showcase for urban stormwater management. The small site includes three bioretention cells that accept all of the roof run-off and there is no stormwater connection to the City's sewer system. Indigenous plant material that require low-maintenance have been integrated throughout the site. Hidden from view are the geo-thermal wells and the recycled water storage tanks, which hold filtered water from the showers for use in toilet flushing.

Sustainable design features include:

- Storm runoff from the parking lots drain into two bioswales which collect sediment, clean heavy metals and encourage infiltration into the soils. The bioswales are planted with cordgrass and swichgrass to absorb a high percentage of rainwater.
- The central courtyard is planted with buffalo grass and requires no chemicals and only a couple of lawn mowings per year.
- Shortgrass prairie grasses are planted on side lots to improve the environmental quality of the area.
- Pervious limestone screening walk connects the building to adjacent streets.
- Steep slopes west of the building are stabilized with a geo-web stabilization product to resist the potential erosion of the hillside into parking lot.
- Building orientation with courtyard configuration maximizes natural daylight into the facility. Every room has access to daylight. - Custom insulated windows for high thermal capacity were installed.
- A geo-thermal heat pump system was installed.
- Passive solar techniques were utilized, including the thermal floor mass in the south-facing multi-purpose room.
- Fritted glass in the south curtain wall helps to manage heat gain.

MANUFACTURERS/SUPPLIERS	
DIV 04	*Brick:* ACME, Midwest.
DIV 07:	*Membrane:* Carlisle.
DIV 08:	*Entrances & Storefronts:* **Kawneer**; *Windows:* **VOS.**
DIV 14:	*Elevators:* **Schindler** .

COST PER SQUARE FOOT $ 305.99

ARCHITECT

Chatham Design Group
116 Fairhope Avenue, Fairhope, AL 36532
www.allplans.com

FILE UNDER

RESIDENTIAL
Fairhope, Alabama

CONSTRUCTION TEAM

STRUCTURAL ENGINEER:

J. Martin Pitts, LLC

116 Fairhope Avenue, Fairhope, AL 36532

GENERAL CONTRACTOR:

Jeremy Friedman Kaloosa Builders

P.O. Box 1184, Fairhope, AL 36533

GENERAL DESCRIPTION

SITE: .35 Acre.
NUMBER OF BUILDINGS: Two.
BUILDING SIZE: Detached garage, 528; first floor, 2,324; second floor, 1,286; porches, 475; total, 4,613* square feet.
BUILDING HEIGHT: Garage, 22'; first floor, 10'; second floor, 9'; total, 34'10".
BASIC CONSTRUCTION TYPE: New/Wood Frame.
FOUNDATION: Slab-on-grade.
EXTERIOR WALLS: Concrete siding.
ROOF: Asphalt shingles.
FLOORS: Wood.
INTERIOR WALLS: Wood stud drywall.

FAIRHOPE GREEN HOME

Construction Period: Apr 2008 to Oct 2008 • Total Square Feet: 4,613

C.S.I. Divisions (1 through 16)			COST	% OF COST	SQ.FT. COST	SPECIFICATIONS
1.	1.	GENERAL REQUIREMENTS	32,366	7.61	8.97	Energy Star® option, engineering fees, permits, insurance, interest, miscellaneous.
3.	3.	CONCRETE	45,000	6.74	7.94	Cast-in-place, reinforcing, formwork (Concrete breakdown: 91 cubic yards foundation).
4.	4.	MASONRY	6,635	1.56	1.84	—
6.	6.	WOOD/PLASTICS/COMPOSITE	110,698	29.88	35.19	Rough carpentry, finish carpentry, architectural woodwork, Zip wall & roofing system.
7.	7.	THERMAL & MOIST. PROTECTION	20,246	4.76	5.61	Waterproofing & dampproofing, insulation, asphalt shingles.
8.	8.	OPENINGS	36,246	8.52	10.04	Doors & frames, entrances, hardware, windows, garage doors.
9.	9.	FINISHES	77,293	18.18	21.41	Plaster & gypsum board, ceilings, flooring, wall finishes, painting & coating.
10.	10.	SPECIALTIES	16,006	3.76	4.43	Columns, mirrors, mailbox, shutters.
14.	14.	EQUIPMENT	12,835	3.02	3.56	Appliances
15.	22.	PLUMBING	21,951	5.16	6.08	—
15.	23.	HVAC	18,535	4.36	5.13	—
16.	26.	ELECTRICAL	27,426	6.45	7.60	—
TOTAL BUILDING COST			**425,237**	**100%**	**$117.79**	
2.	31.	EARTHWORK	3,975			Earth moving.
2.	32.	EXTERIOR IMPROVEMENTS	12,880			Landscaping.
TOTAL PROJECT COST			**442,092**	*(Excluding architectural and engineering fees)*		

UPDATED ESTIMATE TO JAN 2011: $162.49 SF

DCD Subscribers: Access this case study and hundreds more for instant date and location calculations at www.dcd.com.

Fairhope Green Home
Fairhope, Alabama

Architect: Chatham Design Group

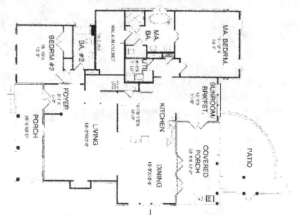

The first NAHB Green Certified home in Alabama under the National Association of Home Builders (NAHB) National Green Building Program', the Fairhope Green Home is a stunning home with detached garage. The home is Green Certified for Energy Efficiency, Water Conservation, Indoor Air Quality, and Resource Efficiency.

The home designed by Chatham Home Planning and built by Jeremy Friedman, a Certified Green Professional, includes exposed rafter tails, Cypress board and batten siding with Cedar shake details. The home includes 5 bedrooms, 4 bathrooms, a gourmet Thermad® or kitchen, luxury master suite, keeping room, outdoor kitchen and detached garage with covered breeze way.

The Fairhope Green Home included energy saving features such as the ZIP System® roof and wall sheathing. The Zip System® roof and wall sheathing by Huber Engineered Woods eliminates the need for housewrap and felt. The ZIP system® structural roof and wall panels also have built-in protective barriers. When used with the ZIP System® tape, the ZIP System® Wall provides a code recognized, water resistive barrier and air barrier for all-in-one for superior moisture protection and enhanced energy efficiency.

The home was so successful it earned 439 points and a gold rating from the NAHB National Green Building Program.

Sustainable design features include:

- Waterfurnace Geothermal Heating & Cooling
- Icynene® Spray Foam Insulation
- Dual Flush Toilets & Low Flow Fixtures
- Advanced Framing Techniques
- Native Draught Tolerant Landscaping
- Reclaimed Heat Pine Flooring
- Reclaimed Pecan Mantel and Beams

LEED® GOLD PENDING

COST PER SQUARE FOOT $162.49

ARCHITECT
Krista Atkins Nutter, ASID, LLC
1288 Baldwin Road, Milford, OH 45150

FILE UNDER
RESIDENTIAL
Milford, Ohio

CONSTRUCTION TEAM

STRUCTURAL ENGINEER:
Advantage Group Engineers
660 Lincoln Avenue #305, Cincinnati, OH 45206

GENERAL DESCRIPTION

SITE: 1.75 Acres.
NUMBER OF BUILDINGS: One.
BUILDING SIZE: Garage (unfinished), 728; basement (unfinished), 1,295; first floor, 1,323; second floor, 966; total, 2,289 (finished).
BUILDING HEIGHT: Basement, 8'5"; first floor, 8'; second floor, 8'; floor to floor, 9'4"; total, 19' (total height is 19' above grade on south side and 27' above grade on north side).
BASIC CONSTRUCTION TYPE: New/ICF/Stick Frame.
FOUNDATION: Insulated Concrete Form (ICF).
EXTERIOR WALLS: Stick frame, metal & composite siding.
ROOF: Metal.
FLOORS: Concrete, wood.
INTERIOR WALLS: Wood stud drywall.

NUTTER GREEN HOME
Construction Period: Apr 2006 to June 2007 • Total Square Feet: 2,289

C.S.I. Divisions (1 through 16)			COST	% OF COST	SQ.FT. COST	SPECIFICATIONS
		PROCUREMENT & CONTRACTING	5,100	2.34	2.23	Permits, fees.
3.	3.	CONCRETE	47,891	22.02	20.92	Reinforcing, cast-in-place (Concrete breakdown: 60 cubic yards foundation walls; 21 cubic yards first floor, 17 cubic yards basement floor).
5.	5.	METALS	1,200	0.55	0.52	Fabrications.
6.	6.	WOOD/PLASTICS/COMPOSITE	39,882	18.34	17.42	Rough carpentry, finish carpentry, architectural woodwork.
7.	7.	THERMAL & MOIST. PROTECTION	19,131	8.79	8.36	Thermal protection, roofing & siding panels, membrane roofing, roof & wall specialties & accessories.
8.	8.	OPENINGS	16,801	7.72	7.34	Specialty doors & frames, windows, hardware.
9.	9.	FINISHES	14,795	6.80	6.47	Plaster & gypsum board, tiling, flooring, painting & coating.
10.	10.	SPECIALTIES	3,269	1.50	1.43	Information, fireplaces & stoves.
14.	14.	EQUIPMENT	4,200	1.93	1.84	Appliances.
15.	22.	PLUMBING	14,400	6.62	6.29	Piping & pumps, equipment, fixtures.
15.	23.	HVAC	6,900	3.17	3.02	Facility fuel systems, central HVAC equipment.
16.	26.	ELECTRICAL	6,024	2.77	2.63	Medium voltage distribution, lighting.
16.	28.	ELECTRONIC SAFETY & SECURITY	1,200	0.55	0.52	Detection & alarm.
16.	29.	ELECTRICAL POWER GENERATION	36,768	16.90	16.06	Power generation.
TOTAL BUILDING COST			**217,561**	**100%**	**$95.05**	
2.	2.	EXISTING CONDITIONS	2,930			Assessment.
2.	31.	EARTHWORK	22,000			Earth moving.
2.	32.	EXTERIOR IMPROVEMENTS	12,200			Bases, bollards, & paving, improvements, wetlands, planting.
2.	33.	UTILITIES	18,900			Water, sanitary.
TOTAL PROJECT COST			**273,591**	*(Excluding architectural and engineering fees)*		

UPDATED ESTIMATE TO JAN 2011: $128.32 SF

DCD Subscribers: Access this case study and hundreds more for instant date and location calculations at www.dcd.com.

Nutter Green Home
Milford, Ohio

Architect: Krista Atkins Nutter, ASID, LLC

Photos Courtesy of Krista Atkins Nutter, ASID,LLC

Krista Atkins Nutter, designer and owner of the Nutter Green Home, is a certified interior designer with an MS Architecture degree and is a Professor of Design. She and her husband, Kenneth have built a super-insulated, energy-efficient, passive-solar, solar-electric home with solar hot water heating and a rainwater catchment system to provide water for toilet-flushing, clothes-washing, and outdoor uses. The home is designed for the local climate, and has achieved an Energy Star Rating of 5+ Stars, a 2007 Cincinnati Sustainability Award and a 2008 Excellence in Design Award Runner-Up selection in Environmental Design + Construction Magazine. One of the main goals of the home is to serve as an educational model for the community. The Nutters have given seminars on the Nutter Green Home to various groups, including students of all ages, the US EPA, the NeoCon World's Trade Fair in Chicago, and Green Energy Ohio.

The site for the project is a sub-urban infill lot with surrounding homes 25 years of age. The site was selected for its unobstructed southern exposure and close proximity to services such as retail, churches, healthcare, schools, ect. The home is a 2,300-square-foot modern-industrial, environmentally-sensitive dwelling utilizing strategies such as an insulated concrete form foundation, spray foam insulation, value engineered framing, suspended concrete floors, environmentally-sensitive materials such as recycled glass counter tops and bamboo stair treads, water-conserving low flow faucets and dual flush toilets, two high-efficiency heat pumps for back up heating and cooling, a heat recovery ventilator to supply the home with fresh air, and a 4.4 kW photovoltaic roof-mounted array. The exterior of the home features recycleable Galvalume and fiber composite siding. All concrete used in the project utilized 20% fly-ash. The home with its grid-tied solar PV system, installed in March 2007, recorded its first positive electric bill in Aug./Sept. 2007 (which was quite a hot month in Cincinnati where temperatures consistently reached the 90s Fahrenheit). With both heat pump condensing units running and the thermostats set to 74-degrees, the electric bill from mid-August to mid-September was $8. All bills

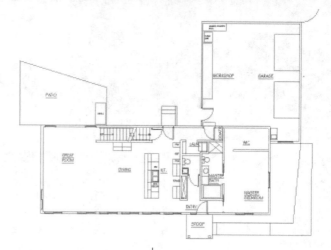

prior to that were credits ranging from $40 to $60. Superior indoor air quality is achieved through zero-voc finishes, low or no-formaldehyde construction products, a high-efficiency electronic air filtration system, a heat recovery ventilation system, and by sealing all metal ductwork with mastic.

A grant from the Ohio Department of Development was awarded to the owners to help offset the cost of the PV and solar hot water heating systems.

The home also features native landscaping including a rain garden, a constructed wetland, a native prairie, a vegetable garden, and other plants native to southwestern Ohio. Like the home, the gardens will be open to the public for tours and seminars which discuss the importance of native landscapes, habitat preservation, and stormwater runoff retention. The National Wildlife Federation has designated the property as a "Certified Backyard Wildlife Habitat."

ENERGY STAR 5+STARS

COST PER SQUARE FOOT $128.32

ARCHITECT
Rob Paulus Architect
990 E. 17th Street, Tucson, AZ 85719
www.robpaulus.com

CONSTRUCTION TEAM

STRUCTURAL ENGINEER:
  Grenier Engineering, Inc.
  5515 E. 5th Street, Tucson, AZ 85711
GENERAL CONTRACTOR:
  Epstein/Fenton LLC
  4855 N. Shamrock Place, #109, Tucson, AZ 85705
ELECTRICAL & MECHANICAL ENGINEER:
  GLHN Architects & Engineers, Inc.
  2939 E. Broadway Blvd., Tucson, AZ 85716
LANDSCAPE ARCHITECT:
  Chris Winters + Associates
  820 N. 3rd Street, Phoenix, AZ 85004

GENERAL DESCRIPTION

SITE: 0.86 Acres.
NUMBER OF BUILDINGS: Three.
BUILDING SIZE: Building 1 First floor, 4,490; Building 2 first floor, 3,810; second floor, 4,100; Building 3, first floor, 100; total, 12,500 sf.
BUILDING HEIGHT: Building 1, low end, 9'; Building 1, high end, 14'8"; Building 2, first floor, 8'4"; Building 2, low end, 19'2'" Building 2, high end, 23'6"; Building 3, 10'4'.
BASIC CONSTRUCTION TYPE: Wood Frame/New.
FOUNDATION: Cast-in-place, slab-on-grade.
EXTERIOR WALLS: CMU, metal panel over wood frame.
ROOF: Metal, membrane.
FLOORS: Concrete, wood.
INTERIOR WALLS: Wood stud drywall.

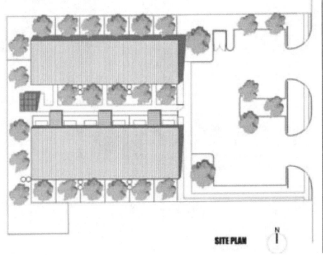

SITE PLAN

SILVERADO FLATS
Construction Period: Aug 2006 to Mar 2007 • Total Square Feet: 12,500

C.S.I. Divisions (1 through 16)			COST	% OF COST	SQ.FT. COST	SPECIFICATIONS
1.	1.	GENERAL REQUIREMENTS	183,474	17.05	14.68	Temporary utilities & facilities, clean-up, safety, supervision, fee, sales tax, termite protection.
3.	3.	CONCRETE	83,360	7.75	6.67	Concrete (Foundation 140 cubic yards, floors, 240 cubic yards), accessories.
4.	4.	MASONRY	18,570	1.73	1.49	—
5.	5.	METALS	35,000	3.25	2.80	Stairs, canopies, trellis, PV structure.
6.	6.	WOOD/PLASTICS/COMPOSITE	166,219	15.45	13.30	Rough carpentry, finish carpentry, (cabinetry provided & installed by owner).
7.	7.	THERMAL & MOIST. PROTECTION	109,266	10.15	8.74	Roofing & siding, insulation.
8.	8.	OPENINGS	61,094	5.68	4.89	Doors, hardware, windows.
9.	9.	FINISHES	140,155	13.02	11.21	Stucco, plaster, drywall, ceramic tile, flooring, paint.
14.	14.	EQUIPMENT	24,000	2.23	1.92	Appliances.
15.	21.	FIRE SUPRESSION	13,500	1.25	1.08	—
15.	22.	PLUMBING	73,605	6.84	5.89	—
15.	23.	HVAC	60,260	5.60	4.82	—
16.	26.	ELECTRICAL	107,576	10.00	8.60	Electrical, fixtures, photovoltaic panels.
TOTAL BUILDING COST			**1,076,079**	**100%**	**$86.09**	
2.	2.	EXISTING CONDITIONS	9,000			Demolition.
2.	31.	EARTHWORK	46,787			Earthwork.
2.	32.	EXTERIOR IMPROVEMENTS	19,900			Landscaping, improvements.
2.	33.	UTILITIES	45,475			—
TOTAL PROJECT COST			**1,197,241**			*(Excluding architectural and engineering fees)*

UPDATED ESTIMATE TO JAN 2011: $126.72 SF

Silverado Flats
Tucson, Arizona

Architect: Rob Paulus Architect

Photos Courtesy of Ross Cooperthwaite/Cooperthwaite Photography

Silverado Flats provides a sophisticated alternative to the ubiquitous stucco-box apartment building. The project utilizes an infill site to create twelve apartments with multiple outdoor spaces, an abundance of light, and a shared entry court to foster a sense of community.

Program requirements informed two distinct sizes of units: single story one-bedroom and double story two-bedroom apartments. The concept evolved into a metal-clad wedge shape, reaching towards the Santa Catalina Mountains, with the middle chopped out to create the central courtyard between. Steel trellis structures and frequent trees create rhythmic interruptions along the processional entry path. Unique fencing materials surround private yards for each unit while vertical road culverts become rainwater harvesting tanks. A small laundry room at the end of the courtyard plays host to an array of photovoltaic cells, providing a focal point and renewable energy for the site.

The design emphasizes a pared-down aesthetic as a contrast to the complexity of the outside desert landscape. Exposed concrete floors, industrial light fixtures, dramatic ceilings and large window openings create loft-style living space. One-bedrooms feature open, flexible living spaces and two private patios. Two-bedroom units boast mountain views and are designed to accommodate either couples or roommates by providing private lavatories off each bedroom outside the shared bath. Interior openings allow slivers of light and snapshot views as operable panels slide past each other, while accent colors highlight the circulation core.

Sustainable principals influenced every step of the design process. Careful site planning optimizes allowable density while maintaining open space. Passive solar orientation and active solar collection use the intense sun to advantage. Durable, integral color materials were chosen for both resource conservation and maintenance ease. High-efficiency insulation keeps units temperate, while abundant patio spaces provide exterior living space to enjoy. The central location near a through-city bike path provides viable transportation options. Rainwater harvesting and native, low water use vegetation provides visible reminders of the importance of water conservation in the desert climate.

What helped make this project special was the enthusiasm of the client. They were expressly interested in creating unique, modern spaces with energy-conscious ideals at the fore. Their involvement in every step of the project enriched the design process, allowing a pure artistic expression that provides functional resolution of the needs of the owner.

MANUFACTURERS/SUPPLIERS	
DIV 07:	*Manufactured Roofing & Siding:* **MBCI.**
DIV 08	*Windows:* **International Window Corporation;** *Entrances:* **Therma-Tru.**
DIV 09:	*Flooring:* **Tarkett.**
DIV 26:	*Photovoltaic Collectors:* **Kyocera** *KC175 (12); Mounting System:* **Unirac Solar Mount.**

COST PER SQUARE FOOT $ 126.72

DESIGNER
JOHN KOSMER
P.O. Box 92,
Fly Creek, NY 13337
www.solarhouseproject.com

FILE UNDER
RESIDENTIAL
Fly Creek, New York

CONSTRUCTION TEAM

GENERAL CONTRACTOR:
Building with Integrity, Inc.
228 Poolbrook Road, Laurens, NY 13796
SOLAR ENGINEER:
Bruce Brownell
Adirondack Alternative Energy
98 Northville Road, Edinburg, NY 12134

GENERAL DESCRIPTION

SITE: 26 acres.
NUMBER OF BUILDINGS: One.(house and attached garage)
BUILDING SIZE: Garage, 1,008; First floor, 1,400; Second floor 1,400; total, 3,808 square feet.
BUILDING HEIGHT: First floor, 9'6"; Second floor, 8'-6" total, 28'6"
BASIC CONSTRUCTION TYPE: New/Wood Frame.
FOUNDATION: Cast-in-place.
EXTERIOR WALLS: Pre-finished concrete siding.
ROOF: Asphalt shingles.
FLOORS: Concrete.
INTERIOR WALLS: Wood stud drywall.

SOLAR HOUSE

Construction Period: May 2006 to May 2007 Total Square Feet 3,808

C.S.I. Divisions (1 through 16)			COST	% OF COST	SQ.FT. COST	SPECIFICATIONS
		PROCUREMENT & CONT. REQ.	2,500	0.59	0.76	Solicitation, contracting forms & supplements, project forms, conditions of the contract, revisions, clarifications, & modifications.
1.	1.	GENERAL REQUIREMENTS	-	-	-	-
3.	3.	CONCRETE	28,600	6.73	8.66	Forming & accessories, reinforcing, mass, cutting & boring.
6.	6.	WOOD/PLASTICS/COMPOSITE	-	-	-	
7.	7.	THERMAL & MOIST. PROTECT	85,000	20.00	25.73	Dampproofing & waterproofing, thermal protection, weather barriers, steep slope roofing, roofing & siding panels, membrane roofing, flashing & sheet metal, roof & wall
8.	8.	OPENINGS	42,000	9.88	12.71	Doors & frames, specialty doors & frames, windows, roof windows & skylights, hardware.
9.	9.	FINISHES	71,000	16.70	21.49	Plaster & gypsum board, tiling, ceilings, flooring, wall finishes, acoustic treatment, painting & coating.
10.	10.	SPECIALTIES	15,000	3.53	4.54	Fireplaces & stoves, solar collectors.
11.	11.	EQUIPMENT	20,000	4.70	6.05	Residential.
13.	13.	SPECIAL CONSTRUCTIONS	70,000	16.47	21.19	Garage.
15.	22.	PLUMBING	37,000	8.70	11.20	Equipment, fixtures.
15.	23.	HVAC	32,000	7.53	9.68	Facility fuel systems, piping & pumps, air cleaning devices, central heating, central HVAC equipment.
16.	26.	ELECTRICAL	22,000	5.17	6.65	Medium-voltage distribution, lighting.
TOTAL BUILDING COST			**425,100**	**100.00**	**$128.66**	
2.	2.	EXISTING CONDITIONS	1,500			Assessment, subsurface investigation.
2.	31.	EARTHWORK	20,000			Site clearing, earth moving, excavation support & protection.
2.	32.	EXTERIOR IMROVEMENTS	5,000			Improvements.
2.	33.	UTILITIES	14,500			Wells, sanitary sewerage utilities, electrical.
TOTAL PROJECT COST			**466,100**	*(Excluding architectural and engineering fees)*		

UPDATED ESTIMATE TO JAN 2011: $164.56 SF

DCD Subscribers: Access this case study and hundreds more for instant date and location calculations at www.dcd.com.

Solar House
Fly Creek, New York

Designer: John Kosmer

Photos Courtesy of John Kosmer and Simonton®

In the 21st century, any home design that does not provide maximum annual energy savings is a nonstarter. In this new epoch, if you don't incorporate super energy efficiency into a new home plan, you simply become a 21st Century Nero: fiddling with home design as America burns fuel," stated John Kosmer owner and designer of a passive solar home.

Kosmer designed a 4,000-square-foot traditional style passive solar home located in cold upstate New York that uses $2.50 a day for heat - between $900 and $1,200 a year. The house was built at the same cost as a comparable size Energy Star® qualified home, but uses 70% less energy to heat. This savings would be even greater if the house was not in the Snow Belt and had regular height ceilings.

Bruce Brownell, a nationally recognized solar engineer for over 30 years, tailored the passive solar system that was incorporated in the house. Four inch thick rigid polyurethane was used on the exterior walls, under the roof and beneath the one-foot concrete slab. A state-of-the-art boiler provides supplemental heat during the coldest weather. And, the exterior finish is a pre-finished concrete siding with a 50-year warranty.

Simonton windows were used throughout the home and garage. "My research firmly led me to Simonton Windows®. The Energy Star compliant windows from Simonton with a glass package that includes Argon gas filling, double glazing and Low E Softcoat to retain the sun's energy," Kosmer states.

Over 80 casement windows in the home pour sunlight into the white interior and on the natural blonde bamboo floors throughout the house. Air heated by the sun rises several stories and is drawn into air ducts. A large squirrel cage fan in the main duct channels the air into the concrete slab and then redistributes it into the home to keep it at a comfortable temperature.

The price per square foot was kept low using several strategies such as a simple structure of a 30 x 50-footprint with one dormer; using sheet rock for staircases instead of a traditional banister; marble tile for the window sills; and a modified truss in the garage for a 20-foot clear span that netted a 11 x 30-foot deep storage area on the second floor. A comprehensive description of how the home was built and functions can be seen at:

www.SolarHouseProject.com.

FIRST FLOOR

SUSTAINABLE	
MANUFACTURERS/SUPPLIERS	
DIV 07:	*Shingles:* **Certtainteed Presidential. Shake**; *Siding:* **Hardie Plank**; *Skylights:* Velux; *Insulation;* **Dow Turff-R®** Rigid Insulation; **Certainteed Fiberglass Insulation.**
DIV 08:	*Doors:* **Therma Thru**; *Windows:* **Simonton.**
DIV 09:	*Tile:* **Crosville Ceramic Eco-Tile**; *Bamboo Flooring:* **Mannington.**
DIV 12:	*Appliances:* **GE Profile.**
DIV 23:	*Water Heating:* **Baxi Energy Star Qualified**; *Panels:* **Luna 3** .

COST PER SQUARE FOOT $164.56

Part Three

Unit-In-Place Costs

DIVISION	1	GENERAL REQUIREMENTS
DIVISION	2	SITEWORK & DEMOLITION
DIVISION	3	CONCRETE
DIVISION	4	MASONRY
DIVISION	5	METALS
DIVISION	6	WOOD & PLASTICS
DIVISION	7	MOISTURE & THERMAL PROTECTION
DIVISION	8	DOORS & WINDOWS
DIVISION	9	FINISHES
DIVISION	10	SPECIALTIES
DIVISION	11	EQUIPMENT
DIVISION	12	FURNISHINGS
DIVISION	13	SPECIAL CONSTRUCTION
DIVISION	15	MECHANICAL
DIVISION	16	ELECTRICAL

	Unit	Total		Unit	Total

01020.10 ALLOWANCES
Overhead, $20,000 project

Minimum	PCT	15.00			
Average	PCT	20.00			
Maximum	PCT	40.00			

$100,000 project

Minimum	PCT	12.00
Average	PCT	15.00
Maximum	PCT	25.00

$500,000 project

Minimum	PCT	10.00
Average	PCT	12.00
Maximum	PCT	20.00

Profit, $20,000 project

Minimum	PCT	10.00
Average	PCT	15.00
Maximum	PCT	25.00

$100,000 project

Minimum	PCT	10.00
Average	PCT	12.00
Maximum	PCT	20.00

$500,000 project

Minimum	PCT	5.00
Average	PCT	10.00
Maximum	PCT	15.00

Professional fees, Architectural
$100,000 project

Minimum	PCT	5.00
Average	PCT	10.00
Maximum	PCT	20.00

$500,000 project

Minimum	PCT	5.00
Average	PCT	8.00
Maximum	PCT	12.00

Structural engineering

Minimum	PCT	2.00
Average	PCT	3.00
Maximum	PCT	5.00

Mechanical engineering

Minimum	PCT	4.00
Average	PCT	5.00
Maximum	PCT	15.00

Taxes, Sales tax

Minimum	PCT	4.00
Average	PCT	5.00
Maximum	PCT	10.00

Unemployment

Minimum	PCT	3.00
Average	PCT	6.50
Maximum	PCT	8.00
Social security (FICA)	PCT	7.85

01050.10 FIELD STAFF
Superintendent

Minimum	YEAR	86,500
Average	YEAR	126,500
Maximum	YEAR	181,500

Foreman

Minimum	YEAR	48,400
Average	YEAR	77,000
Maximum	YEAR	114,400

Bookkeeper/timekeeper

Minimum	YEAR	27,500
Average	YEAR	36,300
Maximum	YEAR	60,500

Watchman

Minimum	YEAR	19,300
Average	YEAR	24,200
Maximum	YEAR	38,500

01330.10 SURVEYING
Surveying

Small crew	DAY	810
Average crew	DAY	1,250
Large crew	DAY	1,600

Lot lines and boundaries

Minimum	ACRE	580
Average	ACRE	1,250
Maximum	ACRE	2,050

01380.10 JOB REQUIREMENTS
Job photographs, small jobs

Minimum	EA.	120
Average	EA.	170
Maximum	EA.	400

Large projects

Minimum	EA.	580
Average	EA.	870
Maximum	EA.	2,900

01410.10 TESTING
Testing concrete, per test

Minimum	EA.	19.00
Average	EA.	31.75
Maximum	EA.	63.75

01500.10 TEMPORARY FACILITIES
Barricades, temporary
Highway

Concrete	L.F.	15.25
Wood	L.F.	5.29
Steel	L.F.	5.29

Pedestrian barricades

Plywood	S.F.	3.97
Chain link fence	S.F.	4.30

Trailers, general office type, per month

Minimum	EA.	200
Average	EA.	330
Maximum	EA.	660

Crew change trailers, per month

Minimum	EA.	120
Average	EA.	130
Maximum	EA.	200

01505.10 MOBILIZATION
Equipment mobilization
Bulldozer

Minimum	EA.	190
Average	EA.	390
Maximum	EA.	650

Backhoe/front-end loader

Minimum	EA.	110
Average	EA.	190
Maximum	EA.	430

Truck crane

Minimum	EA.	470
Average	EA.	730
Maximum	EA.	1,250

01525.10 CONSTRUCTION AIDS
Scaffolding/staging, rent per month
Measured by lineal feet of base

10' high	L.F.	11.75
20' high	L.F.	21.50
30' high	L.F.	30.00

Measured by square foot of surface

Minimum	S.F.	.53
Average	S.F.	.90
Maximum	S.F.	1.62

Tarpaulins, fabric, per job

Minimum	S.F.	.24
Average	S.F.	.42
Maximum	S.F.	1.07

01570.10 SIGNS
Construction signs, temporary
Signs, 2' x 4'

Minimum	EA.	34.25
Average	EA.	82.00
Maximum	EA.	290

Signs, 4' x 8'

Minimum	EA.	71.75
Average	EA.	190
Maximum	EA.	800

Signs, 8' x 8'

Minimum	EA.	92.25
Average	EA.	290
Maximum	EA.	2,900

	Unit	Total		Unit	Total

01600.10 EQUIPMENT

Air compressor
60 cfm

By day	EA	90.50
By week	EA	270
By month	EA	820

300 cfm

By day	EA	190
By week	EA	570
By month	EA	1,700

Air tools, per compressor, per day

Minimum	EA	34.75
Average	EA	43.50
Maximum	EA	61.00

Generators, 5 kw

By day	EA	87.00
By week	EA	260
By month	EA	800

Heaters, salamander type, per week

Minimum	EA	100.00
Average	EA	150
Maximum	EA	310

Pumps, submersible
50 gpm

By day	EA	69.75
By week	EA	210
By month	EA	620

Pickup truck

By day	EA	130
By week	EA	380
By month	EA	1,200

Dump truck
6 cy truck

By day	EA	350
By week	EA	1,050
By month	EA	3,150

10 cy truck

By day	EA	430
By week	EA	1,300
By month	EA	3,900

16 cy truck

By day	EA	700
By week	EA	2,100
By month	EA	6,250

01600.10 EQUIPMENT (Cont.)

Backhoe, track mounted
1/2 cy capacity

By day	EA	710
By week	EA	2,150
By month	EA	6,450

Backhoe/loader, rubber tired
1/2 cy capacity

By day	EA	430
By week	EA	1,300
By month	EA	3,900

3/4 cy capacity

By day	EA	520
By week	EA	1,550
By month	EA	4,700

Bulldozer
75 hp

By day	EA	610
By week	EA	1,850
By month	EA	5,500

Cranes, crawler type
15 ton capacity

By day	EA	780
By week	EA	2,350
By month	EA	7,050

Truck mounted, hydraulic
15 ton capacity

By day	EA	740
By week	EA	2,200
By month	EA	6,400

Loader, rubber tired
1 cy capacity

By day	EA	520
By week	EA	1,550
By month	EA	4,700

	Unit	Total

02115.66 SEPTIC TANKS
Remove septic tank

	Unit	Total
1000 gals	EA.	190
2000 gals	EA.	220

02210.10 SOIL BORING
Borings, uncased, stable earth

	Unit	Total
2-1/2" dia.	L.F.	28.00
4" dia.	L.F.	31.75
Cased, including samples		
2-1/2" dia.	L.F.	37.25
4" dia.	L.F.	63.75
Drilling in rock		
No sampling	L.F.	58.75
With casing and sampling	L.F.	74.25
Test pits		
Light soil	EA.	370
Heavy soil	EA.	560

02220.10 BUILDING DEMOLITION
Building, complete with disposal

	Unit	Total
Wood frame	C.F.	32

02220.15 BUILDING DEMOLITION, SELECTIVE
Partition removal

	Unit	Total
Concrete block partitions		
8" thick	S.F.	2.71
Brick masonry partitions		
4" thick	S.F.	2.03
8" thick	S.F.	2.54
Stud partitions		
Metal or wood, with drywall both sides	S.F.	2.03
Door and frame removal		
Wood in framed wall		
2'6"x6'8"	EA.	29.00
3'x6'8"	EA.	33.75
Ceiling removal		
Acoustical tile ceiling		
Adhesive fastened	S.F.	.41
Furred and glued	S.F.	.34
Suspended grid	S.F.	.25
Drywall ceiling		
Furred and nailed	S.F.	.45
Nailed to framing	S.F.	.41
Window removal		
Metal windows, trim included		
2'x3'	EA.	40.50
3'x4'	EA.	50.75
4'x8'	EA.	100.00
Wood windows, trim included		
2'x3'	EA.	22.50
3'x4'	EA.	27.00
6'x8'	EA.	40.50
Concrete block walls, not including toothing		
4" thick	S.F.	2.26
6" thick	S.F.	2.39
8" thick	S.F.	2.54
Rubbish handling		
Load in dumpster or truck		
Minimum	C.F.	.90
Maximum	C.F.	1.35
Rubbish hauling		
Hand loaded on trucks, 2 mile trip	C.Y.	34.50
Machine loaded on trucks, 2 mile trip	C.Y.	22.25

02225.13 CORE DRILLING
Concrete

	Unit	Total
6" thick		
3" dia.	EA.	38.00
8" thick		
3" dia.	EA.	53.00

02225.15 CURB & GUTTER

	Unit	Total
Removal, plain concrete curb	L.F.	5.58
Plain concrete curb and 2' gutter	L.F.	7.69

02225.20 FENCE
Remove fencing

	Unit	Total
Chain link, 8' high		
For disposal	L.F.	2.03
For reuse	L.F.	5.08
Wood		
4' high	S.F.	1.35
Masonry		

02225.20 FENCE (Cont.)

	Unit	Total
8" thick		
4' high	S.F.	4.06
6' high	S.F.	5.08

02225.25 GUARDRAIL
Remove standard guardrail

	Unit	Total
Steel	L.F.	7.44
Wood	L.F.	5.72

02225.30 HYDRANT

	Unit	Total
Remove and reset fire hydrant	EA.	1,100

02225.40 PAVEMENT DEMOLITION
Concrete pavement, 6" thick

	Unit	Total
No reinforcement	S.Y.	14.75
With wire mesh	S.Y.	22.25
With rebars	S.Y.	28.00
Sidewalk, 4" thick, with disposal	S.Y.	7.44

02225.42 DRAINAGE PIPING
Remove drainage pipe, not including excavation

	Unit	Total
12" dia.	L.F.	9.30
18" dia.	L.F.	11.75

02225.43 GAS PIPING
Remove welded steel pipe, not including excavation

	Unit	Total
4" dia.	L.F.	14.00
5" dia.	L.F.	22.25

02225.45 SANITARY PIPING
Remove sewer pipe, not including excavation

	Unit	Total
4" dia.	L.F.	8.92

02225.48 WATER PIPING
Remove water pipe, not including excavation

	Unit	Total
4" dia.	L.F.	10.25

02225.50 SAW CUTTING PAVEMENT
Pavement, bituminous

	Unit	Total
2" thick	L.F.	1.72
3" thick	L.F.	2.15
Concrete pavement, with wire mesh		
4" thick	L.F.	3.32
5" thick	L.F.	3.59
Plain concrete, unreinforced		
4" thick	L.F.	2.87
5" thick	L.F.	3.32

02230.10 CLEARING AND GRUBBING
Clear wooded area

	Unit	Total
Light density	ACRE	5,600
Medium density	ACRE	7,450
Heavy density	ACRE	8,900

02230.50 TREE CUTTING & CLEARING
Cut trees and clear out stumps

	Unit	Total
9" to 12" dia.	EA.	450
To 24" dia.	EA.	560
24" dia. and up	EA.	740
Loading and trucking		
For machine load, per load, round trip		
1 mile	EA.	89.25
3 mile	EA.	100.00
5 mile	EA.	110
10 mile	EA.	150
20 mile	EA.	220
Hand loaded, round trip		
1 mile	EA.	220
3 mile	EA.	250
5 mile	EA.	290
10 mile	EA.	350
20 mile	EA.	430

02315.10 BASE COURSE
Base course, crushed stone

	Unit	Total
3" thick	S.Y.	3.75
4" thick	S.Y.	4.85
6" thick	S.Y.	7.02
Base course, bank run gravel		
4" deep	S.Y.	3.19
6" deep	S.Y.	4.53
Prepare and roll sub base		
Minimum	S.Y.	.60
Average	S.Y.	.74
Maximum	S.Y.	.99

DIVISION 2 SITE CONSTRUCTION

	Unit	Total
02315.20 BORROW		
Borrow fill, F.O.B. at pit		
Sand, haul to site, round trip		
10 mile	C.Y.	28.50
20 mile	C.Y.	36.25
30 mile	C.Y.	46.25
Place borrow fill and compact		
Less than 1 in 4 slope	C.Y.	22.50
Greater than 1 in 4 slope	C.Y.	24.50
02315.30 BULK EXCAVATION		
Excavation, by small dozer		
Large areas	C.Y.	1.72
Small areas	C.Y.	2.87
Trim banks	C.Y.	4.31
Hydraulic excavator		
1 cy capacity		
Light material	C.Y.	3.72
Medium material	C.Y.	4.46
Wet material	C.Y.	5.58
Blasted rock	C.Y.	6.37
1-1/2 cy capacity		
Light material	C.Y.	1.49
Medium material	C.Y.	1.99
Wet material	C.Y.	2.38
Wheel mounted front-end loader		
7/8 cy capacity		
Light material	C.Y.	2.98
Medium material	C.Y.	3.40
Wet material	C.Y.	3.97
Blasted rock	C.Y.	4.76
1-1/2 cy capacity		
Light material	C.Y.	1.70
Medium material	C.Y.	1.83
Wet material	C.Y.	1.99
Blasted rock	C.Y.	2.17
2-1/2 cy capacity		
Light material	C.Y.	1.40
Medium material	C.Y.	1.49
Wet material	C.Y.	1.59
Blasted rock	C.Y.	1.70
Track mounted front-end loader		
1-1/2 cy capacity		
Light material	C.Y.	1.99
Medium material	C.Y.	2.17
Wet material	C.Y.	2.38
Blasted rock	C.Y.	2.65
2-3/4 cy capacity		
Light material	C.Y.	1.19
Medium material	C.Y.	1.32
Wet material	C.Y.	1.49
Blasted rock	C.Y.	1.70
02315.40 BUILDING EXCAVATION		
Structural excavation, unclassified earth		
3/8 cy backhoe	C.Y.	16.00
3/4 cy backhoe	C.Y.	12.00
1 cy backhoe	C.Y.	9.93
Foundation backfill and compaction		
by machine	C.Y.	23.75
02315.45 HAND EXCAVATION		
Excavation		
To 2' deep		
Normal soil	C.Y.	45.00
Sand and gravel	C.Y.	40.50
Medium clay	C.Y.	50.75
Heavy clay	C.Y.	58.00
Loose rock	C.Y.	67.75
To 6' deep		
Normal soil	C.Y.	58.00
Sand and gravel	C.Y.	50.75
Medium clay	C.Y.	67.75
Heavy clay	C.Y.	81.25
Loose rock	C.Y.	100.00
Backfilling foundation without compaction,		
6" lifts	C.Y.	25.50
Compaction of backfill around structures or in trench		
By hand with air tamper	C.Y.	29.00
By hand with vibrating plate tamper	C.Y.	27.00
1 ton roller	C.Y.	43.00

	Unit	Total
02315.45 HAND EXCAVATION (Cont.)		
Miscellaneous hand labor		
Trim slopes, sides of excavation	S.F.	.07
Trim bottom of excavation	S.F.	.08
Excavation around obstructions and services	C.Y.	140
02315.50 ROADWAY EXCAVATION		
Roadway excavation		
1/4 mile haul	C.Y.	2.38
2 mile haul	C.Y.	3.97
5 mile haul	C.Y.	5.96
Spread base course	C.Y.	2.98
Roll and compact	C.Y.	3.97
02315.60 TRENCHING		
Trenching and continuous footing excavation		
By gradall, 1 cy capacity		
Light soil	C.Y.	3.40
Medium soil	C.Y.	3.67
Heavy/wet soil	C.Y.	3.97
Loose rock	C.Y.	4.33
Blasted rock	C.Y.	4.58
By hydraulic excavator		
1/2 cy capacity		
Light soil	C.Y.	3.97
Medium soil	C.Y.	4.33
Heavy/wet soil	C.Y.	4.76
Loose rock	C.Y.	5.29
Blasted rock	C.Y.	5.96
1 cy capacity		
Light soil	C.Y.	2.80
Medium soil	C.Y.	2.98
Heavy/wet soil	C.Y.	3.18
Loose rock	C.Y.	3.40
Blasted rock	C.Y.	3.67
1-1/2 cy capacity		
Light soil	C.Y.	2.51
Medium soil	C.Y.	2.65
Heavy/wet soil	C.Y.	2.80
Loose rock	C.Y.	2.98
Blasted rock	C.Y.	3.18
2 cy capacity		
Light soil	C.Y.	2.38
Medium soil	C.Y.	2.51
Heavy/wet soil	C.Y.	2.65
Loose rock	C.Y.	2.80
Blasted rock	C.Y.	2.98
Hand excavation		
Bulk, wheeled 100'		
Normal soil	C.Y.	45.00
Sand or gravel	C.Y.	40.50
Medium clay	C.Y.	58.00
Heavy clay	C.Y.	81.25
Loose rock	C.Y.	100.00
Trenches, up to 2' deep		
Normal soil	C.Y.	50.75
Sand or gravel	C.Y.	45.00
Medium clay	C.Y.	67.75
Heavy clay	C.Y.	100.00
Loose rock	C.Y.	140
Trenches, to 6' deep		
Normal soil	C.Y.	58.00
Sand or gravel	C.Y.	50.75
Medium clay	C.Y.	81.25
Heavy clay	C.Y.	140
Loose rock	C.Y.	200
Backfill trenches		
With compaction		
By hand	C.Y.	33.75
By 60 hp tracked dozer	C.Y.	2.15
02315.70 UTILITY EXCAVATION		
Trencher, sandy clay, 8" wide trench		
18" deep	L.F.	1.92
24" deep	L.F.	2.15
36" deep	L.F.	2.46
Trench backfill, 95% compaction		
Tamp by hand	C.Y.	25.50
Vibratory compaction	C.Y.	20.25
Trench backfilling, with borrow sand,		
place & compact	C.Y.	36.25

	Unit	Total
02315.75 GRAVEL AND STONE		
F.O.B. PLANT		
No. 21 crusher run stone	C.Y.	43.75
No. 26 crusher run stone	C.Y.	43.75
No. 57 stone	C.Y.	43.75
No. 67 gravel	C.Y.	31.50
No. 68 stone	C.Y.	43.75
No. 78 stone	C.Y.	43.75
No. 78 gravel, (pea gravel)	C.Y.	31.50
No. 357 or B-3 stone	C.Y.	43.75
Structural & foundation backfill		
No. 21 crusher run stone	TON	35.00
No. 26 crusher run stone	TON	35.00
No. 57 stone	TON	35.00
No. 67 gravel	TON	24.25
No. 68 stone	TON	35.00
No. 78 stone	TON	35.00
No. 78 gravel, (pea gravel)	TON	24.25
No. 357 or B-3 stone	TON	35.00
02315.80 HAULING MATERIAL		
Haul material by 10 cy dump truck, round trip distance		
1 mile	C.Y.	4.79
2 mile	C.Y.	5.75
5 mile	C.Y.	7.84
10 mile	C.Y.	8.62
20 mile	C.Y.	9.58
30 mile	C.Y.	11.50
Site grading, cut & fill, sandy clay, 200' haul,		
75 hp dozer	C.Y.	3.45
Spread topsoil by equipment on site	C.Y.	3.83
Site grading (cut and fill to 6") less than 1 acre		
75 hp dozer	C.Y.	5.75
1.5 cy backhoe/loader	C.Y.	8.62
02340.05 SOIL STABILIZATION		
Straw bale secured with rebar	L.F.	8.90
Filter barrier, 18" high filter fabric	L.F.	5.89
Sediment fence, 36" fabric with 6" mesh	L.F.	9.40
Soil stabilization with tar paper, burlap,		
straw and stakes	S.F.	.42
02360.20 SOIL TREATMENT		
Soil treatment, termite control pretreatment		
Under slabs	S.F.	.61
By walls	S.F.	.65
02370.40 RIPRAP		
Riprap		
Crushed stone blanket, max size 2-1/2"	TON	99.75
Stone, quarry run, 300 lb. stones	TON	100
400 lb. stones	TON	100
500 lb. stones	TON	99.75
750 lb. stones	TON	98.25
Dry concrete riprap in bags 3" thick,		
80 lb. per bag	BAG	9.19
02455.60 STEEL PILES		
H-section piles		
8x8, 36 lb/ft		
30' long	L.F.	26.00
40' long	L.F.	23.75
Tapered friction piles, fluted casing, up to 50'		
With 4000 psi concrete no reinforcing		
12" dia.	L.F.	24.75
14" dia.	L.F.	27.50
02455.65 STEEL PIPE PILES		
Concrete filled, 3000# concrete, up to 40'		
8" dia.	L.F.	28.75
10" dia.	L.F.	33.75
12" dia.	L.F.	39.25
Pipe piles, non-filled		
8" dia.	L.F.	24.50
10" dia.	L.F.	27.75
12" dia.	L.F.	32.25
Splice		
8" dia.	EA.	150
10" dia.	EA.	150
12" dia.	EA.	180
Standard point		
8" dia.	EA.	160
10" dia.	EA.	170
12" dia.	EA.	240

	Unit	Total
02455.65 STEEL PIPE PILES (Cont.)		
Heavy duty point		
8" dia.	EA.	170
10" dia.	EA.	190
12" dia.	EA.	260
02455.80 WOOD AND TIMBER PILES		
Treated wood piles, 12" butt, 8" tip		
25' long	L.F.	21.75
30' long	L.F.	20.25
35' long	L.F.	18.50
40' long	L.F.	17.25
02465.50 PRESTRESSED PILING		
Prestressed concrete piling, less than 60' long		
10" sq.	L.F.	17.50
12" sq.	L.F.	22.25
Straight cylinder, less than 60' long		
12" dia.	L.F.	21.50
14" dia.	L.F.	27.00
02510.10 WELLS		
Domestic water, drilled and cased		
4" dia.	L.F.	94.50
6" dia.	L.F.	110
02510.40 DUCTILE IRON PIPE		
Ductile iron pipe, cement lined, slip-on joints		
4"	L.F.	19.75
6"	L.F.	22.75
8"	L.F.	28.25
Mechanical joint pipe		
4"	L.F.	24.50
6"	L.F.	28.25
8"	L.F.	35.25
Fittings, mechanical joint, 90 degree elbow		
4"	EA.	220
6"	EA.	270
8"	EA.	390
45 degree elbow		
4"	EA.	190
6"	EA.	250
8"	EA.	340
02510.60 PLASTIC PIPE		
PVC, class 150 pipe		
4" dia.	L.F.	9.10
6" dia.	L.F.	12.75
8" dia.	L.F.	16.75
Schedule 40 pipe		
1-1/2" dia.	L.F.	3.26
2" dia.	L.F.	3.83
2-1/2" dia.	L.F.	4.67
3" dia.	L.F.	5.56
4" dia.	L.F.	7.14
6" dia.	L.F.	11.25
90 degree elbows		
1"	EA.	7.62
1-1/2"	EA.	8.38
2"	EA.	9.90
2-1/2"	EA.	15.75
3"	EA.	18.25
4"	EA.	26.75
6"	EA.	65.50
45 degree elbows		
1"	EA.	8.07
1-1/2"	EA.	9.04
2"	EA.	10.25
2-1/2"	EA.	15.75
3"	EA.	21.00
4"	EA.	31.50
6"	EA.	66.25
Tees		
1"	EA.	9.24
1-1/2"	EA.	10.25
2"	EA.	12.00
2-1/2"	EA.	20.50
3"	EA.	25.00
4"	EA.	37.75
6"	EA.	98.25
Couplings		
1"	EA.	7.45
1-1/2"	EA.	7.75

02510.60 PLASTIC PIPE (Cont.)	Unit	Total
2"	EA.	8.89
2-1/2"	EA.	11.50
3"	EA.	14.25
4"	EA.	17.75
6"	EA.	37.25
Drainage pipe		
PVC schedule 80		
1" dia.	L.F.	4.47
1-1/2" dia.	L.F.	4.90
ABS, 2" dia.	L.F.	5.76
2-1/2" dia.	L.F.	7.28
3" dia.	L.F.	8.28
4" dia.	L.F.	10.75
6" dia.	L.F.	16.00
8" dia.	L.F.	22.00
90 degree elbows		
1"	EA.	10.50
1-1/2"	EA.	11.25
2"	EA.	13.00
2-1/2"	EA.	21.25
3"	EA.	22.50
4"	EA.	34.25
6"	EA.	66.00
45 degree elbows		
1"	EA.	12.75
1-1/2"	EA.	14.25
2"	EA.	16.75
2-1/2"	EA.	25.50
3"	EA.	27.75
4"	EA.	45.75
6"	EA.	95.75

02530.20 VITRIFIED CLAY PIPE	Unit	Total
Vitrified clay pipe, extra strength		
6" dia.	L.F.	15.50
8" dia.	L.F.	16.75
10" dia.	L.F.	20.75

02530.30 MANHOLES	Unit	Total
Precast sections, 48" dia.		
Base section	EA.	520
1'0" riser	EA.	240
1'4" riser	EA.	270
2'8" riser	EA.	340
4'0" riser	EA.	500
2'8" cone top	EA.	420
Precast manholes, 48" dia.		
4' deep	EA.	1,050
6' deep	EA.	1,500
7' deep	EA.	1,700
8' deep	EA.	1,950
10' deep	EA.	2,250
Cast-in-place, 48" dia., with frame and cover		
5' deep	EA.	1,650
6' deep	EA.	2,050
8' deep	EA.	2,600
10' deep	EA.	3,100
Brick manholes, 48" dia. with cover, 8" thick		
4' deep	EA.	1,100
6' deep	EA.	1,300
8' deep	EA.	1,600
10' deep	EA.	1,900
Frames and covers, 24" diameter		
300 lb	EA.	400
400 lb	EA.	420
Steps for manholes		
7" x 9"	EA.	20.00
8" x 9"	EA.	24.25

02530.40 SANITARY SEWERS	Unit	Total
Clay		
6" pipe	L.F.	13.00
PVC		
4" pipe	L.F.	8.08
6" pipe	L.F.	10.75

02540.10 DRAINAGE FIELDS	Unit	Total
Perforated PVC pipe, for drain field		
4" pipe	L.F.	6.84
6" pipe	L.F.	8.84

02540.50 SEPTIC TANKS	Unit	Total
Septic tank, precast concrete		
1000 gals	EA.	1,100
2000 gals	EA.	2,950
Leaching pit, precast concrete, 72" diameter		
3' deep	EA.	970
6' deep	EA.	1,500
8' deep	EA.	1,900

02630.70 UNDERDRAIN	Unit	Total
Drain tile, clay		
6" pipe	L.F.	9.49
8" pipe	L.F.	12.50
Porous concrete, standard strength		
6" pipe	L.F.	9.03
8" pipe	L.F.	9.59
Corrugated metal pipe, perforated type		
6" pipe	L.F.	11.25
8" pipe	L.F.	12.50
Perforated clay pipe		
6" pipe	L.F.	11.75
8" pipe	L.F.	13.75
Drain tile, concrete		
6" pipe	L.F.	8.19
8" pipe	L.F.	10.25
Perforated rigid PVC underdrain pipe		
4" pipe	L.F.	5.51
6" pipe	L.F.	7.75
8" pipe	L.F.	9.62
Underslab drainage, crushed stone		
3" thick	S.F.	.98
4" thick	S.F.	1.17
6" thick	S.F.	1.30
Plastic filter fabric for drain lines	S.F.	.56

02740.20 ASPHALT SURFACES	Unit	Total
Asphalt wearing surface, flexible pavement		
1" thick	S.Y.	5.78
1-1/2" thick	S.Y.	8.04
Binder course		
1-1/2" thick	S.Y.	7.55
2" thick	S.Y.	9.79
Bituminous sidewalk, no base		
2" thick	S.Y.	10.25
3" thick	S.Y.	14.25

02750.10 CONCRETE PAVING	Unit	Total
Concrete paving, reinforced, 5000 psi concrete		
6" thick	S.Y.	44.75
7" thick	S.Y.	50.00
8" thick	S.Y.	55.50

02810.40 LAWN IRRIGATION	Unit	Total
Residential system, complete		
Minimum	ACRE	15,900
Maximum	ACRE	30,300

02820.10 CHAIN LINK FENCE	Unit	Total
Chain link fence, 9 ga., galvanized, with posts 10' o.c.		
4' high	L.F.	9.61
5' high	L.F.	12.75
6' high	L.F.	15.25
Corner or gate post, 3" post		
4' high	EA.	91.75
5' high	EA.	100
6' high	EA.	110
Gate with gate posts, galvanized, 3' wide		
4' high	EA.	190
5' high	EA.	250
6' high	EA.	270
Fabric, galvanized chain link, 2" mesh, 9 ga.		
4' high	L.F.	5.01
5' high	L.F.	6.08
6' high	L.F.	8.28
Line post, no rail fitting, galvanized, 2-1/2" dia.		
4' high	EA.	37.00
5' high	EA.	40.50
6' high	EA.	43.75
Vinyl coated, 9 ga., with posts 10' o.c.		
4' high	L.F.	10.25
5' high	L.F.	12.25
6' high	L.F.	15.25

	Unit	Total
02820.10 CHAIN LINK FENCE (Cont.)		
Gate, with posts, 3' wide		
4' high	EA.	210
5' high	EA.	260
6' high	EA.	280
Fabric, vinyl, chain link, 2" mesh, 9 ga.		
4' high	L.F.	5.01
5' high	L.F.	6.08
6' high	L.F.	8.28
Swing gates, galvanized, 4' high		
Single gate		
3' wide	EA.	300
4' wide	EA.	320
6' high		
Single gate		
3' wide	EA.	400
4' wide	EA.	420
02880.70 RECREATIONAL COURTS		
Walls, galvanized steel		
8' high	L.F.	22.50
10' high	L.F.	26.25
12' high	L.F.	30.50
Vinyl coated		
8' high	L.F.	21.75
10' high	L.F.	26.00
12' high	L.F.	29.50
Gates, galvanized steel		
Single, 3' transom		
3'x7'	EA.	540
4'x7'	EA.	590
5'x7'	EA.	760
6'x7'	EA.	860
Vinyl coated		
Single, 3' transom		
3'x7'	EA.	860
4'x7'	EA.	950
5'x7'	EA.	990
6'x7'	EA.	1,050
02910.10 TOPSOIL		
Spread topsoil, with equipment		
Minimum	C.Y.	12.00
Maximum	C.Y.	15.00
By hand		
Minimum	C.Y.	40.50
Maximum	C.Y.	50.75
Area prep. seeding (grade, rake and clean)		
Square yard	S.Y.	32
By acre	ACRE	1,600
Remove topsoil and stockpile on site		
4" deep	C.Y.	9.93
6" deep	C.Y.	9.16
Spreading topsoil from stock pile		
By loader	C.Y.	10.75
By hand	C.Y.	120
Top dress by hand	S.Y.	1.19
Place imported top soil		
By loader		
4" deep	S.Y.	1.19
6" deep	S.Y.	1.32
By hand		
4" deep	S.Y.	4.51
6" deep	S.Y.	5.08
Plant bed preparation, 18" deep		
With backhoe/loader	S.Y.	2.98
By hand	S.Y.	6.77
02920.10 FERTILIZING		
Fertilizing (23#/1000 sf)		
By square yard	S.Y.	16
By acre	ACRE	820
Liming (70#/1000 sf)		
By square yard	S.Y.	21
By acre	ACRE	1,050
02920.30 SEEDING		
Mechanical seeding, 175 lb/acre		
By square yard	S.Y.	31
By acre	ACRE	1,300
450 lb/acre		
By square yard	S.Y.	63

	Unit	Total
02920.30 SEEDING (Cont.)		
By acre	ACRE	2,550
Seeding by hand, 10 lb per 100 s.y.		
By square yard	S.Y.	69
By acre	ACRE	2,850
Reseed disturbed areas	S.F.	26
02930.10 PLANTS		
Euonymus coloratus, 18" (Purple Wintercreeper)	EA.	9.14
Hedera Helix, 2-1/4" pot (English ivy)	EA.	7.76
Liriope muscari, 2" clumps	EA.	8.18
Santolina, 12"	EA.	8.79
Vinca major or minor, 3" pot	EA.	4.83
Cortaderia argentia, 2 gallon (Pampas Grass)	EA.	19.00
Ophiopogan japonicus, 1 quart (4" pot)	EA.	8.18
Ajuga reptans, 2-3/4" pot (carpet bugle)	EA.	4.83
Pachysandra terminalis, 2-3/4" pot (Japanese Spurge)	EA.	5.10
02930.30 SHRUBS		
Juniperus conferia litoralis, 18"-24" (Shore Juniper)	EA.	51.00
Horizontalis plumosa, 18"-24" (Andorra Juniper)	EA.	53.00
Sabina tamar-iscfolia-tamarix juniper, 18"-24"	EA.	53.00
Chin San Jose, 18"-24" (San Jose Juniper)	EA.	53.00
Sargenti, 18"-24" (Sargent's Juniper)	EA.	51.00
Nandina domestica, 18"-24" (Heavenly Bamboo)	EA.	39.50
Raphiolepis Indica Springtime, 18"-24"	EA.	41.25
Osmanthus Heterophyllus Gulftide, 18"-24"	EA.	43.00
Ilex Cornuta Burfordi Nana, 18"-24"	EA.	46.75
Glabra, 18"-24" (Inkberry Holly)	EA.	45.00
Azalea, Indica types, 18"-24"	EA.	48.50
Kurume types, 18"-24"	EA.	52.50
Berberis Julianae, 18"-24" (Wintergreen Barberry)	EA.	37.50
Pieris Japonica Japanese, 18"-24"	EA.	37.50
Ilex Cornuta Rotunda, 18"-24"	EA.	41.25
Juniperus Horiz. Plumosa, 24"-30"	EA.	43.50
Rhodopendrow Hybrids, 24"-30"	EA.	82.00
Aucuba Japonica Varigata, 24"-30"	EA.	41.25
Ilex Crenata Willow Leaf, 24"-30"	EA.	43.50
Cleyera Japonica, 30"-36"	EA.	52.50
Pittosporum Tobira, 30"-36"	EA.	56.50
Prumus Laurocerasus, 30"-36"	EA.	83.50
Ilex Cornuta Burfordi, 30"-36" (Burford Holly)	EA.	56.50
Abelia Grandiflora, 24"-36" (Yew Podocarpus)	EA.	41.50
Podocarpos Macrophylla, 24"-36"	EA.	55.00
Pyracantha Coccinea Lalandi, 3'-4' (Firethorn)	EA.	45.25
Photinia Frazieri, 3'-4' (Red Photinia)	EA.	57.00
Forsythia Suspensa, 3'-4' (Weeping Forsythia)	EA.	45.25
Camellia Japonica, 3'-4' (Common Camellia)	EA.	60.25
Juniperus Chin Torulosa, 3'-4' (Hollywood Juniper)	EA.	62.50
Cupressocyparis Leylandi, 3'-4'	EA.	56.75
Ilex Opaca Fosteri, 5'-6' (Foster's Holly)	EA.	160
Opaca, 5'-6' (American Holly)	EA.	210
Nyrica Cerifera, 4'-5' (Southern Wax Myrtles)	EA.	68.25
Ligustrum Japonicum, 4'-5' (Japanese Privet)	EA.	60.00
02930.60 TREES		
Cornus Florida, 5'-6' (White flowering Dogwood)	EA.	120
Prunus Serrulata Kwanzan, 6'-8' (Kwanzan Cherry)	EA.	140
Caroliniana, 6'-8' (Carolina Cherry Laurel)	EA.	160
Cercis Canadensis, 6'-8' (Eastern Redbud)	EA.	120
Koelreuteria Paniculata, 8'-10' (Goldenrain Tree)	EA.	190
Acer Platanoides, 1-3/4"-2" (11'-13')	EA.	260
Rubrum, 1-3/4"-2" (11'-13') (Red Maple)	EA.	210
Saccharum, 1-3/4"-2" (Sugar Maple)	EA.	320
Fraxinus Pennsylvanica, 1-3/4"-2"	EA.	190
Celtis Occidentalis, 1-3/4"-2"	EA.	250
Glenditsia Triacantos Inermis, 2"	EA.	240
Prunus Cerasifera 'Thundercloud', 6'-8'	EA.	140
Yeodensis, 6'-8' (Yoshino Cherry)	EA.	140
Lagerstroemia Indica, 8'-10' (Crapemyrtle)	EA.	210
Crataegus Phaenopyrum, 8'-10'	EA.	300

02930.60 TREES (Cont.)

	Unit	Total
Quercus Borealis, 1-3/4"-2" (Northern Red Oak)	EA	220
Quercus Acutissima, 1-3/4"-2" (8'-10')	EA	210
Saliz Babylonica, 1-3/4"-2" (Weeping Willow)	EA	140
Tilia Cordata Greenspire, 1-3/4"-2" (10'-12')	EA	380
Malus, 2"-2-1/2" (8'-10') (Flowering Crabapple)	EA	220
Platanus Occidentalis, (12'-14')	EA	310
Pyrus Calleryana Bradford, 2"-2-1/2"	EA	250
Quercus Palustris, 2"-2-1/2" (12'-14') (Pin Oak)	EA	270
Phellos, 2-1/2"-3" (Willow Oak)	EA	300
Nigra, 2"-2-1/2" (Water Oak)	EA	260
Magnolia Soulangeana, 4'-5' (Saucer Magnolia)	EA	140
Grandiflora, 6'-8' (Southern Magnolia)	EA	190
Cedrus Deodara, 10'-12' (Deodare Cedar)	EA	320
Gingko Biloba, 10'-12' (2"-2-1/2")	EA	300
Pinus Thunbergi, 5'-6' (Japanese Black Pine)	EA	130
Strobus, 6'-8' (White Pine)	EA	140
Taeda, 6'-8' (Loblolly Pine)	EA	130
Quercus Virginiana, 2"-2-1/2" (Live Oak)	EA	300

02935.10 SHRUB & TREE MAINTENANCE

	Unit	Total
Moving shrubs on site		
3' high	EA	40.50
4' high	EA	45.00
Moving trees on site		
6' high	EA	49.50
8' high	EA	55.75
10' high	EA	74.25
Palm trees		
10' high	EA	74.25
40' high	EA	450

02935.30 WEED CONTROL

	Unit	Total
Weed control, bromicil, 15 lb./acre, wettable powder	ACRE	490
Vegetation control, by application of plant killer	S.Y	.18
Weed killer, lawns and fields	S.Y	.32

02945.20 LANDSCAPE ACCESSORIES

	Unit	Total
Steel edging, 3/16" x 4"	L.F.	1.16
Landscaping stepping stones, 15"x15", white	EA	7.86
Wood chip mulch	C.Y.	67.75
2" thick	S.Y.	3.31
4" thick	S.Y.	5.87
6" thick	S.Y.	8.52
Gravel mulch, 3/4" stone	C.Y.	72.75
White marble chips, 1" deep	S.F.	1.05
Peat moss		
2" thick	S.Y.	4.36
4" thick	S.Y.	8.02
6" thick	S.Y.	12.00
Landscaping timbers, treated lumber		
4" x 4"	L.F.	2.70
6" x 6"	L.F.	4.15
8" x 8"	L.F.	6.09

	Unit	Total
03110.05 BEAM FORMWORK		
Beam forms, job built		
Beam bottoms		
1 use	S.F.	12.75
4 uses	S.F.	9.22
5 uses	S.F.	8.88
Beam sides		
1 use	S.F.	8.76
5 uses	S.F.	6.00
03110.15 COLUMN FORMWORK		
Column, square forms, job built		
8" x 8" columns		
1 use	S.F.	13.75
5 uses	S.F.	10.25
12" x 12" columns		
1 use	S.F.	12.50
5 uses	S.F.	9.26
Round fiber forms, 1 use		
10" dia.	L.F.	14.75
12" dia.	L.F.	16.00
03110.18 CURB FORMWORK		
Curb forms, Straight, 6" high		
1 use	L.F.	7.12
5 uses	L.F.	5.04
Curved, 6" high		
1 use	L.F.	8.58
5 uses	L.F.	6.17
03110.25 EQUIPMENT PAD FORMWORK		
Equipment pad, job built		
1 use	S.F.	9.82
3 uses	S.F.	7.40
5 uses	S.F.	6.21
03110.35 FOOTING FORMWORK		
Wall footings, job built, continuous		
1 use	S.F.	6.78
3 uses	S.F.	5.65
5 uses	S.F.	5.04
03110.50 GRADE BEAM FORMWORK		
Grade beams, job built		
1 use	S.F.	7.67
3 uses	S.F.	5.83
5 uses	S.F.	5.10
03110.53 PILE CAP FORMWORK		
Pile cap forms, job built, Square		
1 use	S.F.	9.31
5 uses	S.F.	6.16
03110.55 SLAB / MAT FORMWORK		
Mat foundations, job built		
1 use	S.F.	8.96
3 uses	S.F.	6.83
5 uses	S.F.	5.92
Edge forms		
6" high		
1 use	L.F.	7.20
3 uses	L.F.	5.38
5 uses	L.F.	4.72
03110.65 WALL FORMWORK		
Wall forms, exterior, job built, Up to 8' high wall		
1 use	S.F.	7.93
3 uses	S.F.	6.07
5 uses	S.F.	5.34
Retaining wall forms		
1 use	S.F.	8.32
3 uses	S.F.	6.38
5 uses	S.F.	5.61
Column pier and pilaster		
1 use	S.F.	13.50
5 uses	S.F.	8.80
03110.90 MISCELLANEOUS FORMWORK		
Keyway forms (5 uses)		
2 x 4	L.F.	2.83
2 x 6	L.F.	3.23
Bulkheads		
Walls, with keyways		
3 piece	L.F.	9.90
Ground slab, with keyway		
2 piece	L.F.	8.11
3 piece	L.F.	9.37

	Unit	Total
03110.90 MISCELLANEOUS FORMWORK (Cont.)		
Chamfer strips		
Wood		
1/2" wide	L.F.	1.37
3/4" wide	L.F.	1.44
1" wide	L.F.	1.52
PVC		
1/2" wide	L.F.	2.10
3/4" wide	L.F.	2.18
1" wide	L.F.	2.63
03210.05 BEAM REINFORCING		
Beam-girders		
#3 - #4	TON	2,700
#5 - #6	TON	2,250
03210.15 COLUMN REINFORCING		
Columns		
#3 - #4	TON	2,900
#5 - #6	TON	2,400
03210.20 ELEVATED SLAB REINFORCING		
Elevated slab		
#3 - #4	TON	2,000
#5 - #6	TON	1,750
03210.25 EQUIP. PAD REINFORCING		
Equipment pad		
#3 - #4	TON	2,400
#5 - #6	TON	2,150
03210.35 FOOTING REINFORCING		
Footings		
#3 - #4	TON	2,200
#5 - #6	TON	1,950
#7 - #8	TON	1,800
Straight dowels, 24" long		
3/4" dia. (#6)	EA.	9.36
5/8" dia. (#5)	EA.	7.91
1/2" dia. (#4)	EA.	6.46
03210.45 FOUNDATION REINFORCING		
Foundations		
#3 - #4	TON	2,200
#5 - #6	TON	1,950
#7 - #8	TON	1,800
03210.50 GRADE BEAM REINFORCING		
Grade beams		
#3 - #4	TON	2,150
#5 - #6	TON	1,850
#7 - #8	TON	1,750
03210.53 PILE CAP REINFORCING		
Pile caps		
#3 - #4	TON	2,700
#5 - #6	TON	2,400
#7 - #8	TON	2,200
03210.55 SLAB / MAT REINFORCING		
Bars, slabs		
#3 - #4	TON	2,200
#5 - #6	TON	1,950
Wire mesh, slabs		
Galvanized		
4x4		
W1.4xW1.4	S.F.	.71
W2.0xW2.0	S.F.	.82
W2.9xW2.9	S.F.	1.02
W4.0xW4.0	S.F.	1.33
6x6		
W1.4xW1.4	S.F.	.59
W2.0xW2.0	S.F.	.73
W2.9xW2.9	S.F.	.89
W4.0xW4.0	S.F.	.98
03210.65 WALL REINFORCING		
Walls		
#3 - #4	TON	2,100
#5 - #6	TON	1,850
Masonry wall (horizontal)		
#3 - #4	TON	3,550
#5 - #6	TON	3,050
Galvanized		
#3 - #4	TON	4,450
#5 - #6	TON	4,000
Masonry wall (vertical)		
#3 - #4	TON	4,150

	Unit	Total
03210.65 WALL REINFORCING (Cont.)		
#5 - #6	TON	3,400
Galvanized		
#3 - #4	TON	5,050
#5 - #6	TON	4,350
03250.40 CONCRETE ACCESSORIES		
Expansion joint, poured		
Asphalt		
1/2" x 1"	L.F.	1.60
1" x 2"	L.F.	3.36
Expansion joint, premolded, in slabs		
Asphalt		
1/2" x 6"	L.F.	1.93
1" x 12"	L.F.	2.89
Cork		
1/2" x 6"	L.F.	2.83
1" x 12"	L.F.	8.23
Neoprene sponge		
1/2" x 6"	L.F.	3.71
1" x 12"	L.F.	11.25
Polyethylene foam		
1/2" x 6"	L.F.	2.05
1" x 12"	L.F.	6.13
Polyurethane foam		
1/2" x 6"	L.F.	2.39
1" x 12"	L.F.	4.37
Polyvinyl chloride foam		
1/2" x 6"	L.F.	3.95
1" x 12"	L.F.	7.67
Rubber, gray sponge		
1/2" x 6"	L.F.	5.58
1" x 12"	L.F.	21.00
Asphalt felt control joints or bond breaker, screed joints		
4" slab	L.F.	2.13
6" slab	L.F.	2.55
8" slab	L.F.	3.15
Waterstops, Polyvinyl chloride, Ribbed		
3/16" thick x		
4" wide	L.F.	3.24
6" wide	L.F.	4.18
1/2" thick x		
9" wide	L.F.	8.04
Ribbed with center bulb		
3/16" thick x 9" wide	L.F.	7.22
3/8" thick x 9" wide	L.F.	8.04
Dumbbell type, 3/8" thick x 6" wide	L.F.	7.92
Plain, 3/8" thick x 9" wide	L.F.	9.42
Center bulb, 3/8" thick x 9" wide	L.F.	11.75
Rubber		
Vapor barrier		
4 mil polyethylene	S.F.	.18
6 mil polyethylene	S.F.	.22
Gravel porous fill, under floor slabs, 3/4" stone	C.Y.	89.75
Reinforcing accessories		
Beam bolsters		
1-1/2" high, plain	L.F.	1.20
Galvanized	L.F.	1.96
3" high		
Plain	L.F.	1.61
Galvanized	L.F.	2.77
Slab bolsters		
1" high		
Plain	L.F.	.91
Galvanized	L.F.	1.55
2" high		
Plain	L.F.	1.14
Galvanized	L.F.	1.94
Chairs, high chairs		
3" high		
Plain	EA.	3.10
Galvanized	EA.	3.27
8" high		
Plain	EA.	4.41
Galvanized	EA.	6.55
Continuous, high chair		
3" high		
Plain	L.F.	2.76
Galvanized	L.F.	3.29

	Unit	Total
03300.10 CONCRETE ADMIXTURES		
Concrete admixtures		
Water reducing admixture	GAL	9.35
Set retarder	GAL	20.75
Air entraining agent	GAL	8.25
03350.10 CONCRETE FINISHES		
Floor finishes		
Broom	S.F.	.58
Screed	S.F.	.51
Darby	S.F.	.51
Steel float	S.F.	.68
Wall finishes		
Burlap rub, with cement paste	S.F.	.79
03360.10 PNEUMATIC CONCRETE		
Pneumatic applied concrete (gunite)		
2" thick	S.F.	8.04
3" thick	S.F.	10.25
4" thick	S.F.	12.25
Finish surface		
Minimum	S.F.	2.61
Maximum	S.F.	5.22
03370.10 CURING CONCRETE		
Sprayed membrane		
Slabs	S.F.	.14
Walls	S.F.	.18
Curing paper		
Slabs	S.F.	.19
Walls	S.F.	.20
Burlap		
7.5 oz.	S.F.	.21
12 oz.	S.F.	.23
03380.05 BEAM CONCRETE		
Beams and girders		
2500# or 3000# concrete		
By crane	C.Y.	190
By pump	C.Y.	180
By hand buggy	C.Y.	150
3500# or 4000# concrete		
By crane	C.Y.	190
By pump	C.Y.	180
By hand buggy	C.Y.	150
03380.15 COLUMN CONCRETE		
Columns		
2500# or 3000# concrete		
By crane	C.Y.	180
By pump	C.Y.	180
3500# or 4000# concrete		
By crane	C.Y.	180
By pump	C.Y.	180
03380.20 ELEVATED SLAB CONCRETE		
Elevated slab		
2500# or 3000# concrete		
By crane	C.Y.	150
By pump	C.Y.	140
By hand buggy	C.Y.	150
03380.25 EQUIPMENT PAD CONCRETE		
Equipment pad		
2500# or 3000# concrete		
By chute	C.Y.	120
By pump	C.Y.	170
By crane	C.Y.	180
3500# or 4000# concrete		
By chute	C.Y.	120
By pump	C.Y.	170
03380.35 FOOTING CONCRETE		
Continuous footing		
2500# or 3000# concrete		
By chute	C.Y.	120
By pump	C.Y.	160
By crane	C.Y.	170
Spread footing		
2500# or 3000# concrete		
By chute	C.Y.	110
By pump	C.Y.	150
By crane	C.Y.	160

	Unit	Total		Unit	Total

03380.50 GRADE BEAM CONCRETE
Grade beam
  2500# or 3000# concrete
    By chute — C.Y. 110
    By crane — C.Y. 160
    By pump — C.Y. 150
    By hand buggy — C.Y. 140
  3500# or 4000# concrete
    By chute — C.Y. 120
    By crane — C.Y. 170
    By pump — C.Y. 160
    By hand buggy — C.Y. 150

03380.53 PILE CAP CONCRETE
Pile cap
  2500# or 3000 concrete
    By chute — C.Y. 120
    By crane — C.Y. 180
    By pump — C.Y. 170
    By hand buggy — C.Y. 150

03380.55 SLAB / MAT CONCRETE
Slab on grade
  2500# or 3000# concrete
    By chute — C.Y. 120
    By crane — C.Y. 140
    By pump — C.Y. 140
    By hand buggy — C.Y. 140

03380.58 SIDEWALKS
Walks, cast in place with wire mesh, base not incl.
  4" thick — S.F. 2.88
  5" thick — S.F. 3.69
  6" thick — S.F. 4.57

03380.65 WALL CONCRETE
Walls
  2500# or 3000# concrete
  To 4'
    By chute — C.Y. 120
    By crane — C.Y. 180
    By pump — C.Y. 170
  To 8'
    By crane — C.Y. 180
    By pump — C.Y. 180
  Filled block (CMU)
    3000# concrete, by pump
      4" wide — S.F. 3.29
      6" wide — S.F. 4.28
      8" wide — S.F. 5.48

03400.90 PRECAST SPECIALTIES
Precast concrete, coping, 4' to 8' long
  12" wide — L.F. 13.75
  10" wide — L.F. 13.75
Splash block, 30"x12"x4" — EA. 49.75
Stair unit, per riser — EA. 120
Sun screen and trellis, 8' long, 12" high
  4" thick blades — EA. 120

03600.10 GROUTING
Grouting for bases
  Non-metallic grout
    1" deep — S.F. 15.75
    2" deep — S.F. 21.75
Portland cement grout (1 cement to 3 sand)
  1/2" joint thickness
    6" wide joints — L.F. 1.95
    8" wide joints — L.F. 2.32
  1" joint thickness
    4" wide joints — L.F. 1.86
    6" wide joints — L.F. 2.16

	Unit	Total
04100.10 MASONRY GROUT		
Grout, non shrink, non-metallic, trowelable	C.F.	6.44
Grout door frame, hollow metal		
Single	EA.	68.25
Double	EA.	76.00
Grout-filled concrete block (CMU)		
4" wide	S.F.	2.19
6" wide	S.F.	2.90
8" wide	S.F.	3.51
12" wide	S.F.	4.44
Grout-filled individual CMU cells		
4" wide	L.F.	1.40
6" wide	L.F.	1.49
8" wide	L.F.	1.62
10" wide	L.F.	1.89
12" wide	L.F.	2.02
Bond beams or lintels, 8" deep		
6" thick	L.F.	2.58
8" thick	L.F.	3.01
10" thick	L.F.	3.48
12" thick	L.F.	4.00
Cavity walls		
2" thick	S.F.	3.51
3" thick	S.F.	3.93
4" thick	S.F.	4.53
6" thick	S.F.	5.84
04150.10 MASONRY ACCESSORIES		
Foundation vents	EA.	52.50
Bar reinforcing		
Horizontal		
#3 - #4	Lb.	2.97
#5 - #6	Lb.	2.63
Vertical		
#3 - #4	Lb.	3.48
#5 - #6	Lb.	2.97
Horizontal joint reinforcing		
Truss type		
4" wide, 6" wall	L.F.	.40
6" wide, 8" wall	L.F.	.41
8" wide, 10" wall	L.F.	.46
10" wide, 12" wall	L.F.	.47
12" wide, 14" wall	L.F.	.54
Ladder type		
4" wide, 6" wall	L.F.	.34
6" wide, 8" wall	L.F.	.37
8" wide, 10" wall	L.F.	.40
10" wide, 12" wall	L.F.	.43
Rectangular wall ties		
3/16" dia., galvanized		
2" x 6"	EA.	1.19
2" x 8"	EA.	1.21
2" x 10"	EA.	1.27
2" x 12"	EA.	1.32
4" x 6"	EA.	1.41
4" x 8"	EA.	1.46
4" x 10"	EA.	1.59
4" x 12"	EA.	1.68
1/4" dia., galvanized		
2" x 6"	EA.	1.49
2" x 8"	EA.	1.57
2" x 10"	EA.	1.66
2" x 12"	EA.	1.79
4" x 6"	EA.	1.76
4" x 8"	EA.	1.83
4" x 10"	EA.	1.96
4" x 12"	EA.	2.00
"Z" type wall ties, galvanized		
6" long		
1/8" dia.	EA.	1.15
3/16" dia.	EA.	1.17
1/4" dia.	EA.	1.19
8" long		
1/8" dia.	EA.	1.17
3/16" dia.	EA.	1.19
1/4" dia.	EA.	1.21
10" long		
1/8" dia.	EA.	1.19
3/16" dia.	EA.	1.24

	Unit	Total
04150.10 MASONRY ACCESSORIES (Cont.)		
1/4" dia.	EA.	1.29
Dovetail anchor slots		
Galvanized steel, filled		
24 ga.	L.F.	2.08
20 ga.	L.F.	2.29
16 oz. copper, foam filled	L.F.	3.30
Dovetail anchors		
16 ga.		
3-1/2" long	EA.	1.09
5-1/2" long	EA.	1.15
12 ga.		
3-1/2" long	EA.	1.17
5-1/2" long	EA.	1.39
Dovetail, triangular galvanized ties, 12 ga.		
3" x 3"	EA.	1.40
5" x 5"	EA.	1.44
7" x 7"	EA.	1.52
7" x 9"	EA.	1.57
Brick anchors		
Corrugated, 3-1/2" long		
16 ga.	EA.	1.06
12 ga.	EA.	1.22
Non-corrugated, 3-1/2" long		
16 ga.	EA.	1.15
12 ga.	EA.	1.39
Cavity wall anchors, corrugated, galvanized		
5" long		
16 ga.	EA.	1.52
12 ga.	EA.	1.85
7" long		
28 ga.	EA.	1.59
24 ga.	EA.	1.79
22 ga.	EA.	1.81
16 ga.	EA.	1.93
Mesh ties, 16 ga., 3" wide		
8" long	EA.	1.75
12" long	EA.	1.85
20" long	EA.	2.24
24" long	EA.	2.38
04150.20 MASONRY CONTROL JOINTS		
Control joint, cross shaped PVC	L.F.	3.44
Closed cell joint filler		
1/2"	L.F.	1.64
3/4"	L.F.	2.03
Rubber, for		
4" wall	L.F.	3.77
PVC, for		
4" wall	L.F.	2.57
04150.50 MASONRY FLASHING		
Through-wall flashing		
5 oz. coated copper	S.F.	7.57
0.030" elastomeric	S.F.	4.47
04210.10 BRICK MASONRY		
Standard size brick, running bond		
Face brick, red (6.4/sf)		
Veneer	S.F.	13.75
Cavity wall	S.F.	12.50
9" solid wall	S.F.	25.25
Common brick (6.4/sf)		
Select common for veneers	S.F.	12.25
Back-up		
4" thick	S.F.	9.73
8" thick	S.F.	16.75
Glazed brick (7.4/sf)		
Veneer	S.F.	21.25
Buff or gray face brick (6.4/sf)		
Veneer	S.F.	14.50
Cavity wall	S.F.	13.25
Jumbo or oversize brick (3/sf)		
4" veneer	S.F.	9.17
4" back-up	S.F.	8.33
8" back-up	S.F.	12.00
Norman brick, red face, (4.5/sf)		
4" veneer	S.F.	12.25
Cavity wall	S.F.	11.50
Chimney, standard brick, including flue		
16" x 16"	L.F.	88.25

	Unit	Total

04210.10 BRICK MASONRY (Cont.)

	Unit	Total
16" x 20"	L.F.	93.50
16" x 24"	L.F.	110
20" x 20"	L.F.	130
20" x 24"	L.F.	130
20" x 32"	L.F.	150
Window sill, face brick on edge	L.F.	15.75

04210.60 PAVERS, MASONRY

	Unit	Total
Brick walk laid on sand, sand joints		
Laid flat, (4.5 per sf)	S.F.	9.39
Laid on edge, (7.2 per sf)	S.F.	14.50
Precast concrete patio blocks		
2" thick		
Natural	S.F.	4.36
Colors	S.F.	5.42
Exposed aggregates, local aggregate		
Natural	S.F.	7.84
Colors	S.F.	8.39
Granite or limestone aggregate	S.F.	10.25
White tumblestone aggregate	S.F.	6.85
Stone pavers, set in mortar		
Bluestone		
1" thick		
Irregular	S.F.	19.00
Snapped rectangular	S.F.	19.25
1-1/2" thick, random rectangular	S.F.	23.50
2" thick, random rectangular	S.F.	27.25
Slate		
Natural cleft		
Irregular, 3/4" thick	S.F.	21.25
Random rectangular		
1-1/4" thick	S.F.	27.25
1-1/2" thick	S.F.	30.25
Granite blocks		
3" thick, 3" to 6" wide		
4" to 12" long	S.F.	30.50
6" to 15" long	S.F.	25.50

04220.10 CONCRETE MASONRY UNITS

	Unit	Total
Hollow, load bearing		
4"	S.F.	5.12
6"	S.F.	5.90
8"	S.F.	6.52
10"	S.F.	7.78
12"	S.F.	8.71
Solid, load bearing		
4"	S.F.	5.90
6"	S.F.	6.31
8"	S.F.	7.51
10"	S.F.	8.11
12"	S.F.	10.25
Back-up block, 8" x 16"		
2"	S.F.	4.21
4"	S.F.	4.35
6"	S.F.	5.17
8"	S.F.	5.68
10"	S.F.	6.80
12"	S.F.	7.55
Foundation wall, 8" x 16"		
6"	S.F.	5.62
8"	S.F.	6.20
10"	S.F.	7.40
12"	S.F.	8.25
Solid		
6"	S.F.	6.31
8"	S.F.	7.51
10"	S.F.	8.11
12"	S.F.	10.25
Exterior, styrofoam inserts, std weight, 8" x 16"		
6"	S.F.	7.84
8"	S.F.	8.46
10"	S.F.	10.00
12"	S.F.	12.75
Lightweight		
6"	S.F.	8.30
8"	S.F.	9.17
10"	S.F.	9.87
12"	S.F.	12.00

	Unit	Total

04220.10 CONCRETE MASONRY UNITS (Cont.)

	Unit	Total
Acoustical slotted block		
4"	S.F.	9.34
6"	S.F.	9.57
8"	S.F.	11.25
Filled cavities		
4"	S.F.	10.75
6"	S.F.	11.75
8"	S.F.	13.75
Hollow, split face		
4"	S.F.	7.00
6"	S.F.	7.66
8"	S.F.	8.17
10"	S.F.	9.02
12"	S.F.	9.78
Split rib profile		
4"	S.F.	8.55
6"	S.F.	9.19
8"	S.F.	10.00
10"	S.F.	10.50
12"	S.F.	11.00
Solar screen concrete block		
4" thick		
6" x 6"	S.F.	15.75
8" x 8"	S.F.	15.50
12" x 12"	S.F.	13.50
8" thick		
8" x 16"	S.F.	12.75
Vertical reinforcing		
4' o.c., add 5% to labor		
2'8" o.c., add 15% to labor		
Interior partitions, add 10% to labor		

04220.90 BOND BEAMS & LINTELS

	Unit	Total
Bond beam, no grout or reinforcement		
8" x 16" x		
4" thick	L.F.	5.29
6" thick	L.F.	6.17
8" thick	L.F.	6.65
10" thick	L.F.	7.41
12" thick	L.F.	8.03
Beam lintel, no grout or reinforcement		
8" x 16" x		
10" thick	L.F.	10.75
12" thick	L.F.	12.75
Precast masonry lintel		
6 lf, 8" high x		
4" thick	L.F.	13.75
6" thick	L.F.	15.25
8" thick	L.F.	17.00
10" thick	L.F.	18.50
10 lf, 8" high x		
4" thick	L.F.	11.75
6" thick	L.F.	13.25
8" thick	L.F.	14.75
10" thick	L.F.	18.00
Steel angles and plates		
Minimum	Lb.	1.76
Maximum	Lb.	2.80
Various size angle lintels		
1/4" stock		
3" x 3"	L.F.	8.55
3" x 3-1/2"	L.F.	9.10
3/8" stock		
3" x 4"	L.F.	12.50
3-1/2" x 4"	L.F.	13.00
4" x 4"	L.F.	14.00
5" x 3-1/2"	L.F.	14.75
6" x 3-1/2"	L.F.	16.00
1/2" stock		
6" x 4"	L.F.	17.50

04270.10 GLASS BLOCK

	Unit	Total
Glass block, 4" thick		
6" x 6"	S.F.	42.75
8" x 8"	S.F.	29.25
12" x 12"	S.F.	31.00

DIVISION 4 MASONRY

	Unit	Total
04295.10 PARGING / MASONRY PLASTER		
Parging		
1/2" thick	S.F.	3.67
3/4" thick	S.F.	4.58
1" thick	S.F.	5.56
04400.10 STONE		
Rubble stone		
Walls set in mortar		
8" thick	S.F.	27.75
12" thick	S.F.	38.50
18" thick	S.F.	49.25
24" thick	S.F.	64.00
Dry set wall		
8" thick	S.F.	25.50
12" thick	S.F.	31.75
18" thick	S.F.	43.25
24" thick	S.F.	52.50
Thresholds, 7/8" thick, 3' long, 4" to 6" wide		
Plain	EA.	69.00
Beveled	EA.	72.00
Window sill		
6" wide, 2" thick	L.F.	35.50
Stools		
5" wide, 7/8" thick	L.F.	42.50
Granite veneer facing panels, polished		
7/8" thick		
Black	S.F.	66.50
Gray	S.F.	56.50
Slate, panels		
1" thick	S.F.	45.25
Sills or stools		
1" thick		
6" wide	L.F.	32.00
10" wide	L.F.	41.00

	Unit	Total
04520.10 RESTORATION AND CLEANING		
Masonry cleaning		
Washing brick		
Smooth surface	S.F.	1.05
Rough surface	S.F.	1.42
Steam clean masonry		
Smooth face		
Minimum	S.F.	.66
Maximum	S.F.	.97
Rough face		
Minimum	S.F.	.88
Maximum	S.F.	1.33
04550.10 REFRACTORIES		
Flue liners		
Rectangular		
8" x 12"	L.F.	16.00
12" x 12"	L.F.	18.50
12" x 18"	L.F.	26.50
16" x 16"	L.F.	29.25
18" x 18"	L.F.	34.00
20" x 20"	L.F.	49.50
24" x 24"	L.F.	58.75
Round		
18" dia.	L.F.	46.00
24" dia.	L.F.	81.7

	Unit	Total

05050.10 STRUCTURAL WELDING
Welding
  Single pass

	Unit	Total
1/8"	L.F.	3.24
3/16"	L.F.	4.42
1/4"	L.F.	5.61

05050.90 METAL ANCHORS
Anchor bolts
  3/8" x

	Unit	Total
8" long	EA.	.98
12" long	EA.	1.17

  1/2" x

	Unit	Total
8" long	EA.	1.46
12" long	EA.	1.70

  5/8" x

	Unit	Total
8" long	EA.	1.36
12" long	EA.	1.61

  3/4" x

	Unit	Total
8" long	EA.	1.95
12" long	EA.	2.19

Non-drilling anchor

	Unit	Total
1/4"	EA.	.65
3/8"	EA.	.80
1/2"	EA.	1.23

Self-drilling anchor

	Unit	Total
1/4"	EA.	1.63
3/8"	EA.	2.44
1/2"	EA.	3.26

05050.95 METAL LINTELS
Lintels, steel

	Unit	Total
Plain	Lb.	2.78
Galvanized	Lb.	3.45

05120.10 STRUCTURAL STEEL
Beams and girders, A-36

	Unit	Total
Welded	TON	3,300
Bolted	TON	3,200

Columns
  Pipe

	Unit	Total
6" dia.	Lb.	2.17

Structural tube
  6" square

	Unit	Total
Light sections	TON	4,850

05410.10 METAL FRAMING
Furring channel, galvanized
  Beams and columns, 3/4"

	Unit	Total
12" o.c.	S.F.	6.21
16" o.c.	S.F.	5.59

  Walls, 3/4"

	Unit	Total
12" o.c.	S.F.	3.31
16" o.c.	S.F.	2.73
24" o.c.	S.F.	2.16

  1-1/2"

	Unit	Total
12" o.c.	S.F.	3.57
16" o.c.	S.F.	2.92
24" o.c.	S.F.	2.28

Stud, load bearing
  16" o.c.
   16 ga.

	Unit	Total
2-1/2"	S.F.	3.79
3-5/8"	S.F.	4.01
4"	S.F.	4.06
6"	S.F.	4.78

   18 ga.

	Unit	Total
2-1/2"	S.F.	3.57
3-5/8"	S.F.	3.79
4"	S.F.	3.84
6"	S.F.	4.51
8"	S.F.	4.83

   20 ga.

	Unit	Total
2-1/2"	S.F.	3.13
3-5/8"	S.F.	3.24
4"	S.F.	3.30
6"	S.F.	3.79
8"	S.F.	3.95

  24" o.c.
   16 ga.

	Unit	Total
2-1/2"	S.F.	3.06
3-5/8"	S.F.	3.23

05410.10 METAL FRAMING (Cont.)

	Unit	Total
4"	S.F.	3.28
6"	S.F.	3.68
8"	S.F.	4.02

   18 ga.

	Unit	Total
2-1/2"	S.F.	2.90
3-5/8"	S.F.	3.01
4"	S.F.	3.06
6"	S.F.	3.46
8"	S.F.	3.68

   20 ga.

	Unit	Total
2-1/2"	S.F.	2.68
3-5/8"	S.F.	2.74
4"	S.F.	2.79
6"	S.F.	3.14
8"	S.F.	3.30

05520.10 RAILINGS
Railing, pipe
  1-1/4" diameter, welded steel
   2-rail

	Unit	Total
Primed	L.F.	41.25
Galvanized	L.F.	49.50

   3-rail

	Unit	Total
Primed	L.F.	52.50
Galvanized	L.F.	63.75

  Wall mounted, single rail, welded steel

	Unit	Total
Primed	L.F.	28.75
Galvanized	L.F.	34.75

  Wall mounted, single rail, welded steel

	Unit	Total
Primed	L.F.	30.50
Galvanized	L.F.	37.00

  Wall mounted, single rail, welded steel

	Unit	Total
Primed	L.F.	33.00
Galvanized	L.F.	40.00

05700.10 ORNAMENTAL METAL
Railings, square bars, 6" o.c., shaped top rails

	Unit	Total
Steel	L.F.	120
Aluminum	L.F.	140
Bronze	L.F.	230
Stainless steel	L.F.	220

Laminated metal or wood handrails

	Unit	Total
2-1/2" round or oval shape	L.F.	310

Aluminum louvers
  Residential use, fixed type, with screen

	Unit	Total
8" x 8"	EA.	46.25
12" x 12"	EA.	48.00
12" x 18"	EA.	51.75
14" x 24"	EA.	61.75
18" x 24"	EA.	65.75
30" x 24"	EA.	82.25

06050.10 ACCESSORIES

Column/post base, cast aluminum

	Unit	Total
4" x 4"	EA.	27.25
6" x 6"	EA.	36.00

Bridging, metal, per pair

	Unit	Total
12" o.c.	EA.	7.08
16" o.c.	EA.	6.35

Anchors
Bolts, threaded two ends, with nuts and washers
1/2" dia.

	Unit	Total
4" long	EA.	5.49
7-1/2" long	EA.	5.81

3/4" dia.

	Unit	Total
7-1/2" long	EA.	8.68
15" long	EA.	11.00

Framing anchors

	Unit	Total
10 gauge	EA.	5.29

Bolts, carriage

	Unit	Total
1/4 x 4	EA.	5.73
5/16 x 6	EA.	6.44
3/8 x 6	EA.	7.29
1/2 x 6	EA.	7.41

Joist and beam hangers
18 ga.

	Unit	Total
2 x 4	EA.	6.42
2 x 6	EA.	7.20
2 x 8	EA.	7.46
2 x 10	EA.	8.56
2 x 12	EA.	9.56

16 ga.

	Unit	Total
3 x 6	EA.	9.38
3 x 8	EA.	11.00
3 x 10	EA.	12.75
3 x 12	EA.	14.00
3 x 14	EA.	15.00
4 x 6	EA.	13.00
4 x 8	EA.	13.75
4 x 10	EA.	15.00
4 x 12	EA.	17.75
4 x 14	EA.	19.00

Rafter anchors, 18 ga., 1-1/2" wide

	Unit	Total
5-1/4" long	EA.	5.30
10-3/4" long	EA.	5.82

Shear plates

	Unit	Total
2-5/8" dia.	EA.	6.72
4" dia.	EA.	10.00

Sill anchors

	Unit	Total
Embedded in concrete	EA.	7.59

Split rings

	Unit	Total
2-1/2" dia.	EA.	7.67
4" dia.	EA.	9.88

Strap ties, 14 ga., 1-3/8" wide

	Unit	Total
12" long	EA.	6.33
18" long	EA.	6.88
24" long	EA.	8.46
36" long	EA.	10.25

Toothed rings

	Unit	Total
2-5/8" dia.	EA.	10.75
4" dia.	EA.	12.75

06110.10 BLOCKING

Steel construction
Walls

	Unit	Total
2x4	L.F.	3.90
2x6	L.F.	4.64
2x8	L.F.	5.20
2x10	L.F.	5.77
2x12	L.F.	7.09

Ceilings

	Unit	Total
2x4	L.F.	4.44
2x6	L.F.	5.37
2x8	L.F.	6.07
2x10	L.F.	6.82
2x12	L.F.	8.39

Wood construction
Walls

	Unit	Total
2x4	L.F.	3.32
2x6	L.F.	3.88
2x8	L.F.	4.33

06110.10 BLOCKING (Cont.)

	Unit	Total
2x10	L.F.	4.75
2x12	L.F.	5.89

Ceilings

	Unit	Total
2x4	L.F.	3.68
2x6	L.F.	4.35
2x8	L.F.	4.87
2x10	L.F.	5.37
2x12	L.F.	6.62

06110.20 CEILING FRAMING

Ceiling joists
12" o.c.

	Unit	Total
2x4	S.F.	1.84
2x6	S.F.	2.18
2x8	S.F.	2.63
2x10	S.F.	2.93
2x12	S.F.	4.24

16" o.c.

	Unit	Total
2x4	S.F.	1.50
2x6	S.F.	1.79
2x8	S.F.	2.13
2x10	S.F.	2.34
2x12	S.F.	3.39

24" o.c.

	Unit	Total
2x4	S.F.	1.21
2x6	S.F.	1.47
2x8	S.F.	1.80
2x10	S.F.	2.01
2x12	S.F.	3.66

Headers and nailers

	Unit	Total
2x4	L.F.	2.10
2x6	L.F.	2.36
2x8	L.F.	2.71
2x10	L.F.	3.05
2x12	L.F.	3.43

Sister joists for ceilings

	Unit	Total
2x4	L.F.	4.15
2x6	L.F.	4.97
2x8	L.F.	6.07
2x10	L.F.	7.56
2x12	L.F.	9.96

06110.30 FLOOR FRAMING

Floor joists
12" o.c.

	Unit	Total
2x6	S.F.	1.79
2x8	S.F.	2.17
2x10	S.F.	2.62
2x12	S.F.	3.41
2x14	S.F.	4.54
3x6	S.F.	3.63
3x8	S.F.	4.43
3x10	S.F.	5.32
3x12	S.F.	6.19
3x14	S.F.	6.99
4x6	S.F.	4.41
4x8	S.F.	5.39
4x10	S.F.	6.62
4x12	S.F.	7.85
4x14	S.F.	9.01

16" o.c.

	Unit	Total
2x6	S.F.	1.52
2x8	S.F.	1.78
2x10	S.F.	2.00
2x12	S.F.	2.31
2x14	S.F.	4.10
3x6	S.F.	3.02
3x8	S.F.	3.70
3x10	S.F.	4.42
3x12	S.F.	5.12
3x14	S.F.	5.89
4x6	S.F.	3.65
4x8	S.F.	4.61
4x10	S.F.	5.55
4x12	S.F.	6.43
4x14	S.F.	7.47

	Unit	Total
06110.30 FLOOR FRAMING] (Cont.)		
Sister joists for floors		
2x4	L.F.	3.68
2x6	L.F.	4.35
2x8	L.F.	5.20
2x10	L.F.	6.24
2x12	L.F.	8.39
3x6	L.F.	7.34
3x8	L.F.	8.57
3x10	L.F.	9.97
3x12	L.F.	11.50
4x6	L.F.	7.90
4x8	L.F.	9.48
4x10	L.F.	11.00
4x12	L.F.	13.00
06110.40 FURRING		
Furring, wood strips		
Walls		
On masonry or concrete walls		
1x2 furring		
12" o.c.	S.F.	1.96
16" o.c.	S.F.	1.78
24" o.c.	S.F.	1.63
1x3 furring		
12" o.c.	S.F.	2.06
16" o.c.	S.F.	1.86
24" o.c.	S.F.	1.67
On wood walls		
1x2 furring		
12" o.c.	S.F.	1.49
16" o.c.	S.F.	1.33
24" o.c.	S.F.	1.21
1x3 furring		
12" o.c.	S.F.	1.59
16" o.c.	S.F.	1.41
24" o.c.	S.F.	1.25
Ceilings		
On masonry or concrete ceilings		
1x2 furring		
12" o.c.	S.F.	3.23
16" o.c.	S.F.	2.90
24" o.c.	S.F.	2.63
1x3 furring		
12" o.c.	S.F.	3.33
16" o.c.	S.F.	2.98
24" o.c.	S.F.	2.67
On wood ceilings		
1x2 furring		
12" o.c.	S.F.	2.26
16" o.c.	S.F.	2.03
24" o.c.	S.F.	1.84
1x3		
12" o.c.	S.F.	2.36
16" o.c.	S.F.	2.11
24" o.c.	S.F.	1.88
06110.50 ROOF FRAMING		
Roof framing		
Rafters, gable end		
0-2 pitch (flat to 2-in-12)		
12" o.c.		
2x4	S.F.	1.71
2x6	S.F.	2.01
2x8	S.F.	2.45
2x10	S.F.	2.72
2x12	S.F.	4.00
16" o.c.		
2x6	S.F.	1.68
2x8	S.F.	2.00
2x10	S.F.	2.26
2x12	S.F.	3.24
24" o.c.		
2x6	S.F.	1.22
2x8	S.F.	1.68
2x10	S.F.	1.84
2x12	S.F.	2.69
4-6 pitch (4-in-12 to 6-in-12)		
12" o.c.		
2x4	S.F.	1.77

	Unit	Total
06110.50 ROOF FRAMING (Cont.)		
2x6	S.F.	2.11
2x8	S.F.	2.60
2x10	S.F.	2.87
2x12	S.F.	3.74
16" o.c.		
2x6	S.F.	1.72
2x8	S.F.	2.21
2x10	S.F.	2.36
2x12	S.F.	3.06
24" o.c.		
2x6	S.F.	1.44
2x8	S.F.	1.83
2x10	S.F.	1.99
2x12	S.F.	2.65
8-12 pitch (8-in-12 to 12-in-12)		
12" o.c.		
2x4	S.F.	1.85
2x6	S.F.	2.28
2x8	S.F.	2.75
2x10	S.F.	3.02
2x12	S.F.	3.98
16" o.c.		
2x6	S.F.	1.82
2x8	S.F.	2.36
2x10	S.F.	2.57
2x12	S.F.	3.27
24" o.c.		
2x6	S.F.	1.48
2x8	S.F.	1.91
2x10	S.F.	2.11
2x12	S.F.	2.85
Ridge boards		
2x6	L.F.	3.23
2x8	L.F.	3.72
2x10	L.F.	4.30
2x12	L.F.	5.65
Hip rafters		
2x6	L.F.	2.48
2x8	L.F.	2.75
2x10	L.F.	3.05
2x12	L.F.	4.01
Jack rafters		
4-6 pitch (4-in-12 to 6-in-12)		
16" o.c.		
2x6	S.F.	2.30
2x8	S.F.	2.84
2x10	S.F.	3.08
2x12	S.F.	3.76
24" o.c.		
2x6	S.F.	1.81
2x8	S.F.	2.25
2x10	S.F.	2.53
2x12	S.F.	3.00
8-12 pitch (8-in-12 to 12-in-12)		
16" o.c.		
2x6	S.F.	2.84
2x8	S.F.	3.15
2x10	S.F.	3.94
2x12	S.F.	4.82
24" o.c.		
2x6	S.F.	2.18
2x8	S.F.	2.53
2x10	S.F.	3.22
2x12	S.F.	4.04
Sister rafters		
2x4	L.F.	4.13
2x6	L.F.	4.97
2x8	L.F.	6.07
2x10	L.F.	7.56
2x12	L.F.	10.50
Fascia boards		
2x4	L.F.	3.01
2x6	L.F.	3.23
2x8	L.F.	3.75
2x10	L.F.	3.92
2x12	L.F.	5.13

DIVISION 6 WOOD AND PLASTICS

	Unit	Total
06110.50 ROOF FRAMING (Cont.)		
Cant strips		
Fiber		
3x3	L.F.	1.82
4x4	L.F.	2.01
Wood		
3x3	L.F.	3.27
06110.60 SLEEPERS		
Sleepers, over concrete		
12" o.c.		
1x2	S.F.	1.45
1x3	S.F.	1.60
2x4	S.F.	2.21
2x6	S.F.	2.65
16" o.c.		
1x2	S.F.	1.26
1x3	S.F.	1.35
2x4	S.F.	1.87
2x6	S.F.	2.25
06110.65 SOFFITS		
Soffit framing		
2x3	L.F.	4.06
2x4	L.F.	4.44
2x6	L.F.	4.97
2x8	L.F.	5.60
06110.70 WALL FRAMING		
Framing wall, studs		
12" o.c.		
2x3	S.F.	1.41
2x4	S.F.	1.59
2x6	S.F.	1.92
2x8	S.F.	2.35
16" o.c.		
2x3	S.F.	1.17
2x4	S.F.	1.32
2x6	S.F.	1.59
2x8	S.F.	2.00
24" o.c.		
2x3	S.F.	.99
2x4	S.F.	1.09
2x6	S.F.	1.35
2x8	S.F.	1.59
Plates, top or bottom		
2x3	L.F.	1.87
2x4	L.F.	2.05
2x6	L.F.	2.36
2x8	L.F.	2.71
Headers, door or window		
2x6		
Single		
3' long	EA.	27.75
6' long	EA.	36.50
Double		
3' long	EA.	32.75
6' long	EA.	44.75
2x8		
Single		
4' long	EA.	36.25
8' long	EA.	47.00
Double		
4' long	EA.	44.00
8' long	EA.	61.00
2x10		
Single		
5' long	EA.	45.25
10' long	EA.	62.50
Double		
5' long	EA.	53.75
10' long	EA.	72.75
2x12		
Single		
6' long	EA.	47.75
12' long	EA.	66.75
Double		
6' long	EA.	62.00
12' long	EA.	87.00

	Unit	Total
06115.10 FLOOR SHEATHING		
Sub-flooring, plywood, CDX		
1/2" thick	S.F.	1.34
5/8" thick	S.F.	1.85
3/4" thick	S.F.	2.13
Structural plywood		
1/2" thick	S.F.	1.42
5/8" thick	S.F.	2.00
3/4" thick	S.F.	2.18
Board type subflooring		
1x6		
Minimum	S.F.	2.15
Maximum	S.F.	2.56
1x8		
Minimum	S.F.	2.36
Maximum	S.F.	2.71
1x10		
Minimum	S.F.	2.58
Maximum	S.F.	2.81
Underlayment		
Hardboard, 1/4" tempered	S.F.	1.27
Plywood, CDX		
3/8" thick	S.F.	1.42
1/2" thick	S.F.	1.58
5/8" thick	S.F.	1.79
3/4" thick	S.F.	2.06
06115.20 ROOF SHEATHING		
Sheathing		
Plywood, CDX		
3/8" thick	S.F.	1.44
1/2" thick	S.F.	1.58
5/8" thick	S.F.	1.79
3/4" thick	S.F.	2.06
Structural plywood		
3/8" thick	S.F.	1.22
1/2" thick	S.F.	1.42
5/8" thick	S.F.	1.63
3/4" thick	S.F.	1.82
06115.30 WALL SHEATHING		
Sheathing		
Plywood, CDX		
3/8" thick	S.F.	1.54
1/2" thick	S.F.	1.68
5/8" thick	S.F.	1.91
3/4" thick	S.F.	2.21
Waferboard		
3/8" thick	S.F.	1.49
1/2" thick	S.F.	1.76
5/8" thick	S.F.	2.05
3/4" thick	S.F.	2.16
Structural plywood		
3/8" thick	S.F.	1.32
1/2" thick	S.F.	1.57
5/8" thick	S.F.	2.12
3/4" thick	S.F.	2.33
Gypsum, 1/2" thick	S.F.	1.15
Asphalt impregnated fiberboard, 1/2" thick	S.F.	1.53
06125.10 WOOD DECKING		
Decking, T&G solid		
Cedar		
3" thick	S.F.	10.50
4" thick	S.F.	12.75
Fir		
3" thick	S.F.	5.15
4" thick	S.F.	6.09
Southern yellow pine		
3" thick	S.F.	5.29
4" thick	S.F.	5.62
White pine		
3" thick	S.F.	6.08
4" thick	S.F.	7.70
06130.10 HEAVY TIMBER		
Mill framed structures		
Beams to 20' long		
Douglas fir		
6x8	L.F.	12.75
6x10	L.F.	14.25
6x12	L.F.	16.00

	Unit	Total		Unit	Total

06130.10 HEAVY TIMBER (Cont.)

	Unit	Total
6x14	L.F.	18.25
6x16	L.F.	19.25
8x10	L.F.	16.50
8x12	L.F.	18.75
8x14	L.F.	20.75
8x16	L.F.	22.75
Southern yellow pine		
6x8	L.F.	11.50
6x10	L.F.	12.75
6x12	L.F.	15.00
6x14	L.F.	16.50
6x16	L.F.	17.75
8x10	L.F.	15.00
8x12	L.F.	17.25
8x14	L.F.	19.00
8x16	L.F.	20.75
Columns to 12' high		
Douglas fir		
6x6	L.F.	14.50
8x8	L.F.	17.50
10x10	L.F.	24.50
12x12	L.F.	27.50
Southern yellow pine		
6x6	L.F.	13.75
8x8	L.F.	16.25
10x10	L.F.	21.25
12x12	L.F.	25.00
Posts, treated		
4x4	L.F.	3.61
6x6	L.F.	7.01

06190.20 WOOD TRUSSES

	Unit	Total
Truss, fink, 2x4 members		
3-in-12 slope		
24' span	EA.	150
26' span	EA.	150
28' span	EA.	160
30' span	EA.	160
34' span	EA.	180
38' span	EA.	180
5-in-12 slope		
24' span	EA.	150
28' span	EA.	160
30' span	EA.	170
32' span	EA.	170
40' span	EA.	220
Gable, 2x4 members		
5-in-12 slope		
24' span	EA.	170
26' span	EA.	180
28' span	EA.	190
30' span	EA.	200
32' span	EA.	200
36' span	EA.	220
40' span	EA.	230
King post type, 2x4 members		
4-in-12 slope		
16' span	EA.	120
18' span	EA.	130
24' span	EA.	130
26' span	EA.	140
30' span	EA.	160
34' span	EA.	170
38' span	EA.	200
42' span	EA.	230

06200.10 FINISH CARPENTRY

	Unit	Total
Mouldings and trim		
Apron, flat		
9/16 x 2	L.F.	3.99
9/16 x 3-1/2	L.F.	5.45
Base		
Colonial		
7/16 x 2-1/4	L.F.	4.26
7/16 x 3	L.F.	4.65
7/16 x 3-1/4	L.F.	4.81
9/16 x 3	L.F.	4.89
9/16 x 3-1/4	L.F.	5.01
11/16 x 2-1/4	L.F.	5.27

06200.10 FINISH CARPENTRY (Cont.)

	Unit	Total
Ranch		
7/16 x 2-1/4	L.F.	4.43
7/16 x 3-1/4	L.F.	4.81
9/16 x 2-1/4	L.F.	4.73
9/16 x 3	L.F.	4.79
9/16 x 3-1/4	L.F.	4.95
Casing		
11/16 x 2-1/2	L.F.	4.06
11/16 x 3-1/2	L.F.	4.41
Chair rail		
9/16 x 2-1/2	L.F.	4.41
9/16 x 3-1/2	L.F.	5.14
Closet pole		
1-1/8" dia.	L.F.	4.69
1-5/8" dia.	L.F.	5.22
Cove		
9/16 x 1-3/4	L.F.	3.99
11/16 x 2-3/4	L.F.	4.53
Crown		
9/16 x 1-5/8	L.F.	5.30
9/16 x 2-5/8	L.F.	6.00
11/16 x 3-5/8	L.F.	6.49
11/16 x 4-1/4	L.F.	7.89
11/16 x 5-1/4	L.F.	8.80
Drip cap		
1-1/16 x 1-5/8	L.F.	4.53
Glass bead		
3/8 x 3/8	L.F.	3.92
1/2 x 9/16	L.F.	4.08
5/8 x 5/8	L.F.	4.11
3/4 x 3/4	L.F.	4.20
Half round		
1/2	L.F.	2.86
5/8	L.F.	3.13
3/4	L.F.	3.41
Lattice		
1/4 x 7/8	L.F.	2.69
1/4 x 1-1/8	L.F.	2.75
1/4 x 1-3/8	L.F.	2.78
1/4 x 1-3/4	L.F.	2.89
1/4 x 2	L.F.	3.03
Ogee molding		
5/8 x 3/4	L.F.	3.87
11/16 x 1-1/8	L.F.	4.53
11/16 x 1-3/8	L.F.	4.98
Parting bead		
3/8 x 7/8	L.F.	4.30
Quarter round		
1/4 x 1/4	L.F.	2.42
3/8 x 3/8	L.F.	2.59
1/2 x 1/2	L.F.	2.77
11/16 x 11/16	L.F.	2.95
3/4 x 3/4	L.F.	3.53
1-1/16 x 1-1/16	L.F.	3.31
Railings, balusters		
1-1/8 x 1-1/8	L.F.	8.68
1-1/2 x 1-1/2	L.F.	8.76
Screen moldings		
1/4 x 3/4	L.F.	5.17
5/8 x 5/16	L.F.	5.39
Shoe		
7/16 x 11/16	L.F.	3.13
Sash beads		
1/2 x 3/4	L.F.	5.54
1/2 x 7/8	L.F.	5.73
1/2 x 1-1/8	L.F.	6.23
5/8 x 7/8	L.F.	6.23
Stop		
5/8 x 1-5/8		
Colonial	L.F.	3.98
Ranch	L.F.	3.98
Stools		
11/16 x 2-1/4	L.F.	9.10
11/16 x 2-1/2	L.F.	9.21
11/16 x 5-1/4	L.F.	10.00
Exterior trim, casing, select pine, 1x3	L.F.	4.98

DIVISION 6 WOOD AND PLASTICS

	Unit	Total
06200.10 FINISH CARPENTRY (Cont.)		
Douglas fir		
1x3	L.F.	3.71
1x4	L.F.	3.99
1x6	L.F.	4.72
1x8	L.F.	5.74
Cornices, white pine, #2 or better		
1x2	L.F.	3.21
1x4	L.F.	3.43
1x6	L.F.	4.28
1x8	L.F.	4.72
1x10	L.F.	5.46
1x12	L.F.	6.23
Shelving, pine		
1x8	L.F.	5.21
1x10	L.F.	5.79
1x12	L.F.	6.39
Plywood shelf, 3/4", with edge band, 12" wide	L.F.	7.42
Adjustable shelf, and rod, 12" wide		
3' to 4' long	EA.	30.50
5' to 8' long	EA.	50.50
Prefinished wood shelves with brackets and supports		
8" wide		
3' long	EA.	64.75
4' long	EA.	72.50
6' long	EA.	98.75
10" wide		
3' long	EA.	70.25
4' long	EA.	95.50
6' long	EA.	110
06220.10 MILLWORK		
Countertop, laminated plastic		
25" x 7/8" thick		
Minimum	L.F.	26.75
Average	L.F.	43.50
Maximum	L.F.	59.50
25" x 1-1/4" thick		
Minimum	L.F.	34.00
Average	L.F.	54.25
Maximum	L.F.	76.00
Add for cutouts	EA.	32.75
Backsplash, 4" high, 7/8" thick	L.F.	28.75
Plywood, sanded, A-C		
1/4" thick	S.F.	2.93
3/8" thick	S.F.	3.16
1/2" thick	S.F.	3.44
A-D		
1/4" thick	S.F.	2.84
3/8" thick	S.F.	3.10
1/2" thick	S.F.	3.39
Base cab., 34-1/2" high, 24" deep, hardwood		
Minimum	L.F.	200
Average	L.F.	240
Maximum	L.F.	270
Wall cabinets		
Minimum	L.F.	72.50
Average	L.F.	95.25
Maximum	L.F.	120
06300.10 WOOD TREATMENT		
Creosote preservative treatment		
8 lb/cf	B.F.	.59
10 lb/cf	B.F.	.72
Salt preservative treatment		
Oil borne		
Minimum	B.F.	.44
Maximum	B.F.	.72
Water borne		
Minimum	B.F.	.33
Maximum	B.F.	.55
Fire retardant treatment		
Minimum	B.F.	.74
Maximum	B.F.	.88
Kiln dried, softwood, add to framing costs		
1" thick	B.F.	.22
2" thick	B.F.	.32
3" thick	B.F.	.45
4" thick	B.F.	.58

	Unit	Total
06420.10 PANEL WORK		
Hardboard, tempered, 1/4" thick		
Natural faced	S.F.	2.12
Plastic faced	S.F.	2.75
Pegboard, natural	S.F.	2.29
Plastic faced	S.F.	2.75
Untempered, 1/4" thick		
Natural faced	S.F.	2.07
Plastic faced	S.F.	2.85
Pegboard, natural	S.F.	2.10
Plastic faced	S.F.	2.72
Plywood unfinished, 1/4" thick		
Birch		
Natural	S.F.	2.62
Select	S.F.	3.06
Knotty pine	S.F.	3.50
Cedar (closet lining)		
Standard boards T&G	S.F.	3.94
Particle board	S.F.	3.06
Plywood, prefinished, 1/4" thick, premium grade		
Birch veneer	S.F.	5.23
Cherry veneer	S.F.	5.77
Chestnut veneer	S.F.	9.34
Lauan veneer	S.F.	3.47
Mahogany veneer	S.F.	5.77
Oak veneer (red)	S.F.	5.77
Pecan veneer	S.F.	6.77
Rosewood veneer	S.F.	9.34
Teak veneer	S.F.	6.87
Walnut veneer	S.F.	6.21
06430.10 STAIRWORK		
Risers, 1x8, 42" wide		
White oak	EA.	64.50
Pine	EA.	59.00
Treads, 1-1/16" x 9-1/2" x 42"		
White oak	EA.	79.00
06440.10 COLUMNS		
Column, hollow, round wood		
12" diameter		
10' high	EA.	750
12' high	EA.	910
14' high	EA.	1,100
16' high	EA.	1,350
24" diameter		
16' high	EA.	2,950
18' high	EA.	3,300
20' high	EA.	4,050
22' high	EA.	4,250
24' high	EA.	4,650

	Unit	Total

07100.10 WATERPROOFING
Membrane waterproofing, elastomeric
Butyl

	Unit	Total
1/32" thick	S.F.	2.68
1/16" thick	S.F.	3.07

Neoprene

1/32" thick	S.F.	3.42
1/16" thick	S.F.	4.64

Plastic vapor barrier (polyethylene)

4 mil	S.F.	.20
6 mil	S.F.	.23
10 mil	S.F.	.30

Bituminous membrane, asphalt felt, 15 lb.

One ply	S.F.	1.64
Two ply	S.F.	1.97
Three ply	S.F.	2.36

Bentonite waterproofing, panels

3/16" thick	S.F.	2.62
1/4" thick	S.F.	2.84

07150.10 DAMPPROOFING
Silicone dampproofing, sprayed on
Concrete surface

1 coat	S.F.	.88
2 coats	S.F.	1.38

Concrete block

1 coat	S.F.	.92
2 coats	S.F.	1.44

Brick

1 coat	S.F.	1.06
2 coats	S.F.	1.57

07160.10 BITUMINOUS DAMPPROOFING
Building paper, asphalt felt

15 lb	S.F.	1.77
30 lb	S.F.	1.98

Asphalt, troweled, cold, primer plus

1 coat	S.F.	2.03
2 coats	S.F.	3.46
3 coats	S.F.	4.59

Fibrous asphalt, hot troweled, primer plus

1 coat	S.F.	2.30
2 coats	S.F.	3.69
3 coats	S.F.	4.95

Asphaltic paint dampproofing, per coat

Brush on	S.F.	.93
Spray on	S.F.	.95

07190.10 VAPOR BARRIERS
Vapor barrier, polyethylene

2 mil	S.F.	.21
6 mil	S.F.	.26
8 mil	S.F.	.30
10 mil	S.F.	.31

07210.10 BATT INSULATION
Ceiling, fiberglass, unfaced

3-1/2" thick, R11	S.F.	.86
6" thick, R19	S.F.	1.05
9" thick, R30	S.F.	1.62

Suspended ceiling, unfaced

3-1/2" thick, R11	S.F.	.83
6" thick, R19	S.F.	1.02
9" thick, R30	S.F.	1.58

Crawl space, unfaced

3-1/2" thick, R11	S.F.	1.00
6" thick, R19	S.F.	1.19
9" thick, R30	S.F.	1.74

Wall, fiberglass
Paper backed

2" thick, R7	S.F.	.68
3" thick, R8	S.F.	.73
4" thick, R11	S.F.	.93
6" thick, R19	S.F.	1.18

Foil backed, 1 side

2" thick, R7	S.F.	1.01
3" thick, R11	S.F.	1.07
4" thick, R14	S.F.	1.13
6" thick, R21	S.F.	1.36

Foil backed, 2 sides

2" thick, R7	S.F.	1.14
3" thick, R11	S.F.	1.35

07210.10 BATT INSULATION (Cont.)

	Unit	Total
4" thick, R14	S.F.	1.53
6" thick, R21	S.F.	1.65

Unfaced

2" thick, R7	S.F.	.80
3" thick, R9	S.F.	.87
4" thick, R11	S.F.	.93
6" thick, R19	S.F.	1.09

Mineral wool batts
Paper backed

2" thick, R6	S.F.	.67
4" thick, R12	S.F.	.98
6" thick, R19	S.F.	1.18

Fasteners, self adhering, attached to ceiling deck

2-1/2" long	EA.	.88
4-1/2" long	EA.	.96
Capped, self-locking washers	EA.	.61

07210.20 BOARD INSULATION
Perlite board, roof

1.00" thick, R2.78	S.F.	.92
1.50" thick, R4.17	S.F.	1.25

Rigid urethane

1" thick, R6.67	S.F.	1.37
1.50" thick, R11.11	S.F.	1.76

Polystyrene

1.0" thick, R4.17	S.F.	.76
1.5" thick, R6.26	S.F.	.99

07210.60 LOOSE FILL INSULATION
Blown-in type
Fiberglass

5" thick, R11	S.F.	.70
6" thick, R13	S.F.	.83
9" thick, R19	S.F.	1.09

Rockwool, attic application

6" thick, R13	S.F.	.74
8" thick, R19	S.F.	.91
10" thick, R22	S.F.	1.09
12" thick, R26	S.F.	1.27
15" thick, R30	S.F.	1.53

Poured type
Fiberglass

1" thick, R4	S.F.	.58
2" thick, R8	S.F.	.91
3" thick, R12	S.F.	1.24
4" thick, R16	S.F.	1.60

Mineral wool

1" thick, R3	S.F.	.61
2" thick, R6	S.F.	.96
3" thick, R9	S.F.	1.36
4" thick, R12	S.F.	1.60

Vermiculite or perlite

2" thick, R4.8	S.F.	1.06
3" thick, R7.2	S.F.	1.44
4" thick, R9.6	S.F.	1.84

Masonry, poured vermiculite or perlite

4" block	S.F.	.53
6" block	S.F.	.67
8" block	S.F.	.89
10" block	S.F.	1.03
12" block	S.F.	1.22

07210.70 SPRAYED INSULATION
Foam, sprayed on
Polystyrene

1" thick, R4	S.F.	1.03
2" thick, R8	S.F.	1.74

Urethane

1" thick, R4	S.F.	.99
2" thick, R8	S.F.	1.65

07310.10 ASPHALT SHINGLES
Standard asphalt shingles, strip shingles

210 lb/square	SQ.	120
235 lb/square	SQ.	130
240 lb/square	SQ.	140
260 lb/square	SQ.	180
300 lb/square	SQ.	200
385 lb/square	SQ.	260

DIVISION 7 MOISTURE AND THERMAL PROTECTION

	Unit	Total
07310.10 ASPHALT SHINGLES (Cont.)		
Roll roofing, mineral surface		
90 lb	SQ.	77.50
110 lb	SQ.	110
140 lb	SQ.	120
07310.50 METAL SHINGLES		
Aluminum, .020" thick		
Plain	SQ.	360
Colors	SQ.	380
Steel, galvanized		
26 ga.		
Plain	SQ.	360
Colors	SQ.	440
24 ga.		
Plain	SQ.	380
Colors	SQ.	450
Porcelain enamel, 22 ga.		
Minimum	SQ.	930
Average	SQ.	1,050
Maximum	SQ.	1,200
07310.60 SLATE SHINGLES		
Slate shingles		
Pennsylvania		
Ribbon	SQ.	880
Clear	SQ.	1,050
Vermont		
Black	SQ.	930
Gray	SQ.	1000
Green	SQ.	1,000
Red	SQ.	1,650
07310.70 WOOD SHINGLES		
Wood shingles, on roofs		
White cedar, #1 shingles		
4" exposure	SQ.	400
5" exposure	SQ.	340
#2 shingles		
4" exposure	SQ.	340
5" exposure	SQ.	270
Resquared and rebutted		
4" exposure	SQ.	380
5" exposure	SQ.	300
On walls		
White cedar, #1 shingles		
4" exposure	SQ.	480
5" exposure	SQ.	410
6" exposure	SQ.	340
#2 shingles		
4" exposure	SQ.	420
5" exposure	SQ.	340
6" exposure	SQ.	290
Add for fire retarding	SQ.	110
07310.80 WOOD SHAKES		
Shakes, hand split, 24" red cedar, on roofs		
5" exposure	SQ.	510
7" exposure	SQ.	450
9" exposure	SQ.	400
On walls		
6" exposure	SQ.	500
8" exposure	SQ.	440
10" exposure	SQ.	390
Add for fire retarding	SQ.	74.75
07460.10 METAL SIDING PANELS		
Aluminum siding panels		
Corrugated		
Plain finish		
.024"	S.F.	4.09
.032"	S.F.	4.41
Painted finish		
.024"	S.F.	4.53
.032"	S.F.	4.85
Steel siding panels		
Corrugated		
22 ga.	S.F.	6.13
24 ga.	S.F.	5.93

	Unit	Total
07460.50 PLASTIC SIDING		
Horizontal vinyl siding, solid		
8" wide		
Standard	S.F.	3.24
Insulated	S.F.	3.51
10" wide		
Standard	S.F.	3.14
Insulated	S.F.	3.39
Vinyl moldings for doors and windows	L.F.	2.88
07460.60 PLYWOOD SIDING		
Rough sawn cedar, 3/8" thick	S.F.	3.53
Fir, 3/8" thick	S.F.	2.73
Texture 1-11, 5/8" thick		
Cedar	S.F.	4.28
Fir	S.F.	3.55
Redwood	S.F.	4.39
Southern Yellow Pine	S.F.	3.24
07460.70 STEEL SIDING		
Ribbed, sheets, galvanized		
22 ga.	S.F.	4.60
24 ga.	S.F.	4.38
Primed		
24 ga.	S.F.	5.00
26 ga.	S.F.	4.22
07460.80 WOOD SIDING		
Beveled siding, cedar		
A grade		
1/2 x 8	S.F.	6.51
3/4 x 10	S.F.	7.43
Clear		
1/2 x 6	S.F.	7.43
1/2 x 8	S.F.	7.02
3/4 x 10	S.F.	8.34
B grade		
1/2 x 6	S.F.	7.27
1/2 x 8	S.F.	26.25
3/4 x 10	S.F.	6.69
Board and batten		
Cedar		
1x6	S.F.	7.74
1x8	S.F.	6.75
1x10	S.F.	6.07
1x12	S.F.	5.45
Pine		
1x6	S.F.	3.91
1x8	S.F.	3.35
1x10	S.F.	3.07
1x12	S.F.	2.80
Redwood		
1x6	S.F.	8.18
1x8	S.F.	7.27
1x10	S.F.	6.67
1x12	S.F.	6.12
Tongue and groove		
Cedar		
1x4	S.F.	7.71
1x6	S.F.	7.38
1x8	S.F.	6.94
1x10	S.F.	6.76
Pine		
1x4	S.F.	4.34
1x6	S.F.	4.11
1x8	S.F.	3.89
1x10	S.F.	3.70
Redwood		
1x4	S.F.	7.99
1x6	S.F.	7.66
1x8	S.F.	7.34
1x10	S.F.	7.00
07510.10 BUILT-UP ASPHALT ROOFING		
Built-up roofing, asphalt felt, including gravel		
2 ply	SQ.	210
3 ply	SQ.	280
4 ply	SQ.	360
Cant strip, 4" x 4"		
Treated wood	L.F.	3.35
Foamglass	L.F.	2.90
New gravel for built-up roofing, 400 lb/sq	SQ.	130

	Unit	Total

07530.10 SINGLE-PLY ROOFING

Elastic sheet roofing
Neoprene, 1/16" thick	S.F.	3.47
PVC		
45 mil	S.F.	2.67

Flashing
Pipe flashing, 90 mil thick		
1" pipe	EA.	41.00
Neoprene flashing, 60 mil thick strip		
6" wide	L.F.	5.92
12" wide	L.F.	9.69

07610.10 METAL ROOFING

Sheet metal roofing, copper, 16 oz, batten seam	SQ.	1,550
Standing seam	SQ.	1,450

Aluminum roofing, natural finish
Corrugated, on steel frame
.0175" thick	SQ.	260
.0215" thick	SQ.	300
.024" thick	SQ.	340
.032" thick	SQ.	380

V-beam, on steel frame
.032" thick	SQ.	390
.040" thick	SQ.	410
.050" thick	SQ.	480

Ridge cap
.019" thick	L.F.	5.52

Corrugated galvanized steel roofing, on steel frame
28 ga.	SQ.	320
26 ga.	SQ.	350
24 ga.	SQ.	380
22 ga.	SQ.	400
26 ga., factory insulated with 1" polystyrene	SQ.	600

Ridge roll
10" wide	L.F.	3.58
20" wide	L.F.	5.90

07620.10 FLASHING AND TRIM

Counter flashing
Aluminum, .032"	S.F.	6.60
Stainless steel, .015"	S.F.	10.00

Copper
16 oz.	S.F.	12.50
20 oz.	S.F.	14.00
24 oz.	S.F.	15.75
32 oz.	S.F.	18.25

Valley flashing
Aluminum, .032"	S.F.	4.72
Stainless steel, .015	S.F.	8.20

Copper
16 oz.	S.F.	10.75
20 oz.	S.F.	13.00
24 oz.	S.F.	14.00
32 oz.	S.F.	16.50

Base flashing
Aluminum, .040"	S.F.	6.79
Stainless steel, .018"	S.F.	10.25

Copper
16 oz.	S.F.	11.75
20 oz.	S.F.	12.00
24 oz.	S.F.	15.00
32 oz.	S.F.	17.50

Flashing and trim, aluminum
.019" thick	S.F.	4.87
.032" thick	S.F.	5.15
.040" thick	S.F.	6.56
Neoprene sheet flashing, .060" thick	S.F.	5.28

Copper, paper backed
2 oz.	S.F.	7.22
5 oz.	S.F.	7.86

07620.20 GUTTERS AND DOWNSPOUTS

Copper gutter and downspout
Downspouts, 16 oz. copper
Round
3" dia.	L.F.	13.75
4" dia.	L.F.	16.25

Rectangular, corrugated
2" x 3"	L.F.	13.50
3" x 4"	L.F.	15.75

	Unit	Total

07620.20 GUTTERS AND DOWNSPOUTS (Cont.)

Rectangular, flat surface
2" x 3"	L.F.	14.75
3" x 4"	L.F.	19.75

Lead-coated copper downspouts
Round
3" dia.	L.F.	17.00
4" dia.	L.F.	20.00

Rectangular, corrugated
2" x 3"	L.F.	17.00
3" x 4"	L.F.	19.75

Rectangular, plain
2" x 3"	L.F.	13.00
3" x 4"	L.F.	14.25

Gutters, 16 oz. copper
Half round
4" wide	L.F.	14.50
5" wide	L.F.	17.00

Type K
4" wide	L.F.	15.50
5" wide	L.F.	16.50

Lead-coated copper gutters
Half round
4" wide	L.F.	16.50
6" wide	L.F.	21.25

Type K
4" wide	L.F.	17.50
5" wide	L.F.	21.75

Aluminum gutter and downspout
Downspouts
2" x 3"	L.F.	4.44
3" x 4"	L.F.	5.09
4" x 5"	L.F.	5.53

Round
3" dia.	L.F.	5.19
4" dia.	L.F.	5.95

Gutters, stock units
4" wide	L.F.	7.07
5" wide	L.F.	7.70

Galvanized steel gutter and downspout
Downspouts, round corrugated
3" dia.	L.F.	5.19
4" dia.	L.F.	5.83
5" dia.	L.F.	7.28
6" dia.	L.F.	8.49

Rectangular
2" x 3"	L.F.	5.01
3" x 4"	L.F.	5.54
4" x 4"	L.F.	6.14

Gutters, stock units
5" wide
Plain	L.F.	7.19
Painted	L.F.	7.33

6" wide
Plain	L.F.	8.16
Painted	L.F.	8.43

07810.10 PLASTIC SKYLIGHTS

Single thickness, not including mounting curb
2' x 4'	EA.	420
4' x 4'	EA.	570
5' x 5'	EA.	780
6' x 8'	EA.	1,550

Double thickness, not including mounting curb
2' x 4'	EA.	540
4' x 4'	EA.	680
5' x 5'	EA.	1,000
6' x 8'	EA.	1,700

07920.10 CAULKING

Caulk exterior, two component
1/4 x 1/2	L.F.	3.01
3/8 x 1/2	L.F.	3.50
1/2 x 1/2	L.F.	4.08

Caulk interior, single component
1/4 x 1/2	L.F.	2.75
3/8 x 1/2	L.F.	3.12
1/2 x 1/2	L.F.	3.57

DIVISION 8 DOORS AND WINDOWS

	Unit	Total		Unit	Total
08110.10 METAL DOORS			**08210.10 WOOD DOORS (Cont.)**		
Flush hollow metal, std. duty, 20 ga., 1-3/8" thick			Lauan faced		
2-6 x 6-8	EA	360	2-4 x 6-8	EA	120
2-8 x 6-8	EA	400	2-6 x 6-8	EA	130
3-0 x 6-8	EA	420	2-8 x 6-8	EA	140
1-3/4" thick			3-0 x 6-8	EA	150
2-6 x 6-8	EA	420	3-4 x 6-8	EA	160
2-8 x 6-8	EA	440	Tempered hardboard faced		
3-0 x 6-8	EA	470	2-4 x 7-0	EA	140
2-6 x 7-0	EA	450	2-6 x 7-0	EA	140
2-8 x 7-0	EA	470	2-8 x 7-0	EA	150
3-0 x 7-0	EA	490	3-0 x 7-0	EA	160
Heavy duty, 20 ga., unrated, 1-3/4"			3-4 x 7-0	EA	160
2-8 x 6-8	EA	450	Solid core, 1-3/4" thick		
3-0 x 6-8	EA	480	Birch faced		
2-8 x 7-0	EA	510	2-4 x 7-0	EA	300
3-0 x 7-0	EA	490	2-6 x 7-0	EA	310
3-4 x 7-0	EA	510	2-8 x 7-0	EA	320
18 ga., 1-3/4", unrated door			3-0 x 7-0	EA	300
2-0 x 7-0	EA	480	3-4 x 7-0	EA	310
2-4 x 7-0	EA	480	Lauan faced		
2-6 x 7-0	EA	480	2-4 x 7-0	EA	230
2-8 x 7-0	EA	520	2-6 x 7-0	EA	250
3-0 x 7-0	EA	530	2-8 x 7-0	EA	270
3-4 x 7-0	EA	540	3-4 x 7-0	EA	270
2", unrated door			3-0 x 7-0	EA	290
2-0 x 7-0	EA	490	Tempered hardboard faced		
2-4 x 7-0	EA	490	2-4 x 7-0	EA	280
2-6 x 7-0	EA	490	2-6 x 7-0	EA	310
2-8 x 7-0	EA	530	2-8 x 7-0	EA	330
3-0 x 7-0	EA	540	3-0 x 7-0	EA	340
3-4 x 7-0	EA	550	3-4 x 7-0	EA	360
08110.40 METAL DOOR FRAMES			Hollow core, 1-3/4" thick		
Hollow metal, stock, 18 ga., 4-3/4" x 1-3/4"			Birch faced		
2-0 x 7-0	EA	210	2-4 x 7-0	EA	230
2-4 x 7-0	EA	220	2-6 x 7-0	EA	230
2-6 x 7-0	EA	220	2-8 x 7-0	EA	240
2-8 x 7-0	EA	220	3-0 x 7-0	EA	240
3-0 x 7-0	EA	230	3-4 x 7-0	EA	250
4-0 x 7-0	EA	260	Lauan faced		
5-0 x 7-0	EA	270	2-4 x 6-8	EA	160
6-0 x 7-0	EA	300	2-6 x 6-8	EA	180
16 ga., 6-3/4" x 1-3/4"			2-8 x 6-8	EA	160
2-0 x 7-0	EA	230	3-0 x 6-8	EA	170
2-4 x 7-0	EA	220	3-4 x 6-8	EA	180
2-6 x 7-0	EA	220	Tempered hardboard		
2-8 x 7-0	EA	220	2-4 x 7-0	EA	150
3-0 x 7-0	EA	230	2-6 x 7-0	EA	160
4-0 x 7-0	EA	290	2-8 x 7-0	EA	160
6-0 x 7-0	EA	320	3-0 x 7-0	EA	170
08210.10 WOOD DOORS			3-4 x 7-0	EA	180
Solid core, 1-3/8" thick			Add-on, louver	EA	87.25
Birch faced			Glass	EA	160
2-4 x 7-0	EA	220	Exterior doors, 3-0 x 7-0 x 2-1/2", solid core		
2-8 x 7-0	EA	230	Carved		
3-0 x 7-0	EA	230	One face	EA	1,300
3-4 x 7-0	EA	380	Two faces	EA	1,750
2-4 x 6-8	EA	220	Closet doors, 1-3/4" thick		
2-6 x 6-8	EA	220	Bi-fold or bi-passing, includes frame and trim		
2-8 x 6-8	EA	230	Paneled		
3-0 x 6-8	EA	230	4-0 x 6-8	EA	550
Lauan faced			6-0 x 6-8	EA	620
2-4 x 6-8	EA	210	Louvered		
2-8 x 6-8	EA	220	4-0 x 6-8	EA	410
3-0 x 6-8	EA	220	6-0 x 6-8	EA	470
3-4 x 6-8	EA	230	Flush		
Tempered hardboard faced			4-0 x 6-8	EA	320
2-4 x 7-0	EA	240	6-0 x 6-8	EA	390
2-8 x 7-0	EA	250	Primed		
3-0 x 7-0	EA	280	4-0 x 6-8	EA	350
3-4 x 7-0	EA	280	6-0 x 6-8	EA	380
Hollow core, 1-3/8" thick			**08210.90 WOOD FRAMES**		
Birch faced			Frame, interior, pine		
2-4 x 7-0	EA	210	2-6 x 6-8	EA	150
2-8 x 7-0	EA	210	2-8 x 6-8	EA	150
3-0 x 7-0	EA	220	3-0 x 6-8	EA	150
3-4 x 7-0	EA	230	5-0 x 6-8	EA	160
			6-0 x 6-8	EA	160

08210.90 WOOD FRAMES (Cont.)

	Unit	Total
2-6 x 7-0	EA.	160
2-8 x 7-0	EA.	170
3-0 x 7-0	EA.	170
5-0 x 7-0	EA.	210
6-0 x 7-0	EA.	210
Exterior, custom, with threshold, including trim		
Walnut		
3-0 x 7-0	EA.	470
6-0 x 7-0	EA.	520
Oak		
3-0 x 7-0	EA.	440
6-0 x 7-0	EA.	490
Pine		
2-4 x 7-0	EA.	230
2-6 x 7-0	EA.	230
2-8 x 7-0	EA.	240
3-0 x 7-0	EA.	240
3-4 x 7-0	EA.	260
6-0 x 7-0	EA.	340

08300.10 SPECIAL DOORS

	Unit	Total
Sliding glass doors		
Tempered plate glass, 1/4" thick		
6' wide		
Economy grade	EA.	1,150
Premium grade	EA.	1,300
12' wide		
Economy grade	EA.	1,700
Premium grade	EA.	2,400
Insulating glass, 5/8" thick		
6' wide		
Economy grade	EA.	1,400
Premium grade	EA.	1,750
12' wide		
Economy grade	EA.	1,800
Premium grade	EA.	2,750
1" thick		
6' wide		
Economy grade	EA.	1,750
Premium grade	EA.	1,950
12' wide		
Economy grade	EA.	2,700
Premium grade	EA.	3,850
Added costs		
Custom quality, add to material, 30%		
Tempered glass, 6' wide, add	S.F.	4.40
Residential storm door		
Minimum	EA.	260
Average	EA.	310
Maximum	EA.	610

08520.10 ALUMINUM WINDOWS

	Unit	Total
Jalousie		
3-0 x 4-0	EA.	390
3-0 x 5-0	EA.	440
Fixed window		
6 sf to 8 sf	S.F.	24.00
12 sf to 16 sf	S.F.	20.50
Projecting window		
6 sf to 8 sf	S.F.	49.50
12 sf to 16 sf	S.F.	41.50
Horizontal sliding		
6 sf to 8 sf	S.F.	30.00
12 sf to 16 sf	S.F.	26.75
Double hung		
6 sf to 8 sf	S.F.	43.25
10 sf to 12 sf	S.F.	37.75
Storm window, 0.5 cfm, up to		
60 u.i. (united inches)	EA.	100
70 u.i.	EA.	100
80 u.i.	EA.	110
90 u.i.	EA.	120
100 u.i.	EA.	120
2.0 cfm, up to		
60 u.i.	EA.	120
70 u.i.	EA.	130
80 u.i.	EA.	130
90 u.i.	EA.	140
100 u.i.	EA.	140

08600.10 WOOD WINDOWS

	Unit	Total
Double hung		
24" x 36"		
Minimum	EA.	250
Average	EA.	360
Maximum	EA.	480
24" x 48"		
Minimum	EA.	280
Average	EA.	410
Maximum	EA.	560
30" x 48"		
Minimum	EA.	300
Average	EA.	410
Maximum	EA.	590
30" x 60"		
Minimum	EA.	320
Average	EA.	490
Maximum	EA.	620
Casement		
1 leaf, 22" x 38" high		
Minimum	EA.	340
Average	EA.	420
Maximum	EA.	500
2 leaf, 50" x 50" high		
Minimum	EA.	850
Average	EA.	1,100
Maximum	EA.	1,300
3 leaf, 71" x 62" high		
Minimum	EA.	1,350
Average	EA.	1,400
Maximum	EA.	1,700
4 leaf, 95" x 75" high		
Minimum	EA.	1,750
Average	EA.	2,050
Maximum	EA.	2,650
5 leaf, 119" x 75" high		
Minimum	EA.	2,250
Average	EA.	2,450
Maximum	EA.	3,200
Picture window, fixed glass, 54" x 54" high		
Minimum	EA.	530
Average	EA.	580
Maximum	EA.	1000
68" x 55" high		
Minimum	EA.	890
Average	EA.	1,000
Maximum	EA.	1,350
Sliding, 40" x 31" high		
Minimum	EA.	320
Average	EA.	480
Maximum	EA.	590
52" x 39" high		
Minimum	EA.	410
Average	EA.	570
Maximum	EA.	650
64" x 72" high		
Minimum	EA.	590
Average	EA.	920
Maximum	EA.	1,000
Awning windows		
34" x 21" high		
Minimum	EA.	340
Average	EA.	400
Maximum	EA.	480
40" x 21" high		
Minimum	EA.	400
Average	EA.	440
Maximum	EA.	520
60" x 36" high		
Minimum	EA.	440
Average	EA.	750
Maximum	EA.	850
Window frame, milled		
Minimum	L.F.	15.75
Average	L.F.	19.00
Maximum	L.F.	26.25

DIVISION 8 DOORS AND WINDOWS

	Unit	Total
08710.10 HINGES		
Hinges		
3 x 3 butts, steel, interior, plain bearing	PAIR	20.75
4 x 4 butts, steel, standard	PAIR	30.50
5 x 4-1/2 butts, bronze/s. steel, heavy duty	PAIR	79.25
08710.20 LOCKSETS		
Latchset, heavy duty		
Cylindrical	EA.	190
Mortise	EA.	210
Lockset, heavy duty		
Cylindrical	EA.	280
Mortise	EA.	330
Lockset		
Privacy (bath or bedroom)	EA.	230
Entry lock	EA.	260
08710.30 CLOSERS		
Door closers		
Standard	EA.	290
Heavy duty	EA.	330
08710.40 DOOR TRIM		
Panic device		
Mortise	EA.	910
Vertical rod	EA.	1,300
Labeled, rim type	EA.	940
Mortise	EA.	1,200
Vertical rod	EA.	1,300
08710.60 WEATHERSTRIPPING		
Weatherstrip, head and jamb, metal strip, neoprene bulb		
Standard duty	L.F.	7.85
Heavy duty	L.F.	8.76
Spring type		
Metal doors	EA.	180
Wood doors	EA.	220
Sponge type with adhesive backing	EA.	100
Thresholds		
Bronze	L.F.	66.25
Aluminum		
Plain	L.F.	49.50
Vinyl insert	L.F.	52.25
Aluminum with grit	L.F.	50.50
Steel		
Plain	L.F.	42.75
Interlocking	L.F.	83.00
08810.10 GLAZING		
Sheet glass, 1/8" thick	S.F.	11.00
Plate glass, bronze or grey, 1/4" thick	S.F.	16.50
Clear	S.F.	14.00
Polished	S.F.	15.75
Plexiglass		
1/8" thick	S.F.	10.25
1/4" thick	S.F.	12.25
Float glass, clear		
3/16" thick	S.F.	10.75
1/4" thick	S.F.	11.50
3/8" thick	S.F.	19.50
Tinted glass, polished plate, twin ground		
3/16" thick	S.F.	13.00
1/4" thick	S.F.	13.50
3/8" thick	S.F.	20.50

	Unit	Total
08810.10 GLAZING (Cont.)		
Insulating glass, two lites, clear float glass		
1/2" thick	S.F.	21.75
5/8" thick	S.F.	25.25
3/4" thick	S.F.	29.75
7/8" thick	S.F.	32.50
1" thick	S.F.	36.25
Glass seal edge		
3/8" thick	S.F.	19.75
Tinted glass		
1/2" thick	S.F.	30.25
1" thick	S.F.	41.25
Tempered, clear		
1" thick	S.F.	59.50
Plate mirror glass		
1/4" thick		
15 sf	S.F.	16.25
Over 15 sf	S.F.	15.00
08910.10 GLAZED CURTAIN WALLS		
Curtain wall, aluminum system, framing sections		
2" x 3"		
Jamb	L.F.	15.75
Horizontal	L.F.	16.00
Mullion	L.F.	19.75
2" x 4"		
Jamb	L.F.	22.25
Horizontal	L.F.	22.75
Mullion	L.F.	22.25
3" x 5-1/2"		
Jamb	L.F.	27.00
Horizontal	L.F.	29.25
Mullion	L.F.	27.25
4" corner mullion	L.F.	36.25
Coping sections		
1/8" x 8"	L.F.	37.25
1/8" x 9"	L.F.	37.50
1/8" x 12-1/2"	L.F.	40.00
Sill section		
1/8" x 6"	L.F.	33.00
1/8" x 7"	L.F.	33.25
1/8" x 8-1/2"	L.F.	33.75
Column covers, aluminum		
1/8" x 26"	L.F.	41.75
1/8" x 34"	L.F.	42.75
1/8" x 38"	L.F.	43.00
Doors		
Aluminum framed, standard hardware		
Narrow stile		
2-6 x 7-0	EA.	910
3-0 x 7-0	EA.	910
3-6 x 7-0	EA.	930

	Unit	Total		Unit	Total

09110.10 METAL STUDS

Studs, non load bearing, galvanized
2-1/2", 20 ga.
	Unit	Total
12" o.c.	S.F.	1.81
16" o.c.	S.F.	1.42

25 ga.
	Unit	Total
12" o.c.	S.F.	1.57
16" o.c.	S.F.	1.25
24" o.c.	S.F.	1.02

3-5/8", 20 ga.
	Unit	Total
12" o.c.	S.F.	2.16
16" o.c.	S.F.	1.70
24" o.c.	S.F.	1.37

25 ga.
	Unit	Total
12" o.c.	S.F.	1.87
16" o.c.	S.F.	1.50
24" o.c.	S.F.	1.22

4", 20 ga.
	Unit	Total
12" o.c.	S.F.	2.24
16" o.c.	S.F.	1.76
24" o.c.	S.F.	1.42

25 ga.
	Unit	Total
12" o.c.	S.F.	1.94
16" o.c.	S.F.	1.54
24" o.c.	S.F.	1.24

6", 20 ga.
	Unit	Total
12" o.c.	S.F.	2.83
16" o.c.	S.F.	2.18
24" o.c.	S.F.	1.81

25 ga.
	Unit	Total
12" o.c.	S.F.	2.41
16" o.c.	S.F.	1.92
24" o.c.	S.F.	1.55

Load bearing studs, galvanized
3-5/8", 16 ga.
	Unit	Total
12" o.c.	S.F.	2.85
16" o.c.	S.F.	2.47

18 ga.
	Unit	Total
12" o.c.	S.F.	1.98
16" o.c.	S.F.	2.25

4", 16 ga.
	Unit	Total
12" o.c.	S.F.	2.93
16" o.c.	S.F.	2.52

6", 16 ga.
	Unit	Total
12" o.c.	S.F.	3.71
16" o.c.	S.F.	3.17

Furring
On beams and columns
	Unit	Total
7/8" channel	L.F.	4.03
1-1/2" channel	L.F.	4.68

On ceilings
3/4" furring channels
	Unit	Total
12" o.c.	S.F.	2.57
16" o.c.	S.F.	2.40
24" o.c.	S.F.	2.08

1-1/2" furring channels
	Unit	Total
12" o.c.	S.F.	3.03
16" o.c.	S.F.	2.67
24" o.c.	S.F.	2.35

On walls
3/4" furring channels
	Unit	Total
12" o.c.	S.F.	2.14
16" o.c.	S.F.	1.94
24" o.c.	S.F.	1.76

1-1/2" furring channels
	Unit	Total
12" o.c.	S.F.	2.52
16" o.c.	S.F.	2.24
24" o.c.	S.F.	1.97

09205.10 GYPSUM LATH

Gypsum lath, 1/2" thick
	Unit	Total
Clipped	S.Y.	10.00
Nailed	S.Y.	10.50

09205.20 METAL LATH

Diamond expanded, galvanized
2.5 lb., on walls
	Unit	Total
Nailed	S.Y.	10.75
Wired	S.Y.	11.75
On ceilings		

09205.20 METAL LATH (Cont.)

	Unit	Total
Nailed	S.Y.	11.75
Wired	S.Y.	13.00

3.4 lb., on walls
	Unit	Total
Nailed	S.Y.	12.25
Wired	S.Y.	13.25

On ceilings
	Unit	Total
Nailed	S.Y.	13.25
Wired	S.Y.	14.50

Flat rib
2.75 lb., on walls
	Unit	Total
Nailed	S.Y.	10.50
Wired	S.Y.	11.50

On ceilings
	Unit	Total
Nailed	S.Y.	11.50
Wired	S.Y.	12.75

3.4 lb., on walls
	Unit	Total
Nailed	S.Y.	11.25
Wired	S.Y.	12.25

On ceilings
	Unit	Total
Nailed	S.Y.	12.25
Wired	S.Y.	13.50

Stucco lath
	Unit	Total
1.8 lb.	S.Y.	11.50
3.6 lb.	S.Y.	12.00

Paper backed
	Unit	Total
Minimum	S.Y.	9.07
Maximum	S.Y.	13.75

09205.60 PLASTER ACCESSORIES

	Unit	Total
Expansion joint, 3/4", 26 ga., galv.	L.F.	2.78
Plaster corner beads, 3/4", galvanized	L.F.	1.91
Casing bead, expanded flange, galvanized	L.F.	1.86
Expanded wing, 1-1/4" wide, galvanized	L.F.	1.96
Joint clips for lath	EA.	.44
Metal base, galvanized, 2-1/2" high	L.F.	2.50
Stud clips for gypsum lath	EA.	.44
Tie wire galvanized, 18 ga., 25 lb. hank	EA.	47.00
Sound deadening board, 1/4"	S.F.	1.19

09210.10 PLASTER

Gypsum plaster, trowel finish, 2 coats
	Unit	Total
Ceilings	S.Y.	22.25
Walls	S.Y.	21.25

3 coats
	Unit	Total
Ceilings	S.Y.	30.75
Walls	S.Y.	28.25

Vermiculite plaster
2 coats
	Unit	Total
Ceilings	S.Y.	31.50
Walls	S.Y.	29.25

3 coats
	Unit	Total
Ceilings	S.Y.	41.25
Walls	S.Y.	38.00

Keenes cement plaster
2 coats
	Unit	Total
Ceilings	S.Y.	22.25
Walls	S.Y.	19.75

3 coats
	Unit	Total
Ceilings	S.Y.	24.75
Walls	S.Y.	22.25

	Unit	Total
On columns, add to installation, 50%	S.Y.	
Chases, fascia, and soffits, add to installation, 50%		S.Y.
Beams, add to installation, 50%	S.Y.	

Patch holes, average size holes
1 sf to 5 sf
	Unit	Total
Minimum	S.F.	11.00
Average	S.F.	12.75
Maximum	S.F.	15.25

Over 5 sf
	Unit	Total
Minimum	S.F.	7.62
Average	S.F.	9.82
Maximum	S.F.	11.00

Patch cracks
	Unit	Total
Minimum	S.F.	4.19
average	S.F.	5.05
Maximum	S.F.	7.62

	Unit	Total
09220.10 PORTLAND CEMENT PLASTER		
Stucco, portland, gray, 3 coat, 1" thick		
Sand finish	S.Y.	30.00
Trowel finish	S.Y.	31.00
White cement		
Sand finish	S.Y.	32.00
Trowel finish	S.Y.	34.50
Scratch coat		
For ceramic tile	S.Y.	7.96
For quarry tile	S.Y.	7.96
Portland cement plaster		
2 coats, 1/2"	S.Y.	15.75
3 coats, 7/8"	S.Y.	19.50
09250.10 GYPSUM BOARD		
Drywall, plasterboard, 3/8" clipped to		
Metal furred ceiling	S.F.	.93
Columns and beams	S.F.	1.65
Walls	S.F.	.87
Nailed or screwed to		
Wood framed ceiling	S.F.	.87
Columns and beams	S.F.	1.51
Walls	S.F.	.82
1/2", clipped to		
Metal furred ceiling	S.F.	.91
Columns and beams	S.F.	1.63
Walls	S.F.	.85
Nailed or screwed to		
Wood framed ceiling	S.F.	.85
Columns and beams	S.F.	1.49
Walls	S.F.	.80
5/8", clipped to		
Metal furred ceiling	S.F.	1.01
Columns and beams	S.F.	1.81
Walls	S.F.	.94
Nailed or screwed to		
Wood framed ceiling	S.F.	1.01
Columns and beams	S.F.	1.81
Walls	S.F.	.94
Vinyl faced, clipped to metal studs		
1/2"	S.F.	1.53
5/8"	S.F.	1.59
Add for		
Fire resistant	S.F.	.11
Water resistant	S.F.	.18
Water and fire resistant	S.F.	.22
Taping and finishing joints		
Minimum	S.F.	.39
Average	S.F.	.51
Maximum	S.F.	.62
Casing bead		
Minimum	L.F.	1.64
Average	L.F.	1.90
Maximum	L.F.	2.82
Corner bead		
Minimum	L.F.	1.65
Average	L.F.	1.95
Maximum	L.F.	2.86
09310.10 CERAMIC TILE		
Glazed wall tile, 4-1/4" x 4-1/4"		
Minimum	S.F.	5.93
Average	S.F.	7.89
Maximum	S.F.	18.25
Base, 4-1/4" high		
Minimum	L.F.	10.25
Average	L.F.	10.75
Maximum	L.F.	12.25
Unglazed floor tile		
Portland cem., cushion edge, face mtd		
1" x 1"	S.F.	11.25
2" x 2"	S.F.	11.25
4" x 4"	S.F.	10.75
Adhesive bed, with white grout		
1" x 1"	S.F.	11.25
4" x 4"	S.F.	11.25
Organic adhesive bed, thin set, back mounted		
1" x 1"	S.F.	11.25
2" x 2"	S.F.	11.25

	Unit	Total
09310.10 CERAMIC TILE (Cont.)		
For group 2 colors, add to material, 10%		
For group 3 colors, add to material, 20%		
For abrasive surface, add to material, 25%		
Unglazed wall tile		
Organic adhesive, face mounted cushion edge		
1" x 1"		
Minimum	S.F.	7.79
Average	S.F.	9.27
Maximum	S.F.	12.00
2" x 2"		
Minimum	S.F.	8.01
Average	S.F.	8.89
Maximum	S.F.	12.25
Back mounted		
1" x 1"		
Minimum	S.F.	7.79
Average	S.F.	9.27
Maximum	S.F.	12.00
2" x 2"		
Minimum	S.F.	8.01
Average	S.F.	8.89
Maximum	S.F.	12.25
For glazed finish, add to material, 25%		
For glazed mosaic, add to material, 100%		
For metallic colors, add to material, 125%		
For exterior wall use, add to total, 25%		
For exterior soffit, add to total, 25%		
For portland cement bed, add to total, 25%		
For dry set portland cement bed, add to total, 10%		
Ceramic accessories		
Towel bar, 24" long		
Minimum	EA.	33.50
Average	EA.	41.50
Maximum	EA.	87.75
Soap dish		
Minimum	EA.	41.50
Average	EA.	52.50
Maximum	EA.	78.00
09330.10 QUARRY TILE		
Floor		
4 x 4 x 1/2"	S.F.	12.50
6 x 6 x 1/2"	S.F.	12.00
6 x 6 x 3/4"	S.F.	13.25
Wall, applied to 3/4" portland cement bed		
4 x 4 x 1/2"	S.F.	14.25
6 x 6 x 3/4"	S.F.	13.00
Cove base		
5 x 6 x 1/2" straight top	L.F.	14.00
6 x 6 x 3/4" round top	L.F.	13.50
Stair treads 6 x 6 x 3/4"	L.F.	20.25
Window sill 6 x 8 x 3/4"	L.F.	17.00
For abrasive surface, add to material, 25%		
09410.10 TERRAZZO		
Floors on concrete, 1-3/4" thick, 5/8" topping		
Gray cement	S.F.	11.50
White cement	S.F.	12.00
Sand cushion, 3" thick, 5/8" top, 1/4"		
Gray cement	S.F.	13.50
White cement	S.F.	14.25
Monolithic terrazzo, 3-1/2" base slab,		
5/8" topping	S.F.	10.50
Terrazzo wainscot, cast-in-place, 1/2" thick	S.F.	20.25
Base, cast in place, terrazzo cove type, 6" high	L.F.	16.25
Curb, cast in place, 6" wide x 6" high,		
polished top	L.F.	35.50
For venetian type terrazzo, add to material, 10%		
For abrasive heavy duty terrazzo, add to material, 15%		
Divider strips		
Zinc	L.F.	1.51
Brass	L.F.	2.80
Stairs, cast-in-place, topping on concrete or metal		
1-1/2" thick treads, 12" wide	L.F.	31.75
Combined tread and riser	L.F.	73.00
Precast terrazzo, thin set		
Terrazzo tiles, non-slip surface		
9" x 9" x 1" thick	S.F.	26.00
12" x 12"		

DIVISION 9 FINISHES

	Unit	Total
09410.10 TERRAZZO (Cont.)		
1" thick	S.F.	27.00
1-1/2" thick	S.F.	28.25
18" x 18" x 1-1/2" thick	S.F.	34.75
24" x 24" x 1-1/2" thick	S.F.	41.50
For white cement, add to material, 10%		
For venetian type terrazzo, add to material, 25%		
Terrazzo wainscot		
12" x 12" x 1" thick	S.F.	22.00
18" x 18" x 1-1/2" thick	S.F.	30.00
Base		
6" high		
Straight	L.F.	17.25
Coved	L.F.	19.75
8" high		
Straight	L.F.	19.25
Coved	L.F.	21.75
Terrazzo curbs		
8" wide x 8" high	L.F.	55.25
6" wide x 6" high	L.F.	48.50
Precast terrazzo stair treads, 12" wide		
1-1/2" thick		
Diamond pattern	L.F.	51.00
Non-slip surface	L.F.	53.25
2" thick		
Diamond pattern	L.F.	53.25
Non-slip surface	L.F.	56.50
Stair risers, 1" thick to 6" high		
Straight sections	L.F.	19.25
Cove sections	L.F.	21.75
Combined tread and riser		
Straight sections		
1-1/2" tread, 3/4" riser	L.F.	75.25
3" tread, 1" riser	L.F.	87.00
Curved sections		
2" tread, 1" riser	L.F.	94.50
3" tread, 1" riser	L.F.	97.00
Stair stringers, notched for treads and risers		
1" thick	L.F.	49.25
2" thick	L.F.	55.00
Landings, structural, nonslip		
1-1/2" thick	S.F.	43.00
3" thick	S.F.	58.25
09510.10 CEILINGS AND WALLS		
Acoustical panels, suspension system not included		
Fiberglass panels		
5/8" thick		
2' x 2'	S.F.	2.28
2' x 4'	S.F.	2.02
3/4" thick		
2' x 2'	S.F.	3.47
2' x 4'	S.F.	2.99
Glass cloth faced fiberglass panels		
3/4" thick	S.F.	3.38
1" thick	S.F.	3.66
Mineral fiber panels		
5/8" thick		
2' x 2'	S.F.	1.87
2' x 4'	S.F.	1.70
3/4" thick		
2' x 2'	S.F.	2.27
2' x 4'	S.F.	2.10
Wood fiber panels		
1/2" thick		
2' x 2'	S.F.	2.43
2' x 4'	S.F.	2.26
5/8" thick		
2' x 2'	S.F.	2.60
2' x 4'	S.F.	2.43
Acoustical tiles, suspension system not included		
Fiberglass tile, 12" x 12"		
5/8" thick	S.F.	2.35
3/4" thick	S.F.	2.78
Glass cloth faced fiberglass tile		
3/4" thick	S.F.	4.40
3" thick	S.F.	4.93
Mineral fiber tile, 12" x 12"		
5/8" thick		
09510.10 CEILINGS AND WALLS (Cont.)		
Standard	S.F.	1.86
Vinyl faced	S.F.	2.69
3/4" thick		
Standard	S.F.	2.25
Vinyl faced	S.F.	3.15
Ceiling suspension systems		
T bar system		
2' x 4'	S.F.	1.68
2' x 2'	S.F.	1.83
Concealed Z bar suspension system, 12" module	S.F.	1.95
For 1-1/2" carrier channels, 4' o.c., add	S.F.	.38
Carrier channel for recessed light fixtures	S.F.	.69
09550.10 WOOD FLOORING		
Wood strip flooring, unfinished		
Fir floor		
C and better		
Vertical grain	S.F.	5.26
Flat grain	S.F.	5.06
Oak floor		
Minimum	S.F.	6.21
Average	S.F.	7.62
Maximum	S.F.	9.91
Maple floor		
25/32" x 2-1/4"		
Minimum	S.F.	5.97
Maximum	S.F.	7.56
33/32" x 3-1/4"		
Minimum	S.F.	8.06
Maximum	S.F.	8.78
Added costs		
For factory finish, add to material, 10%		
For random width floor, add to total, 20%		
For simulated pegs, add to total, 10%		
Wood block industrial flooring		
Creosoted		
2" thick	S.F.	5.55
2-1/2" thick	S.F.	5.97
3" thick	S.F.	6.25
Parquet, 5/16", white oak		
Finished	S.F.	11.75
Unfinished	S.F.	7.01
Gym floor, 2 ply felt, 25/32" maple, finished, in mastic	S.F.	10.75
Over wood sleepers	S.F.	12.00
Finishing, sand, fill, finish, and wax	S.F.	1.96
Refinish sand, seal, and 2 coats of polyurethane	S.F.	2.90
Clean and wax floors	S.F.	.46
09630.10 UNIT MASONRY FLOORING		
Clay brick		
9 x 4-1/2 x 3" thick		
Glazed	S.F.	12.50
Unglazed	S.F.	12.00
8 x 4 x 3/4" thick		
Glazed	S.F.	11.75
Unglazed	S.F.	11.75
For herringbone pattern, add to labor, 15%		
09660.10 RESILIENT TILE FLOORING		
Solid vinyl tile, 1/8" thick, 12" x 12"		
Marble patterns	S.F.	5.76
Solid colors	S.F.	7.08
Travertine patterns	S.F.	7.79
Conductive resilient flooring, vinyl tile		
1/8" thick, 12" x 12"	S.F.	8.20
09665.10 RESILIENT SHEET FLOORING		
Vinyl sheet flooring		
Minimum	S.F.	4.36
Average	S.F.	6.82
Maximum	S.F.	11.25
Cove, to 6"	L.F.	3.33
Fluid applied resilient flooring		
Polyurethane, poured in place, 3/8" thick	S.F.	14.75
Vinyl sheet goods, backed		
0.070" thick	S.F.	4.55
0.093" thick	S.F.	6.70
0.250" thick	S.F.	8.68

DIVISION 9 FINISHES

	Unit	Total
09678.10 RESILIENT BASE AND ACCESSORIES		
Wall base, vinyl		
4" high	L.F.	2.75
6" high	L.F.	3.12
09682.10 CARPET PADDING		
Carpet padding		
Foam rubber, waffle type, 0.3" thick	S.Y.	8.77
Jute padding		
Minimum	S.Y.	6.55
Average	S.Y.	8.05
Maximum	S.Y.	11.00
Sponge rubber cushion		
Minimum	S.Y.	7.32
Average	S.Y.	9.21
Maximum	S.Y.	12.25
Urethane cushion, 3/8" thick		
Minimum	S.Y.	7.32
Average	S.Y.	8.39
Maximum	S.Y.	10.50
09685.10 CARPET		
Carpet, acrylic		
24 oz., light traffic	S.Y.	20.25
28 oz., medium traffic	S.Y.	23.25
Nylon		
15 oz., light traffic	S.Y.	26.00
28 oz., medium traffic	S.Y.	32.25
Nylon		
28 oz., medium traffic	S.Y.	31.00
35 oz., heavy traffic	S.Y.	36.50
Wool		
30 oz., medium traffic	S.Y.	47.50
36 oz., medium traffic	S.Y.	49.75
42 oz., heavy traffic	S.Y.	64.00
Carpet tile		
Foam backed		
Minimum	S.F.	4.25
Average	S.F.	4.88
Maximum	S.F.	7.18
Tufted loop or shag		
Minimum	S.F.	4.52
Average	S.F.	5.35
Maximum	S.F.	8.04
Clean and vacuum carpet		
Minimum	S.Y.	.50
Average	S.Y.	.81
Maximum	S.Y.	1.16
09905.10 PAINTING PREPARATION		
Dropcloths		
Minimum	S.F.	.05
Average	S.F.	.07
Maximum	S.F.	.10
Masking		
Paper and tape		
Minimum	L.F.	.54
Average	L.F.	.68
Maximum	L.F.	.91
Doors		
Minimum	EA.	6.57
Average	EA.	8.77
Maximum	EA.	11.50
Windows		
Minimum	EA.	6.57
Average	EA.	8.77
Maximum	EA.	11.50
Sanding		
Walls and flat surfaces		
Minimum	S.F.	.35
Average	S.F.	.44
Maximum	S.F.	.52
Doors and windows		
Minimum	EA.	8.71
Average	EA.	13.00
Maximum	EA.	17.50
Trim		
Minimum	L.F.	.65
Average	L.F.	.87
Maximum	L.F.	1.16

	Unit	Total
09905.10 PAINTING PREPARATION (Cont.)		
Puttying		
Minimum	S.F.	.81
Average	S.F.	1.06
Maximum	S.F.	1.34
09910.05 EXT PAINTING, SITEWORK		
Concrete Block		
Roller		
First Coat		
Minimum	S.F.	.42
Average	S.F.	.51
Maximum	S.F.	.68
Second Coat		
Minimum	S.F.	.38
Average	S.F.	.45
Maximum	S.F.	.60
Spray		
First Coat		
Minimum	S.F.	.28
Average	S.F.	.30
Maximum	S.F.	.33
Second Coat		
Minimum	S.F.	.22
Average	S.F.	.25
Maximum	S.F.	.29
Fences, Chain Link		
Roller		
First Coat		
Minimum	S.F.	.48
Average	S.F.	.55
Maximum	S.F.	.61
Second Coat		
Minimum	S.F.	.33
Average	S.F.	.37
Maximum	S.F.	.44
Spray		
First Coat		
Minimum	S.F.	.25
Average	S.F.	.28
Maximum	S.F.	.31
Second Coat		
Minimum	S.F.	.21
Average	S.F.	.24
Maximum	S.F.	.25
Fences, Wood or Masonry		
Brush		
First Coat		
Minimum	S.F.	.71
Average	S.F.	.81
Maximum	S.F.	1.03
Second Coat		
Minimum	S.F.	.49
Average	S.F.	.56
Maximum	S.F.	.68
Roller		
First Coat		
Minimum	S.F.	.45
Average	S.F.	.51
Maximum	S.F.	.56
Second Coat		
Minimum	S.F.	.36
Average	S.F.	.41
Maximum	S.F.	.49
Spray		
First Coat		
Minimum	S.F.	.32
Average	S.F.	.37
Maximum	S.F.	.46
Second Coat		
Minimum	S.F.	.26
Average	S.F.	.29
Maximum	S.F.	.35
09910.15 EXT PAINTING, BUILDINGS		
Decks, Wood, Stained		
Brush		
First Coat		
Minimum	S.F.	.39
Average	S.F.	.42
Maximum	S.F.	.46

	Unit	Total

09910.15 EXT PAINTING, BUILDINGS (Cont.)

	Unit	Total
Second Coat		
Minimum	S.F.	.32
Average	S.F.	.33
Maximum	S.F.	.35
Roller		
First Coat		
Minimum	S.F.	.32
Average	S.F.	.33
Maximum	S.F.	.35
Second Coat		
Minimum	S.F.	.29
Average	S.F.	.30
Maximum	S.F.	.33
Spray		
First Coat		
Minimum	S.F.	.27
Average	S.F.	.28
Maximum	S.F.	.31
Second Coat		
Minimum	S.F.	.26
Average	S.F.	.27
Maximum	S.F.	.28
Doors, Wood		
Brush		
First Coat		
Minimum	S.F.	.93
Average	S.F.	1.17
Maximum	S.F.	1.44
Second Coat		
Minimum	S.F.	.78
Average	S.F.	.88
Maximum	S.F.	1.00
Roller		
First Coat		
Minimum	S.F.	.48
Average	S.F.	.57
Maximum	S.F.	.78
Second Coat		
Minimum	S.F.	.39
Average	S.F.	.42
Maximum	S.F.	.57
Spray		
First Coat		
Minimum	S.F.	.27
Average	S.F.	.31
Maximum	S.F.	.37
Second Coat		
Minimum	S.F.	.24
Average	S.F.	.26
Maximum	S.F.	.28
Gutters and Downspouts		
Brush		
First Coat		
Minimum	L.F.	.81
Average	L.F.	.91
Maximum	L.F.	1.03
Second Coat		
Minimum	L.F.	.60
Average	L.F.	.68
Maximum	L.F.	.81
Siding, Wood		
Roller		
First Coat		
Minimum	S.F.	.30
Average	S.F.	.33
Maximum	S.F.	.35
Second Coat		
Minimum	S.F.	.33
Average	S.F.	.35
Maximum	S.F.	.37
Spray		
First Coat		
Minimum	S.F.	.28
Average	S.F.	.30
Maximum	S.F.	.31
Second Coat		
Minimum	S.F.	.24

09910.15 EXT PAINTING, BUILDINGS (Cont.)

	Unit	Total
Average	S.F.	.28
Maximum	S.F.	.37
Stucco		
Roller		
First Coat		
Minimum	S.F.	.40
Average	S.F.	.43
Maximum	S.F.	.49
Second Coat		
Minimum	S.F.	.35
Average	S.F.	.38
Maximum	S.F.	.42
Spray		
First Coat		
Minimum	S.F.	.29
Average	S.F.	.32
Maximum	S.F.	.35
Second Coat		
Minimum	S.F.	.26
Average	S.F.	.28
Maximum	S.F.	.30
Trim		
Brush		
First Coat		
Minimum	L.F.	.38
Average	L.F.	.42
Maximum	L.F.	.49
Second Coat		
Minimum	L.F.	.32
Average	L.F.	.38
Maximum	L.F.	.49
Walls		
Roller		
First Coat		
Minimum	S.F.	.32
Average	S.F.	.32
Maximum	S.F.	.34
Second Coat		
Minimum	S.F.	.29
Average	S.F.	.30
Maximum	S.F.	.33
Spray		
First Coat		
Minimum	S.F.	.18
Average	S.F.	.20
Maximum	S.F.	.23
Second Coat		
Minimum	S.F.	.17
Average	S.F.	.19
Maximum	S.F.	.22
Windows		
Brush		
First Coat		
Minimum	S.F.	.98
Average	S.F.	1.15
Maximum	S.F.	1.42
Second Coat		
Minimum	S.F.	.86
Average	S.F.	.98
Maximum	S.F.	1.15

09910.25 EXT PAINTING, MISC.

	Unit	Total
Shakes		
Spray		
First Coat		
Minimum	S.F.	.34
Average	S.F.	.36
Maximum	S.F.	.38
Second Coat		
Minimum	S.F.	.32
Average	S.F.	.34
Maximum	S.F.	.36
Shingles, Wood		
Roller		
First Coat		
Minimum	S.F.	.42
Average	S.F.	.46
Maximum	S.F.	.50

	Unit	Total		Unit	Total

09910.25 EXT PAINTING, MISC. (Cont.)

	Unit	Total
Second Coat		
Minimum	S.F.	.33
Average	S.F.	.35
Maximum	S.F.	.37
Spray		
First Coat		
Minimum	L.F.	.31
Average	L.F.	.33
Maximum	L.F.	.35
Second Coat		
Minimum	L.F.	.26
Average	L.F.	.27
Maximum	L.F.	.28
Shutters and Louvres		
Brush		
First Coat		
Minimum	EA.	10.75
Average	EA.	13.25
Maximum	EA.	17.75
Second Coat		
Minimum	EA.	6.69
Average	EA.	8.20
Maximum	EA.	10.75
Spray		
First Coat		
Minimum	EA.	3.60
Average	EA.	4.30
Maximum	EA.	5.34
Second Coat		
Minimum	EA.	2.73
Average	EA.	3.60
Maximum	EA.	4.30
Stairs, metal		
Brush		
First Coat		
Minimum	S.F.	.74
Average	S.F.	.81
Maximum	S.F.	.91
Second Coat		
Minimum	S.F.	.49
Average	S.F.	.53
Maximum	S.F.	.60
Spray		
First Coat		
Minimum	S.F.	.41
Average	S.F.	.49
Maximum	S.F.	.52
Second Coat		
Minimum	S.F.	.34
Average	S.F.	.38
Maximum	S.F.	.45

09910.35 INT PAINTING, BUILDINGS

	Unit	Total
Acoustical Ceiling		
Roller		
First Coat		
Minimum	S.F.	.49
Average	S.F.	.60
Maximum	S.F.	.81
Second Coat		
Minimum	S.F.	.42
Average	S.F.	.49
Maximum	S.F.	.60
Spray		
First Coat		
Minimum	S.F.	.28
Average	S.F.	.30
Maximum	S.F.	.35
Second Coat		
Minimum	S.F.	.25
Average	S.F.	.26
Maximum	S.F.	.28
Cabinets and Casework		
Brush		
First Coat		
Minimum	S.F.	.68
Average	S.F.	.74
Maximum	S.F.	.81

09910.35 INT PAINTING, BUILDINGS (Cont.)

	Unit	Total
Second Coat		
Minimum	S.F.	.60
Average	S.F.	.63
Maximum	S.F.	.68
Spray		
First Coat		
Minimum	S.F.	.39
Average	S.F.	.44
Maximum	S.F.	.50
Second Coat		
Minimum	S.F.	.34
Average	S.F.	.36
Maximum	S.F.	.42
Ceilings		
Roller		
First Coat		
Minimum	S.F.	.35
Average	S.F.	.37
Maximum	S.F.	.39
Second Coat		
Minimum	S.F.	.30
Average	S.F.	.33
Maximum	S.F.	.35
Spray		
First Coat		
Minimum	S.F.	.24
Average	S.F.	.26
Maximum	S.F.	.27
Second Coat		
Minimum	S.F.	.21
Average	S.F.	.22
Maximum	S.F.	.24
Doors, Wood		
Brush		
First Coat		
Minimum	S.F.	.91
Average	S.F.	1.11
Maximum	S.F.	1.32
Second Coat		
Minimum	S.F.	.70
Average	S.F.	.77
Maximum	S.F.	.87
Spray		
First Coat		
Minimum	S.F.	.27
Average	S.F.	.31
Maximum	S.F.	.36
Second Coat		
Minimum	S.F.	.24
Average	S.F.	.26
Maximum	S.F.	.28
Trim		
Brush		
First Coat		
Minimum	L.F.	.37
Average	L.F.	.40
Maximum	L.F.	.45
Second Coat		
Minimum	L.F.	.31
Average	L.F.	.36
Maximum	L.F.	.45
Walls		
Roller		
First Coat		
Minimum	S.F.	.32
Average	S.F.	.32
Maximum	S.F.	.35
Second Coat		
Minimum	S.F.	.29
Average	S.F.	.30
Maximum	S.F.	.33
Spray		
First Coat		
Minimum	S.F.	.19
Average	S.F.	.21
Maximum	S.F.	.24

DIVISION 9 FINISHES

	Unit	Total		Unit	Total

09955.10 WALL COVERING (Cont.)

	Unit	Total
Second Coat		
Minimum	S.F.	.18
Average	S.F.	.20
Maximum	S.F.	.23

09955.10 WALL COVERING

	Unit	Total
Vinyl wall covering		
Medium duty	S.F.	1.57
Heavy duty	S.F.	2.57
Over pipes and irregular shapes		
Lightweight, 13 oz.	S.F.	2.47
Medium weight, 25 oz.	S.F.	2.86
Heavy weight, 34 oz.	S.F.	3.40
Cork wall covering		
1' x 1' squares		
1/4" thick	S.F.	5.77
1/2" thick	S.F.	6.97
3/4" thick	S.F.	7.69
Wall fabrics		
Natural fabrics, grass cloths		
Minimum	S.F.	2.23
Average	S.F.	2.47
Maximum	S.F.	6.38
Flexible gypsum coated wall fabric, fire resistant	S.F.	2.13
Vinyl corner guards		
3/4" x 3/4" x 8'	EA.	14.00
2-3/4" x 2-3/4" x 4'	EA.	11.00

09980.15 PAINT

	Unit	Total
Paint, enamel		
600 sf per gal.	GAL	49.50
550 sf per gal.	GAL	46.25
500 sf per gal.	GAL	33.00
450 sf per gal.	GAL	30.75
350 sf per gal.	GAL	29.75
Filler, 60 sf per gal.	GAL	35.25
Latex, 400 sf per gal.	GAL	33.00
Aluminum		
400 sf per gal.	GAL	44.00
500 sf per gal.	GAL	70.50
Red lead, 350 sf per gal.	GAL	61.50
Primer		
400 sf per gal.	GAL	29.75
300 sf per gal.	GAL	29.75
Latex base, interior, white	GAL	33.00
Sealer and varnish		
400 sf per gal.	GAL	30.75
425 sf per gal.	GAL	44.00
600 sf per gal.	GAL	57.25

DIVISION 10 SPECIALTIES

	Unit	Total
10185.10 SHOWER STALLS		
Shower receptors		
Precast, terrazzo		
32" x 32"	EA	550
32" x 48"	EA	720
Concrete		
32" x 32"	EA	300
48" x 48"	EA	360
Shower door, trim and hardware		
Economy, 24" wide, chrome, tempered glass	EA	330
Porcelain enameled steel, flush	EA	560
Baked enameled steel, flush	EA	350
Aluminum, tempered glass, 48" wide, sliding	EA	690
Folding	EA	660
Aluminum and tempered glass, molded plastic		
Complete with receptor and door		
32" x 32"	EA	890
36" x 36"	EA	990
40" x 40"	EA	1,150
10210.10 VENTS AND WALL LOUVERS		
Block vent, 8"x16"x4" alum., w/screen,		
mill finish	EA	170
Standard	EA	99.00
Vents w/screen, 4" deep, 8" wide, 5" high		
Modular	EA	110
Aluminum gable louvers	S.F.	27.00
Vent screen aluminum, 4" wide, continuous	L.F.	6.74
10290.10 PEST CONTROL		
Termite control, Under slab spraying		
Minimum	S.F.	1.18
Average	S.F.	1.30
Maximum	S.F.	1.89
10350.10 FLAGPOLES		
Installed in concrete base, Fiberglass		
25' high	EA	1,600
50' high	EA	5,350
Aluminum		
25' high	EA	1,600
50' high	EA	4,700
Bonderized steel		
25' high	EA	1,800
50' high	EA	3,900
Freestanding tapered, fiberglass		
30' high	EA	1,900
40' high	EA	2,650

	Unit	Total
10800.10 BATH ACCESSORIES		
Grab bar, 1-1/2" dia., stainless steel, wall mounted		
24" long	EA	94.50
36" long	EA	110
1" dia., stainless steel		
12" long	EA	65.75
24" long	EA	76.50
36" long	EA	88.50
Medicine cabinet, 16 x 22, baked enamel, lighted	EA	130
With mirror, lighted	EA	230
Mirror, 1/4" plate glass, up to 10 sf	S.F.	15.00
Mirror, stainless steel frame		
18"x24"	EA	93.00
18"x32"	EA	110
24"x30"	EA	120
24"x60"	EA	390
Soap dish, stainless steel, wall mounted	EA	170
Toilet tissue dispenser, stainless, wall mounted		
Single roll	EA	73.00
Towel bar, stainless steel		
18" long	EA	95.75
24" long	EA	120
30" long	EA	140
36" long	EA	150
Toothbrush and tumbler holder	EA	64.75

	Unit	Total
11010.10 MAINTENANCE EQUIPMENT		
Vacuum cleaning system		
3 valves		
1.5 hp	EA.	2,600
2.5 hp	EA.	2,850
5 valves	EA.	3,600
7 valves	EA.	4,550
11450.10 RESIDENTIAL EQUIPMENT		
Compactor, 4 to 1 compaction	EA.	920
Dishwasher, built-in		
2 cycles	EA.	920
4 or more cycles	EA.	2,400
Disposal		
Garbage disposer	EA.	330
Heaters, electric, built-in		
Ceiling type	EA.	580
Wall type		
Minimum	EA.	340
Maximum	EA.	650
Hood for range, 2-speed, vented		
30" wide	EA.	600
42" wide	EA.	1,100
Ice maker, automatic		
30 lb per day	EA.	1,650
50 lb per day	EA.	2,550
Folding access stairs, disappearing metal stair		
8' long	EA.	840
11' long	EA.	940
12' long	EA.	1,150
Wood frame, wood stair		
22" x 54" x 8'9" long	EA.	210
25" x 54" x 10' long	EA.	240
Ranges electric		
Built-in, 30", 1 oven	EA.	2,100
2 oven	EA.	2,300
Counter top, 4 burner, standard	EA.	1,200
With grill	EA.	2,900
Free standing, 21", 1 oven	EA.	1,250
30", 1 oven	EA.	1,850
2 oven	EA.	3,600
Water softener		
30 grains per gallon	EA.	1,200
70 grains per gallon	EA.	1,500
12302.10 CASEWORK		
Kitchen base cabinet, prefinished, 24" deep, 35" high		
12" wide	EA.	260
18" wide	EA.	320
24" wide	EA.	340
27" wide	EA.	390
36" wide	EA.	450
48" wide	EA.	500
Corner cabinet, 36" wide	EA.	520
Wall cabinet, 12" deep, 12" high		
30" wide	EA.	230
36" wide	EA.	270
15" high		
30" wide	EA.	280
36" wide	EA.	330
24" high		
30" wide	EA.	300
36" wide	EA.	330
30" high		
12" wide	EA.	230
18" wide	EA.	250
24" wide	EA.	260
27" wide	EA.	300
30" wide	EA.	330
36" wide	EA.	330
Corner cabinet, 30" high		
24" wide	EA.	340
30" wide	EA.	400
36" wide	EA.	430
Wardrobe	EA.	900
Vanity with top, laminated plastic		
24" wide	EA.	610
30" wide	EA.	720
36" wide	EA.	780
48" wide	EA.	910

	Unit	Total
12390.10 COUNTER TOPS		
Stainless steel, counter top, with backsplash	S.F.	130
Acid-proof, kemrock surface	S.F.	50.50
12500.10 WINDOW TREATMENT		
Drapery tracks, wall or ceiling mounted		
Basic traverse rod		
50 to 90"	EA.	75.75
84 to 156"	EA.	100
136 to 250"	EA.	120
165 to 312"	EA.	170
Traverse rod with stationary curtain rod		
30 to 50"	EA.	92.50
50 to 90"	EA.	100
84 to 156"	EA.	140
136 to 250"	EA.	160
Double traverse rod		
30 to 50"	EA.	100
50 to 84"	EA.	120
84 to 156"	EA.	150
136 to 250"	EA.	170
12510.10 BLINDS		
Venetian blinds		
2" slats	S.F.	31.00
1" slats	S.F.	32.75
13056.10 VAULTS		
Floor safes		
1.0 cf	EA.	820
1.3 cf	EA.	930
13121.10 PRE-ENGINEERED BUILDINGS		
Pre-engineered metal building, 40'x100'		
14' eave height	S.F.	11.00
16' eave height	S.F.	12.50
13200.10 STORAGE TANKS		
Oil storage tank, underground, single wall, no excv.		
Steel		
500 gals	EA.	3,400
1,000 gals	EA.	4,600
Fiberglass, double wall		
550 gals	EA.	9,150
1,000 gals	EA.	11,700
Above ground		
Steel, single wall		
275 gals	EA.	1,950
500 gals	EA.	4,750
1,000 gals	EA.	6,500
Fill cap	EA.	170
Vent cap	EA.	170
Level indicator	EA.	23

	Unit	Total		Unit	Total
15100.10 SPECIALTIES			**15240.10 VIBRATION CONTROL**		
Wall penetration			Vibration isolator, in-line, stainless connector		
Concrete wall, 6" thick			1/2"	EA...............	130
2" dia.	EA.............	13.50	3/4"	EA...............	140
4" dia.	EA.............	20.25	1"	EA...............	150
12" thick			1-1/4"	EA...............	190
2" dia.	EA.............	18.50	1-1/2"	EA...............	210
4" dia.	EA.............	29.00	2"	EA...............	250
15120.10 BACKFLOW PREVENTERS			2-1/2"	EA...............	360
Backflow preventer, flanged, cast iron, with valves			3"	EA...............	410
3" pipe	EA.............	3,050	4"	EA...............	520
4" pipe	EA.............	3,850	**15290.10 DUCTWORK INSULATION**		
Threaded			Fiberglass duct insulation, plain blanket		
3/4" pipe	EA...............	610	1-1/2" thick	S.F.............	.93
2" pipe	EA.............	1,100	2" thick	S.F.............	1.24
15140.11 PIPE HANGERS, LIGHT			With vapor barrier		
A band, black iron			1-1/2" thick	S.F.............	.96
1/2"	EA...............	5.12	2" thick	S.F.............	1.27
1"	EA...............	5.35	Rigid with vapor barrier		
1-1/4"	EA...............	5.64	2" thick	S.F.............	3.24
1-1/2"	EA...............	6.06	**15410.05 C.I. PIPE, ABOVE GROUND**		
2"	EA...............	6.56	No hub pipe		
2-1/2"	EA...............	7.75	1-1/2" pipe	L.F.............	10.00
3"	EA...............	8.82	2" pipe	L.F.............	10.75
4"	EA.............	10.50	3" pipe	L.F.............	14.25
Copper			4" pipe	L.F.............	20.50
1/2"	EA...............	5.75	No hub fittings, 1-1/2" pipe		
3/4"	EA...............	6.16	1/4 bend	EA.............	27.25
1"	EA...............	6.16	1/8 bend	EA.............	26.00
1-1/4"	EA...............	6.46	Sanitary tee	EA.............	40.00
1-1/2"	EA...............	6.99	Sanitary cross	EA.............	44.00
2"	EA...............	7.55	Plug	EA...............	4.47
2-1/2"	EA.............	10.50	Coupling	EA.............	16.00
3"	EA.............	11.25	Wye	EA.............	42.75
4"	EA.............	12.50	Tapped tee	EA.............	34.00
2 hole clips, galvanized			P-trap	EA.............	32.00
3/4"	EA...............	4.11	Tapped cross	EA.............	36.00
1"	EA...............	4.28	2" pipe		
1-1/4"	EA...............	4.50	1/4 bend	EA.............	32.25
1-1/2"	EA...............	4.74	1/8 bend	EA.............	30.50
2"	EA...............	5.06	Sanitary tee	EA.............	51.25
2-1/2"	EA...............	5.72	Sanitary cross	EA.............	60.00
3"	EA...............	6.43	Plug	EA...............	4.47
4"	EA...............	8.70	Coupling	EA.............	14.00
Perforated strap			Wye	EA.............	60.00
3/4"			Double wye	EA.............	66.50
Galvanized, 20 ga.	L.F...............	3.29	2x1-1/2" wye & 1/8 bend	EA.............	58.25
Copper, 22 ga.	L.F...............	3.52	Double wye & 1/8 bend	EA.............	66.50
J-Hooks			Test tee less 2" plug	EA.............	34.50
1/2"	EA...............	3.35	Tapped tee		
3/4"	EA...............	3.41	2"x2"	EA.............	38.00
1"	EA...............	3.56	2"x1-1/2"	EA.............	37.00
1-1/4"	EA...............	3.66	P-trap		
1-1/2"	EA...............	3.75	2"x2"	EA.............	36.50
2"	EA...............	3.78	Tapped cross		
3"	EA...............	4.08	2"x1-1/2"	EA.............	42.25
4"	EA...............	4.16	3" pipe		
PVC coated hangers, galvanized, 28 ga.			1/4 bend	EA.............	41.50
1-1/2" x 12"	EA...............	5.11	1/8 bend	EA.............	39.50
2" x 12"	EA...............	5.50	Sanitary tee	EA.............	51.75
3" x 12"	EA...............	5.98	3"x2" sanitary tee	EA.............	50.00
4" x 12"	EA...............	6.51	3"x1-1/2" sanitary tee	EA.............	50.75
Copper, 30 ga.			Sanitary cross	EA.............	81.25
1-1/2" x 12"	EA...............	5.61	3x2" sanitary cross	EA.............	77.50
2" x 12"	EA...............	6.21	Plug	EA...............	6.62
3" x 12"	EA...............	6.75	Coupling	EA.............	16.00
4" x 12"	EA...............	7.34	Wye	EA.............	65.00
Wire hook hangers			3x2" wye	EA.............	60.75
Black wire, 1/2" x			Double wye	EA.............	82.00
4"	EA...............	3.31	3x2" double wye	EA.............	77.00
6"	EA...............	3.54	3x2" wye & 1/8 bend	EA.............	57.25
Copper wire hooks			3x1-1/2" wye & 1/8 bend	EA.............	57.25
1/2" x			Double wye & 1/8 bend	EA.............	82.00
4"	EA...............	3.47	3x2" double wye & 1/8 bend	EA.............	77.00
6"	EA...............	3.71	3x2" reducer	EA.............	32.75
			Test tee, less 3" plug	EA.............	46.50
			Plug	EA...............	6.62
			3x3" tapped tee	EA.............	69.50

	Unit	Total		Unit	Total

15410.05 C.I. PIPE, ABOVE GROUND (Cont.)

	Unit	Total
3x2" tapped tee	EA.	50.75
3x1-1/2" tapped tee	EA.	47.50
P-trap	EA.	58.75
3x2" tapped cross	EA.	56.50
3x1-1/2" tapped cross	EA.	54.75
Closet flange, 3-1/2" deep	EA.	33.25
4" pipe		
1/4 bend	EA.	47.25
1/8 bend	EA.	42.25
Sanitary tee	EA.	72.25
4x3" sanitary tee	EA.	70.25
4x2" sanitary tee	EA.	66.50
Sanitary cross	EA.	120
4x3" sanitary cross	EA.	110
4x2" sanitary cross	EA.	100.00
Plug	EA.	10.25
Coupling	EA.	15.75
Wye	EA.	75.75
4x3" wye	EA.	72.25
4x2" wye	EA.	65.75
Double wye	EA.	130
4x3" double wye	EA.	100
4x2" double wye	EA.	96.25
Wye & 1/8 bend	EA.	85.75
4x3" wye & 1/8 bend	EA.	75.50
4x2" wye & 1/8 bend	EA.	69.25
Double wye & 1/8 bend	EA.	150
4x3" double wye & 1/8 bend	EA.	120
4x2" double wye & 1/8 bend	EA.	120
4x3" reducer	EA.	38.75
4x2" reducer	EA.	38.75
Test tee, less 4" plug	EA.	58.75
Plug	EA.	10.25
4x2" tapped tee	EA.	50.75
4x1-1/2" tapped tee	EA.	48.25
P-trap	EA.	80.50
4x2" tapped cross	EA.	68.50
4x1-1/2" tapped cross	EA.	59.75
Closet flange		
3" deep	EA.	49.50
8" deep	EA.	82.00

15410.06 C.I. PIPE, BELOW GROUND

	Unit	Total
No hub pipe		
1-1/2" pipe	L.F.	9.06
2" pipe	L.F.	9.55
3" pipe	L.F.	12.25
4" pipe	L.F.	16.00
Fittings, 1-1/2"		
1/4 bend	EA.	25.00
1/8 bend	EA.	23.50
Plug	EA.	4.47
Wye	EA.	35.00
Wye & 1/8 bend	EA.	29.25
P-trap	EA.	30.50
2"		
1/4 bend	EA.	28.50
1/8 bend	EA.	27.25
Plug	EA.	4.47
Double wye	EA.	54.50
Wye & 1/8 bend	EA.	41.75
Double wye & 1/8 bend	EA.	67.75
P-trap	EA.	32.75
3"		
1/4 bend	EA.	35.75
1/8 bend	EA.	33.75
Plug	EA.	6.62
Wye	EA.	53.00
3x2" wye	EA.	48.75
Wye & 1/8 bend	EA.	56.50
Double wye & 1/8 bend	EA.	85.00
3x2" double wye & 1/8 bend	EA.	73.00
3x2" reducer	EA.	29.50
P-trap	EA.	53.00
4"		
1/4 bend	EA.	41.50
1/8 bend	EA.	36.50
Plug	EA.	10.25

15410.06 C.I. PIPE, BELOW GROUND (Cont.)

	Unit	Total
Wye	EA.	63.75
4x3" wye	EA.	60.25
4x2" wye	EA.	53.75
Double wye	EA.	120
4x3" double wye	EA.	91.50
4x2" double wye	EA.	86.50
Wye & 1/8 bend	EA.	73.75
4x3" wye & 1/8 bend	EA.	63.50
4x2" wye & 1/8 bend	EA.	57.25
Double wye & 1/8 bend	EA.	140
4x3" double wye & 1/8 bend	EA.	110
4x2" double wye & 1/8 bend	EA.	110
4x3" reducer	EA.	33.00
4x2" reducer	EA.	33.00

15410.10 COPPER PIPE

	Unit	Total
Type "K" copper		
1/2"	L.F.	5.56
3/4"	L.F.	8.94
1"	L.F.	11.25
DWV, copper		
1-1/4"	L.F.	11.75
1-1/2"	L.F.	14.50
2"	L.F.	18.25
3"	L.F.	29.50
4"	L.F.	49.25
6"	L.F.	180
Refrigeration tubing, copper, sealed		
1/8"	L.F.	3.05
3/16"	L.F.	3.26
1/4"	L.F.	3.53
Type "L" copper		
1/4"	L.F.	3.22
3/8"	L.F.	4.03
1/2"	L.F.	4.51
3/4"	L.F.	6.24
1"	L.F.	8.56
Type "M" copper		
1/2"	L.F.	3.71
3/4"	L.F.	5.03
1"	L.F.	7.12
Type "K" tube, coil		
1/4" x 60'	EA.	110
1/2" x 60'	EA.	240
1/2" x 100'	EA.	390
3/4" x 60'	EA.	440
3/4" x 100'	EA.	730
1" x 60'	EA.	570
1" x 100'	EA.	950
Type "L" tube, coil		
1/4" x 60'	EA.	120
3/8" x 60'	EA.	190
1/2" x 60'	EA.	250
1/2" x 100'	EA.	420
3/4" x 60'	EA.	400
3/4" x 100'	EA.	670
1" x 60'	EA.	580
1" x 100'	EA.	970

15410.11 COPPER FITTINGS

	Unit	Total
Coupling, with stop		
1/4"	EA.	20.25
3/8"	EA.	24.25
1/2"	EA.	26.00
5/8"	EA.	31.75
3/4"	EA.	34.00
1"	EA.	38.25
Reducing coupling		
1/4" x 1/8"	EA.	25.75
3/8" x 1/4"	EA.	28.00
1/2" x		
3/8"	EA.	31.00
1/4"	EA.	31.50
1/8"	EA.	31.75
3/4" x		
3/8"	EA.	36.75
1/2"	EA.	35.75
1" x		
3/8"	EA.	44.50

	Unit	Total
15410.11 COPPER FITTINGS (Cont.)		
1" x 1/2"	EA.............	44.25
1" x 3/4"	EA.............	43.00
Slip coupling		
1/4"	EA.............	20.00
1/2"	EA.............	24.50
3/4"	EA.............	31.75
1"	EA.............	38.25
Coupling with drain		
1/2"	EA.............	39.50
3/4"	EA.............	47.50
1"	EA.............	55.50
Reducer		
3/8" x 1/4"	EA.............	26.00
1/2" x 3/8"	EA.............	25.50
3/4" x		
1/4"	EA.............	31.25
3/8"	EA.............	31.50
1/2"	EA.............	31.75
1" x		
1/2"	EA.............	36.25
3/4"	EA.............	34.50
Female adapters		
1/4"	EA.............	31.00
3/8"	EA.............	34.25
1/2"	EA.............	32.50
3/4"	EA.............	37.25
1"	EA.............	44.25
Increasing female adapters		
1/8" x		
3/8"	EA.............	30.75
1/2"	EA.............	30.25
1/4" x 1/2"	EA.............	32.50
3/8" x 1/2"	EA.............	34.25
1/2" X		
3/4"	EA.............	37.25
1"	EA.............	46.00
3/4" X		
1"	EA.............	50.50
1-1/4"	EA.............	63.00
1" x		
1-1/4"	EA.............	65.00
1-1/2"	EA.............	68.00
Reducing female adapters		
3/8" x 1/4"	EA.............	33.25
1/2" x		
1/4"	EA.............	34.75
3/8"	EA.............	34.75
3/4" x 1/2"	EA.............	40.25
1" x		
1/2"	EA.............	54.50
3/4"	EA.............	50.00
Female fitting adapters		
1/2"	EA.............	39.50
3/4"	EA.............	42.50
3/4" x 1/2"	EA.............	46.75
1"	EA.............	50.25
Male adapters		
1/4"	EA.............	38.25
3/8"	EA.............	32.25
Increasing male adapters		
3/8" x 1/2"	EA.............	34.50
1/2" x		
3/4"	EA.............	35.75
1"	EA.............	44.75
3/4" x		
1"	EA.............	45.75
1-1/4"	EA.............	50.25
1" x 1-1/4"	EA.............	52.00
Reducing male adapters		
1/2" x		
1/4"	EA.............	39.25
3/8"	EA.............	37.25
3/4" x 1/2"	EA.............	40.00
1" x		
1/2"	EA.............	59.00
3/4"	EA.............	53.75

	Unit	Total
15410.11 COPPER FITTINGS (Cont.)		
Fitting x male adapters		
1/2"	EA.............	43.75
3/4"	EA.............	49.50
1"	EA.............	51.50
90 ells		
1/8"	EA.............	25.25
1/4"	EA.............	26.50
3/8"	EA.............	29.75
1/2"	EA.............	30.00
3/4"	EA.............	32.75
1"	EA.............	38.25
Reducing 90 ell		
3/8" x 1/4"	EA.............	32.50
1/2" x		
1/4"	EA.............	37.50
3/8"	EA.............	37.50
3/4" x 1/2"	EA.............	38.00
1" x		
1/2"	EA.............	45.50
3/4"	EA.............	44.75
Street ells, copper		
1/4"	EA.............	29.50
3/8"	EA.............	30.75
1/2"	EA.............	30.50
3/4"	EA.............	34.00
1"	EA.............	41.75
Female, 90 ell		
1/2"	EA.............	30.00
3/4"	EA.............	33.00
1"	EA.............	38.75
Female increasing, 90 ell		
3/8" x 1/2"	EA.............	39.75
1/2" x		
3/4"	EA.............	38.25
1"	EA.............	48.00
3/4" x 1"	EA.............	47.50
1" x 1-1/4"	EA.............	75.75
Female reducing, 90 ell		
1/2" x 3/8"	EA.............	43.75
3/4" x 1/2"	EA.............	46.75
1" x		
1/2"	EA.............	55.25
3/4"	EA.............	57.25
Male, 90 ell		
1/4"	EA.............	34.50
3/8"	EA.............	38.75
1/2"	EA.............	35.25
3/4"	EA.............	44.75
1"	EA.............	48.75
Male, increasing 90 ell		
1/2" x		
3/4"	EA.............	51.50
1"	EA.............	73.50
3/4" x 1"	EA.............	73.00
1" x 1-1/4"	EA.............	71.50
Male, reducing 90 ell		
1/2" x 3/8"	EA.............	41.75
3/4" x 1/2"	EA.............	53.00
1" x		
1/2"	EA.............	74.75
3/4"	EA.............	72.75
Drop ear ells		
1/2"	EA.............	37.00
Female drop ear ells		
1/2"	EA.............	37.00
1/2" x 3/8"	EA.............	43.00
3/4"	EA.............	54.25
Female flanged sink ell		
1/2"	EA.............	43.75
45 ells		
1/4"	EA.............	29.75
3/8"	EA.............	31.75
45 street ell		
1/4"	EA.............	30.75
3/8"	EA.............	34.75
1/2"	EA.............	31.25
3/4"	EA.............	34.00

	Unit	Total		Unit	Total

15410.11 COPPER FITTINGS (Cont.)

Item	Unit	Total
1"	EA.	42.00
Tee		
1/8"	EA.	28.75
1/4"	EA.	29.25
3/8"	EA.	31.00
Caps		
1/4"	EA.	24.00
3/8"	EA.	28.00
Test caps		
1/2"	EA.	29.75
3/4"	EA.	31.25
1"	EA.	34.00
Flush bushing		
1/4" x 1/8"	EA.	25.00
1/2" x		
1/4"	EA.	31.50
3/8"	EA.	31.25
3/4" x		
3/8"	EA.	35.25
1/2"	EA.	34.75
1" x		
1/2"	EA.	39.75
3/4"	EA.	39.00
Female flush bushing		
1/2" x		
1/2" x 1/8"	EA.	34.75
1/4"	EA.	35.25
Union		
1/4"	EA.	61.00
3/8"	EA.	78.50
Female		
1/2"	EA.	47.00
3/4"	EA.	48.50
Male		
1/2"	EA.	48.50
3/4"	EA.	56.50
1"	EA.	89.00
45 degree wye		
1/2"	EA.	53.25
3/4"	EA.	65.25
1"	EA.	79.25
1" x 3/4" x 3/4"	EA.	97.75
Twin ells		
1" x 3/4" x 3/4"	EA.	49.00
1" x 1" x 1"	EA.	49.00
90 union ells, male		
1/2"	EA.	55.25
3/4"	EA.	74.25
1"	EA.	97.75
DWV fittings, coupling with stop		
1-1/4"	EA.	40.00
1-1/2"	EA.	43.50
1-1/2" x 1-1/4"	EA.	48.25
2"	EA.	49.00
2" x 1-1/4"	EA.	52.75
2" x 1-1/2"	EA.	52.50
3"	EA.	68.25
3" x 1-1/2"	EA.	96.25
3" x 2"	EA.	94.00
4"	EA.	120
Slip coupling		
1-1/2"	EA.	47.75
2"	EA.	52.50
3"	EA.	73.25
90 ells		
1-1/2"	EA.	50.25
1-1/2" x 1-1/4"	EA.	75.00
2"	EA.	64.50
2" x 1-1/2"	EA.	91.25
3"	EA.	120
4"	EA.	290
Street, 90 elbows		
1-1/2"	EA.	54.25
2"	EA.	78.25
3"	EA.	150
4"	EA.	310

15410.11 COPPER FITTINGS (Cont.)

Item	Unit	Total
Female, 90 elbows		
1-1/2"	EA.	54.00
2"	EA.	73.25
Male, 90 elbows		
1-1/2"	EA.	67.75
2"	EA.	100
90 with side inlet		
3" x 3" x 1"	EA.	140
3" x 3" x 1-1/2"	EA.	150
3" x 3" x 2"	EA.	150
45 ells		
1-1/4"	EA.	45.75
1-1/2"	EA.	45.75
2"	EA.	61.00
3"	EA.	95.75
4"	EA.	280
Street, 45 ell		
1-1/2"	EA.	51.75
2"	EA.	66.50
3"	EA.	130
60 ell		
1-1/2"	EA.	60.50
2"	EA.	84.00
3"	EA.	150
22-1/2 ell		
1-1/2"	EA.	66.00
2"	EA.	77.00
3"	EA.	120
11-1/4 ell		
1-1/2"	EA.	69.00
2"	EA.	85.25
3"	EA.	140
Wye		
1-1/4"	EA.	84.25
1-1/2"	EA.	90.75
2"	EA.	110
2" x 1-1/2" x 1-1/2"	EA.	120
2" x 1-1/2" x 2"	EA.	130
2" x 1-1/2" x 2"	EA.	130
3"	EA.	220
3" x 3" x 1-1/2"	EA.	210
3" x 3" x 2"	EA.	210
4"	EA.	410
4" x 4" x 2"	EA.	310
4" x 4" x 3"	EA.	310
Sanitary tee		
1-1/4"	EA.	59.50
1-1/2"	EA.	67.75
2"	EA.	75.50
2" x 1-1/2" x 1-1/2"	EA.	97.50
2" x 1-1/2" x 2"	EA.	98.75
2" x 2" x 1-1/2"	EA.	73.50
3"	EA.	180
3" x 3" x 1-1/2"	EA.	160
3" x 3" x 2"	EA.	160
4"	EA.	410
4" x 4" x 3"	EA.	350
Female sanitary tee		
1-1/2"	EA.	97.50
Long turn tee		
1-1/2"	EA.	97.00
2"	EA.	180
3" x 1-1/2"	EA.	230
Double wye		
1-1/2"	EA.	130
2"	EA.	200
2" x 2" x 1-1/2" x 1-1/2"	EA.	170
3"	EA.	310
3" x 3" x 1-1/2" x 1-1/2"	EA.	310
3" x 3" x 2" x 2"	EA.	310
4" x 4" x 1-1/2" x 1-1/2"	EA.	340
Double sanitary tee		
1-1/2"	EA.	98.25
2"	EA.	180
2" x 2" x 1-1/2"	EA.	170
3"	EA.	220
3" x 3" x 1-1/2" x 1-1/2"	EA.	270

15410.11 COPPER FITTINGS (Cont.)	Unit	Total
3" x 3" x 2" x 2"	EA	230
4" x 4" x 1-1/2" x 1-1/2"	EA	450
Long		
2" x 1-1/2"	EA	200
Twin elbow		
1-1/2"	EA	120
2"	EA	170
2" x 1-1/2" x 1-1/2"	EA	150
Spigot adapter, manoff		
1-1/2" x 2"	EA	89.50
1-1/2" x 3"	EA	100
2"	EA	65.00
2" x 3"	EA	100
2" x 4"	EA	130
3"	EA	140
3" x 4"	EA	220
4"	EA	200
No-hub adapters		
1-1/2" x 2"	EA	68.50
2"	EA	69.00
2" x 3"	EA	110
3"	EA	110
3" x 4"	EA	180
4"	EA	190
Fitting reducers		
1-1/2" x 1-1/4"	EA	47.50
2" x 1-1/2"	EA	56.50
3" x 1-1/2"	EA	98.25
3" x 2"	EA	93.00
Slip joint (Desanco)		
1-1/4"	EA	53.25
1-1/2"	EA	56.00
1-1/2" x 1-1/4"	EA	56.75
Street x slip joint (Desanco)		
1-1/2"	EA	60.75
1-1/2" x 1-1/4"	EA	62.25
Flush bushing		
1-1/2" x 1-1/4"	EA	50.25
2" x 1-1/2"	EA	63.00
3" x 1-1/2"	EA	92.00
3" x 2"	EA	92.00
Male hex trap bushing		
1-1/4" x 1-1/2"	EA	54.75
1-1/2"	EA	51.25
1-1/2" x 2"	EA	59.00
2"	EA	56.50
Round trap bushing		
1-1/2"	EA	53.75
2"	EA	57.25
Female adapter		
1-1/4"	EA	54.50
1-1/2"	EA	68.25
1-1/2" x 2"	EA	120
2"	EA	82.25
2" x 1-1/2"	EA	110
3"	EA	220
Fitting x female adapter		
1-1/2"	EA	79.25
2"	EA	96.00
Male adapters		
1-1/4"	EA	52.00
1-1/4" x 1-1/2"	EA	75.75
1-1/2"	EA	56.75
1-1/2" x 2"	EA	110
2"	EA	73.25
2" x 1-1/2"	EA	120
3"	EA	230
Male x slip joint adapters		
1-1/2" x 1-1/4"	EA	68.75
Dandy cleanout		
1-1/2"	EA	94.00
2"	EA	110
3"	EA	290
End cleanout, flush pattern		
1-1/2" x 1"	EA	71.25
2" x 1-1/2"	EA	80.75
3" x 2-1/2"	EA	140

15410.11 COPPER FITTINGS (Cont.)	Unit	Total
Copper caps		
1-1/2"	EA	48.00
2"	EA	60.50
Closet flanges		
3"	EA	100
4"	EA	150
Drum traps, with cleanout		
1-1/2" x 3" x 6"	EA	240
P-trap, swivel, with cleanout		
1-1/2"	EA	160
P-trap, solder union		
1-1/2"	EA	87.75
2"	EA	130
With cleanout		
1-1/2"	EA	93.00
2"	EA	140
2" x 1-1/2"	EA	140
Swivel joint, with cleanout		
1-1/2" x 1-1/4"	EA	110
1-1/2"	EA	130
2" x 1-1/2"	EA	150
Estabrook TY, with inlets		
3", with 1-1/2" inlet	EA	210
Fine thread adapters		
1/2"	EA	33.00
1/2" x 1/2" IPS	EA	33.50
1/2" x 3/4" IPS	EA	36.50
1/2" x male	EA	31.75
1/2" x female	EA	34.75
Copper pipe fittings		
1/2"		
90 deg ell	EA	14.50
45 deg ell	EA	15.00
Tee	EA	19.50
Cap	EA	7.67
Coupling	EA	14.00
Union	EA	23.75
3/4"		
90 deg ell	EA	18.50
45 deg ell	EA	19.25
Tee	EA	26.00
Cap	EA	9.25
Coupling	EA	17.25
Union	EA	30.25
1"		
90 deg ell	EA	28.50
45 deg ell	EA	31.50
Tee	EA	38.25
Cap	EA	14.25
Coupling	EA	26.00
Union	EA	37.25
15410.14 BRASS I.P.S. FITTINGS		
Fittings, iron pipe size, 45 deg ell		
1/8"	EA	32.75
1/4"	EA	32.75
3/8"	EA	36.00
1/2"	EA	38.50
3/4"	EA	44.25
1"	EA	55.75
90 deg ell		
1/8"	EA	32.25
1/4"	EA	32.25
3/8"	EA	35.50
1/2"	EA	38.00
3/4"	EA	41.25
1"	EA	51.75
90 deg ell, reducing		
1/4" x 1/8"	EA	34.00
3/8" x 1/8"	EA	37.25
3/8" x 1/4"	EA	37.25
1/2" x 1/4"	EA	39.75
1/2" x 3/8"	EA	39.75
3/4" x 1/2"	EA	46.25
1" x 3/8"	EA	54.75
1" x 1/2"	EA	54.75
1" x 3/4"	EA	54.75

	Unit	Total

15410.14 BRASS I.P.S. FITTINGS (Cont.)

	Unit	Total
Street ell, 45 deg		
1/2"	EA.	39.75
3/4"	EA.	46.25
90 deg		
1/8"	EA.	34.00
1/4"	EA.	34.00
3/8"	EA.	37.25
1/2"	EA.	39.75
3/4"	EA.	43.75
1"	EA.	49.75
Tee, 1/8"	EA.	32.50
1/4"	EA.	32.50
3/8"	EA.	35.75
1/2"	EA.	38.25
3/4"	EA.	43.00
1"	EA.	50.75
Tee, reducing, 3/8" x		
1/4"	EA.	39.25
1/2"	EA.	39.25
1/2" x		
1/4"	EA.	41.75
3/8"	EA.	41.75
3/4"	EA.	44.00
3/4" x		
1/4"	EA.	45.50
1/2"	EA.	45.50
1"	EA.	61.00
1" x		
1/2"	EA.	62.50
3/4"	EA.	62.50
Tee, reducing		
1/2" x 3/8" x 1/2"	EA.	40.50
3/4" x 1/2" x 1/2"	EA.	47.25
3/4" x 1/2" x 3/4"	EA.	46.25
1" x 1/2" x 1/2"	EA.	61.50
1" x 1/2" x 3/4"	EA.	61.50
1" x 3/4" x 1/2"	EA.	68.00
1" x 3/4" x 3/4"	EA.	61.50
Union		
1/8"	EA.	47.50
1/4"	EA.	47.50
3/8"	EA.	50.75
1/2"	EA.	53.25
3/4"	EA.	64.00
1"	EA.	77.00
Brass face bushing		
3/8" x 1/4"	EA.	34.75
1/2" x 3/8"	EA.	37.25
3/4" x 1/2"	EA.	40.75
1" x 3/4"	EA.	50.00
Hex bushing, 1/4" x 1/8"	EA.	29.00
1/2" x		
1/4"	EA.	34.25
3/8"	EA.	34.25
5/8" x		
1/8"	EA.	34.25
1/4"	EA.	34.25
3/4" x		
1/8"	EA.	38.25
1/4"	EA.	38.25
3/8"	EA.	37.25
1/2"	EA.	37.25
1" x		
1/4"	EA.	42.50
3/8"	EA.	42.50
1/2"	EA.	41.75
3/4"	EA.	41.75
Caps		
1/8"	EA.	28.50
1/4"	EA.	29.00
3/8"	EA.	32.25
1/2"	EA.	34.75
3/4"	EA.	36.50
1"	EA.	43.25
Couplings		
1/8"	EA.	29.25
1/4"	EA.	29.25

15410.14 BRASS I.P.S. FITTINGS (Cont.)

	Unit	Total
3/8"	EA.	32.50
1/2"	EA.	35.00
3/4"	EA.	39.00
1"	EA.	45.75
Couplings, reducing, 1/4" x 1/8"	EA.	30.50
3/8" x		
1/8"	EA.	33.75
1/4"	EA.	33.75
1/2" x		
1/8"	EA.	38.00
1/4"	EA.	36.75
3/8"	EA.	36.75
3/4" x		
1/4"	EA.	43.00
3/8"	EA.	40.75
1/2"	EA.	40.75
1" x		
1/2"	EA.	45.00
3/4"	EA.	45.00
Square head plug, solid		
1/8"	EA.	29.25
1/4"	EA.	29.25
3/8"	EA.	32.50
1/2"	EA.	35.00
3/4"	EA.	37.75
Cored		
1/2"	EA.	33.75
3/4"	EA.	36.50
1"	EA.	42.00
Countersunk		
1/2"	EA.	35.75
3/4"	EA.	37.50
Locknut		
3/4"	EA.	36.50
1"	EA.	39.75
Close standard red nipple, 1/8"	EA.	25.00
1/8" x		
1-1/2"	EA.	26.75
2"	EA.	27.00
2-1/2"	EA.	27.75
3"	EA.	28.25
3-1/2"	EA.	29.25
4"	EA.	29.75
4-1/2"	EA.	30.00
5"	EA.	30.50
5-1/2"	EA.	31.75
6"	EA.	32.25
1/4" x close	EA.	27.00
1/4" x		
1-1/2"	EA.	29.00
2"	EA.	29.50
2-1/2"	EA.	29.75
3"	EA.	30.00
3-1/2"	EA.	31.00
4"	EA.	31.25
4-1/2"	EA.	31.75
5"	EA.	32.00
5-1/2"	EA.	33.00
6"	EA.	33.25
3/8" x close	EA.	30.25
3/8" x		
1-1/2"	EA.	31.00
2"	EA.	31.50
2-1/2"	EA.	32.75
3"	EA.	34.00
3-1/2"	EA.	34.75
4"	EA.	37.25
4-1/2"	EA.	37.50
5"	EA.	38.25
5-1/2"	EA.	39.00
6"	EA.	40.75
1/2" x close	EA.	34.25
1/2" x		
1-1/2"	EA.	34.75
2"	EA.	36.25
2-1/2"	EA.	37.25
3"	EA.	38.50

DIVISION 15 MECHANICAL

15410.14 BRASS I.P.S. FITTINGS (Cont.)

	Unit	Total
3-1/2"	EA	39.50
4"	EA	40.00
4-1/2"	EA	40.75
5"	EA	41.00
5-1/2"	EA	41.50
6"	EA	42.75
7-1/2"	EA	71.00
8"	EA	71.00
3/4" x close	EA	45.25
3/4" x		
1-1/2"	EA	38.75
2"	EA	40.00
2-1/2"	EA	41.00
3"	EA	41.75
3-1/2"	EA	43.00
4"	EA	43.75
4-1/2"	EA	44.75
5"	EA	45.25
5-1/2"	EA	47.25
6"	EA	48.00
1" x close	EA	44.00
1" x		
2"	EA	49.50
2-1/2"	EA	49.75
3"	EA	50.50
3-1/2"	EA	51.25
4"	EA	53.00
4-1/2"	EA	53.25
5"	EA	56.75
5-1/2"	EA	57.50
6"	EA	60.00

15410.15 BRASS FITTINGS

	Unit	Total
Compression fittings, union		
3/8"	EA	13.25
1/2"	EA	15.50
5/8"	EA	16.25
Union elbow		
3/8"	EA	12.00
1/2"	EA	13.00
5/8"	EA	14.00
Union tee		
3/8"	EA	12.50
1/2"	EA	13.75
5/8"	EA	15.00
Male connector		
3/8"	EA	12.00
1/2"	EA	11.50
5/8"	EA	11.25
Female connector		
3/8"	EA	11.75
1/2"	EA	12.50
5/8"	EA	13.00
Brass flare fittings, union		
3/8"	EA	11.25
1/2"	EA	12.00
5/8"	EA	12.25
90 deg elbow union		
3/8"	EA	13.25
1/2"	EA	15.50
5/8"	EA	18.75
Three way tee		
3/8"	EA	19.75
1/2"	EA	21.25
5/8"	EA	25.00
Cross		
3/8"	EA	29.50
1/2"	EA	40.00
5/8"	EA	61.50
Male connector, half union		
3/8"	EA	10.75
1/2"	EA	11.75
5/8"	EA	12.50
Female connector, half union		
3/8"	EA	11.25
1/2"	EA	11.00
5/8"	EA	12.50

15410.15 BRASS FITTINGS (Cont.)

	Unit	Total
Long forged nut		
3/8"	EA	10.75
1/2"	EA	11.50
5/8"	EA	16.75
Short forged nut		
3/8"	EA	10.50
1/2"	EA	11.00
5/8"	EA	11.25
Nut		
1/8"	EA	.26
1/4"	EA	.26
5/16"	EA	.30
3/8"	EA	.40
1/2"	EA	.58
5/8"	EA	1.21
Sleeve		
1/8"	EA	11.75
1/4"	EA	11.50
5/16"	EA	11.75
3/8"	EA	11.75
1/2"	EA	11.75
5/8"	EA	12.00
Tee		
1/4"	EA	19.50
5/16"	EA	21.00
Male tee		
5/16" x 1/8"	EA	22.75
Female union		
1/8" x 1/8"	EA	16.00
1/4" x 3/8"	EA	17.50
3/8" x 1/4"	EA	17.00
3/8" x 1/2"	EA	17.50
5/8" x 1/2"	EA	21.25
Male union, 1/4"		
1/4" x 1/4"	EA	16.00
3/8"	EA	16.50
1/2"	EA	17.25
5/16" x		
1/8"	EA	16.00
1/4"	EA	16.25
3/8"	EA	17.25
3/8" x		
1/8"	EA	16.25
1/4"	EA	16.50
1/2"	EA	17.00
5/8" x		
3/8"	EA	20.25
1/2"	EA	19.75
Female elbow, 1/4" x 1/4"	EA	20.00
5/16" x		
1/8"	EA	20.50
1/4"	EA	21.50
3/8" x		
3/8"	EA	19.75
1/2"	EA	19.00
Male elbow, 1/8" x 1/8"	EA	20.00
3/16" x 1/4"	EA	19.75
1/4" x		
1/8"	EA	18.50
1/4"	EA	18.75
3/8"	EA	18.50
5/16" x		
1/8"	EA	18.50
1/4"	EA	19.00
3/8"	EA	20.00
3/8" x		
1/8"	EA	18.50
1/4"	EA	19.25
3/8"	EA	18.50
1/2"	EA	19.25
1/2" x		
1/4"	EA	23.75
3/8"	EA	23.25
1/2"	EA	22.75
5/8" x		
3/8"	EA	23.75
1/2"	EA	24.00

DIVISION 15 MECHANICAL

	Unit	Total		Unit	Total
15410.15 BRASS FITTINGS (Cont.)			**15410.30 PVC/CPVC PIPE (Cont.)**		
3/4"	EA.	28.75	Reducing insert	EA.	20.25
Union			Threaded	EA.	21.00
1/8"	EA.	18.50	Female adapter	EA.	20.25
3/16"	EA.	18.50	Coupling	EA.	20.25
1/4"	EA.	18.25	Union	EA.	36.50
5/16"	EA.	18.50	Cap	EA.	20.25
3/8"	EA.	18.75	Flange	EA.	31.50
Reducing union			1-1/2"		
3/8" x 1/4"	EA.	21.75	90 deg elbow	EA.	18.00
5/8" x			45 deg elbow	EA.	18.75
3/8"	EA.	23.00	Tee	EA.	21.50
1/2"	EA.	23.50	Reducing insert	EA.	20.25
15410.17 CHROME PLATED FITTINGS			Threaded	EA.	21.25
Fittings			Male adapter	EA.	20.50
90 ell			Female adapter	EA.	20.50
3/8"	EA.	42.00	Coupling	EA.	20.25
1/2"	EA.	50.00	Union	EA.	47.50
45 ell			Cap	EA.	20.25
3/8"	EA.	50.00	Flange	EA.	43.25
1/2"	EA.	61.25	2"		
Tee			90 deg elbow	EA.	21.75
3/8"	EA.	53.25	45 deg elbow	EA.	22.50
1/2"	EA.	59.75	Tee	EA.	26.25
Coupling			Reducing insert	EA.	25.25
3/8"	EA.	36.00	Threaded	EA.	25.75
1/2"	EA.	36.00	Male adapter	EA.	24.75
Union			Female adapter	EA.	24.75
3/8"	EA.	50.00	Coupling	EA.	24.50
1/2"	EA.	51.25	Union	EA.	61.50
Tee			Cap	EA.	24.50
1/2" x 3/8" x 3/8"	EA.	59.75	Flange	EA.	51.25
1/2" x 3/8" x 1/2"	EA.	60.50	2-1/2"		
15410.30 PVC/CPVC PIPE			90 deg elbow	EA.	43.75
PVC schedule 40			45 deg elbow	EA.	47.00
1/2" pipe	L.F.	2.84	Tee	EA.	48.25
3/4" pipe	L.F.	3.22	Reducing insert	EA.	41.50
1" pipe	L.F.	3.65	Threaded	EA.	43.00
1-1/4" pipe	L.F.	4.19	Male adapter	EA.	43.50
1-1/2" pipe	L.F.	5.08	Female adapter	EA.	42.75
2" pipe	L.F.	5.99	Coupling	EA.	41.50
2-1/2" pipe	L.F.	7.83	Union	EA.	82.00
3" pipe	L.F.	9.62	Cap	EA.	40.00
4" pipe	L.F.	12.75	Flange	EA.	68.50
Fittings, 1/2"			3"		
90 deg ell	EA.	7.69	90 deg elbow	EA.	56.25
45 deg ell	EA.	7.87	45 deg elbow	EA.	58.75
Tee	EA.	8.74	Tee	EA.	65.75
Reducing insert	EA.	10.00	Reducing insert	EA.	52.00
Threaded	EA.	8.40	Threaded	EA.	53.00
Male adapter	EA.	10.00	Male adapter	EA.	54.25
Female adapter	EA.	7.72	Female adapter	EA.	53.00
Coupling	EA.	7.58	Coupling	EA.	52.75
Union	EA.	15.50	Union	EA.	93.50
Cap	EA.	10.00	Cap	EA.	52.00
Flange	EA.	19.75	Flange	EA.	76.50
3/4"			4"		
90 deg elbow	EA.	10.00	90 deg elbow	EA.	72.50
45 deg elbow	EA.	10.75	45 deg elbow	EA.	77.00
Tee	EA.	12.25	Tee	EA.	86.00
Reducing insert	EA.	8.71	Reducing insert	EA.	66.75
Threaded	EA.	10.25	Threaded	EA.	69.25
1"			Male adapter	EA.	65.75
90 deg elbow	EA.	12.25	Female adapter	EA.	66.25
45 deg elbow	EA.	12.75	Coupling	EA.	64.50
Tee	EA.	13.75	Union	EA.	120
Reducing insert	EA.	12.25	Cap	EA.	67.00
Threaded	EA.	13.75	Flange	EA.	97.25
Male adapter	EA.	15.25	PVC schedule 80 pipe		
Female adapter	EA.	15.25	1-1/2" pipe	L.F.	5.47
Coupling	EA.	15.00	2" pipe	L.F.	6.65
Union	EA.	25.25	3" pipe	L.F.	11.00
Cap	EA.	12.25	4" pipe	L.F.	14.00
Flange	EA.	27.75	Fittings, 1-1/2"		
1-1/4"			90 deg elbow	EA.	26.00
90 deg elbow	EA.	18.00	45 deg elbow	EA.	34.00
45 deg elbow	EA.	18.25	Tee	EA.	52.00
Tee	EA.	21.00	Reducing insert	EA.	23.50

15410.30 PVC/CPVC PIPE (Cont.)	Unit	Total
Threaded	EA	24.25
Male adapter	EA	27.50
Female adapter	EA	28.00
Coupling	EA	28.75
Union	EA	45.75
Cap	EA	24.00
Flange	EA	39.75
2"		
90 deg elbow	EA	31.25
45 deg elbow	EA	42.25
Tee	EA	65.00
Reducing insert	EA	29.00
Threaded	EA	29.25
Male adapter	EA	34.25
Female adapter	EA	38.25
2-1/2"		
90 deg elbow	EA	55.00
45 deg elbow	EA	76.50
Tee	EA	79.50
Reducing insert	EA	46.50
Threaded	EA	49.25
Male adapter	EA	49.25
Female adapter	EA	60.25
Coupling	EA	49.00
Union	EA	84.25
Cap	EA	51.25
Flange	EA	67.50
3"		
90 deg elbow	EA	65.25
45 deg elbow	EA	97.25
Tee	EA	97.25
Reducing insert	EA	64.75
Threaded	EA	72.50
Male adapter	EA	62.75
Female adapter	EA	75.50
Coupling	EA	63.00
Union	EA	100
Cap	EA	67.50
Flange	EA	79.75
4"		
90 deg elbow	EA	100
45 deg elbow	EA	150
Tee	EA	120
Reducing insert	EA	81.00
Threaded	EA	95.25
Male adapter	EA	84.00
Coupling	EA	76.50
Union	EA	120
Cap	EA	81.50
Flange	EA	99.25
CPVC schedule 40		
1/2" pipe	L.F.	2.89
3/4" pipe	L.F.	3.27
1" pipe	L.F.	3.83
1-1/4" pipe	L.F.	4.44
1-1/2" pipe	L.F.	5.11
2" pipe	L.F.	6.12
Fittings, CPVC, schedule 80		
1/2", 90 deg ell	EA	9.21
Tee	EA	20.00
3/4", 90 deg ell	EA	10.25
Tee	EA	25.00
1", 90 deg ell	EA	13.50
Tee	EA	27.00
1-1/4", 90 deg ell	EA	19.50
Tee	EA	26.00
1-1/2", 90 deg ell	EA	25.75
Tee	EA	32.00
2", 90 deg ell	EA	27.00
Tee	EA	34.25
15410.33 ABS DWV PIPE		
Schedule 40 ABS		
1-1/2" pipe	L.F.	3.70
2" pipe	L.F.	4.30
3" pipe	L.F.	6.38
4" pipe	L.F.	8.96

15410.33 ABS DWV PIPE (Cont.)	Unit	Total
Fittings		
1/8 bend		
1-1/2"	EA	13.25
2"	EA	17.25
3"	EA	25.75
4"	EA	34.50
Tee, sanitary		
1-1/2"	EA	22.00
2"	EA	27.00
3"	EA	39.75
4"	EA	56.25
Tee, sanitary reducing		
2 x 1-1/2 x 1-1/2	EA	26.75
2 x 1-1/2 x 2	EA	27.75
2 x 2 x 1-1/2	EA	29.75
3 x 3 x 1-1/2	EA	35.25
3 x 3 x 2	EA	40.00
4 x 4 x 1-1/2	EA	56.25
4 x 4 x 2	EA	60.00
4 x 4 x 3	EA	60.75
Wye		
1-1/2"	EA	20.25
2"	EA	28.50
3"	EA	41.00
4"	EA	62.50
Reducer		
2 x 1-1/2	EA	17.00
3 x 1-1/2	EA	26.00
3 x 2	EA	25.00
4 x 2	EA	34.50
4 x 3	EA	34.75
P-trap		
1-1/2"	EA	25.25
2"	EA	29.25
3"	EA	56.00
4"	EA	92.25
Double sanitary, tee		
1-1/2"	EA	28.75
2"	EA	37.25
3"	EA	59.25
4"	EA	85.00
Long sweep, 1/4 bend		
1-1/2"	EA	14.50
2"	EA	18.25
3"	EA	28.50
4"	EA	45.75
Wye, standard		
1-1/2"	EA	23.25
2"	EA	28.50
3"	EA	41.25
4"	EA	62.50
Wye, reducing		
2 x 1-1/2 x 1-1/2	EA	26.50
2 x 2 x 1-1/2	EA	30.00
4 x 4 x 2	EA	50.75
4 x 4 x 3	EA	58.50
Double wye		
1-1/2"	EA	31.75
2"	EA	39.25
3"	EA	63.00
4"	EA	100
2 x 2 x 1-1/2 x 1-1/2	EA	39.25
3 x 3 x 2 x 2	EA	58.00
4 x 4 x 3 x 3	EA	99.50
Combination wye and 1/8 bend		
1-1/2"	EA	25.50
2"	EA	30.50
3"	EA	44.75
4"	EA	68.75
2 x 2 x 1-1/2	EA	29.50
3 x 3 x 1-1/2	EA	44.00
3 x 3 x 2	EA	39.50
4 x 4 x 2	EA	57.00
4 x 4 x 3	EA	61.75

	Unit	Total

15410.80 STEEL PIPE

Black steel, extra heavy pipe, threaded

	Unit	Total
1/2" pipe	L.F.	4.57
3/4" pipe	L.F.	5.23

Fittings, malleable iron, threaded, 1/2" pipe

	Unit	Total
90 deg ell	EA.	22.50
45 deg ell	EA.	23.50
Tee	EA.	32.25
Reducing tee	EA.	36.50
Cap	EA.	14.25
Coupling	EA.	26.50
Union	EA.	34.00
Nipple, 4" long	EA.	22.00

3/4" pipe

	Unit	Total
90 deg ell	EA.	23.00
45 deg ell	EA.	34.50
Tee	EA.	33.75
Reducing tee	EA.	27.75
Cap	EA.	15.00
Coupling	EA.	23.50
Union	EA.	36.00
Nipple, 4" long	EA.	22.50

Cast iron fittings

1/2" pipe

	Unit	Total
90 deg. ell	EA.	24.00
45 deg. ell	EA.	29.00
Reducing tee	EA.	40.50

3/4" pipe

	Unit	Total
90 deg. ell	EA.	24.25
45 deg. ell	EA.	25.50
Tee	EA.	36.50
Reducing tee	EA.	39.00

15410.82 GALVANIZED STEEL PIPE

Galvanized pipe

	Unit	Total
1/2" pipe	L.F.	8.55
3/4" pipe	L.F.	10.75

90 degree ell, 150 lb malleable iron, galvanized

	Unit	Total
1/2"	EA.	13.75
3/4"	EA.	17.25

45 degree ell, 150 lb m.i., galv.

	Unit	Total
1/2"	EA.	15.00
3/4"	EA.	19.25

Tees, straight, 150 lb m.i., galv.

	Unit	Total
1/2"	EA.	17.25
3/4"	EA.	21.25

Tees, reducing, out, 150 lb m.i., galv.

	Unit	Total
1/2"	EA.	19.50
3/4"	EA.	22.25

Couplings, straight, 150 lb m.i., galv.

	Unit	Total
1/2"	EA.	14.25
3/4"	EA.	16.00

Couplings, reducing, 150 lb m.i., galv

	Unit	Total
1/2"	EA.	14.50
3/4"	EA.	16.25

Caps, 150 lb m.i., galv.

	Unit	Total
1/2"	EA.	7.97
3/4"	EA.	8.97

Unions, 150 lb m.i., galv.

	Unit	Total
1/2"	EA.	26.75
3/4"	EA.	30.25

Nipples, galvanized steel, 4" long

	Unit	Total
1/2"	EA.	10.50
3/4"	EA.	12.00

90 degree reducing ell, 150 lb m.i., galv.

	Unit	Total
3/4" x 1/2"	EA.	15.00
1" x 3/4"	EA.	17.50

Square head plug (C.I.)

	Unit	Total
1/2"	EA.	8.59
3/4"	EA.	12.00

15430.23 CLEANOUTS

Cleanout, wall

	Unit	Total
2"	EA.	210
3"	EA.	270
4"	EA.	290

Floor

	Unit	Total
2"	EA.	200
3"	EA.	250
4"	EA.	260

15430.25 HOSE BIBBS

Hose bibb

	Unit	Total
1/2"	EA.	28.25
3/4"	EA.	28.75

15430.60 VALVES

Gate valve, 125 lb, bronze, soldered

	Unit	Total
1/2"	EA.	40.25
3/4"	EA.	45.50

Threaded

	Unit	Total
1/4", 125 lb	EA.	52.75

1/2"

	Unit	Total
125 lb	EA.	51.50
150 lb	EA.	61.00
300 lb	EA.	94.50

3/4"

	Unit	Total
125 lb	EA.	56.25
150 lb	EA.	68.25
300 lb	EA.	110

Check valve, bronze, soldered, 125 lb

	Unit	Total
1/2"	EA.	55.00
3/4"	EA.	64.75

Threaded

1/2"

	Unit	Total
125 lb	EA.	76.00
150 lb	EA.	72.00
200 lb	EA.	74.25

3/4"

	Unit	Total
125 lb	EA.	65.25
150 lb	EA.	89.00
200 lb	EA.	96.25

Vertical check valve, bronze, 125 lb, threaded

	Unit	Total
1/2"	EA.	88.00
3/4"	EA.	120

Globe valve, bronze, soldered, 125 lb

	Unit	Total
1/2"	EA.	76.75
3/4"	EA.	92.50

Threaded

1/2"

	Unit	Total
125 lb	EA.	74.25
150 lb	EA.	91.75
300 lb	EA.	160

3/4"

	Unit	Total
125 lb	EA.	100
150 lb	EA.	110
300 lb	EA.	190

Ball valve, bronze, 250 lb, threaded

	Unit	Total
1/2"	EA.	37.25
3/4"	EA.	44.25

Angle valve, bronze, 150 lb, threaded

	Unit	Total
1/2"	EA.	110
3/4"	EA.	140

Balancing valve, meter connections, circuit setter

	Unit	Total
1/2"	EA.	110
3/4"	EA.	110

Balancing valve, straight type

	Unit	Total
1/2"	EA.	45.25
3/4"	EA.	50.00

Angle type

	Unit	Total
1/2"	EA.	53.00
3/4"	EA.	64.50

Square head cock, 125 lb, bronze body

	Unit	Total
1/2"	EA.	36.75
3/4"	EA.	44.00

Radiator temp control valve, with control and sensor

	Unit	Total
1/2" valve	EA.	160

Pressure relief valve, 1/2", bronze

	Unit	Total
Low pressure	EA.	51.00
High pressure	EA.	55.75

Pressure and temperature relief valve

	Unit	Total
Bronze, 3/4"	EA.	120

Cast iron, 3/4"

	Unit	Total
High pressure	EA.	71.50
Temperature relief	EA.	88.50
Pressure & temp relief valve	EA.	100

Pressure reducing valve, bronze, threaded, 250 lb

	Unit	Total
1/2"	EA.	210
3/4"	EA.	210

	Unit	Total			Unit	Total

15430.60 VALVES (Cont.)

	Unit	Total
Solar water temperature regulating valve		
3/4"	EA	690
Tempering valve, threaded		
3/4"	EA	370
Thermostatic mixing valve, threaded		
1/2"	EA	140
3/4"	EA	140
Sweat connection		
1/2"	EA	150
3/4"	EA	190
Mixing valve, sweat connection		
1/2"	EA	92.75
3/4"	EA	95.25
Liquid level gauge, aluminum body		
3/4"	EA	380
4125 psi, pvc body		
3/4"	EA	450
150 psi, crs body		
3/4"	EA	360
175 psi, bronze body, 1/2"	EA	710

15430.65 VACUUM BREAKERS

	Unit	Total
Vacuum breaker, atmospheric, threaded connection		
3/4"	EA	68.75
Anti-siphon, brass		
3/4"	EA	72.50

15430.68 STRAINERS

	Unit	Total
Strainer, Y pattern, 125 psi, cast iron body, threaded		
3/4"	EA	32.00
250 psi, brass body, threaded		
3/4"	EA	52.75
Cast iron body, threaded		
3/4"	EA	40.50

15430.70 DRAINS, ROOF & FLOOR

	Unit	Total
Floor drain, cast iron, with cast iron top		
2"	EA	180
3"	EA	190
4"	EA	340
Roof drain, cast iron		
2"	EA	260
3"	EA	270
4"	EA	330

15430.80 TRAPS

	Unit	Total
Bucket trap, threaded		
3/4"	EA	230
Inverted bucket steam trap, threaded		
3/4"	EA	270
With stainless interior		
1/2"	EA	180
3/4"	EA	210
Brass interior		
3/4"	EA	290
Cast steel body, threaded, high temperature		
3/4"	EA	700
Float trap, 15 psi		
3/4"	EA	210
Float and thermostatic trap, 15 psi		
3/4"	EA	220
Steam trap, cast iron body, threaded, 125 psi		
3/4"	EA	250
Thermostatic trap, low pressure, angle type, 25 psi		
1/2"	EA	99.75
3/4"	EA	150
Cast iron body, threaded, 125 psi		
3/4"	EA	180

15440.10 BATHS

	Unit	Total
Bath tub, 5' long		
Minimum	EA	720
Average	EA	1,450
Maximum	EA	3,250
6' long		
Minimum	EA	780
Average	EA	1,500
Maximum	EA	4,000
Square tub, whirlpool, 4'x4'		
Minimum	EA	2,100
Average	EA	3,150
Maximum	EA	8,550

15440.10 BATHS (Cont.)

	Unit	Total
5'x5'		
Minimum	EA	2,100
Average	EA	3,150
Maximum	EA	8,700
6'x6'		
Minimum	EA	2,500
Average	EA	3,850
Maximum	EA	9,950
For trim and rough-in		
Minimum	EA	380
Average	EA	570
Maximum	EA	1,350

15440.12 DISPOSALS & ACCESSORIES

	Unit	Total
Continuous feed		
Minimum	EA	190
Average	EA	340
Maximum	EA	580
Batch feed, 1/2 hp		
Minimum	EA	400
Average	EA	690
Maximum	EA	1,150
Hot water dispenser		
Minimum	EA	320
Average	EA	460
Maximum	EA	700
Epoxy finish faucet	EA	410
Lock stop assembly	EA	130
Mounting gasket	EA	55.00
Tailpipe gasket	EA	49.00
Stopper assembly	EA	81.75
Switch assembly, on/off	EA	120
Tailpipe gasket washer	EA	29.75
Stop gasket	EA	34.50
Tailpipe flange	EA	29.00
Tailpipe	EA	39.25

15440.15 FAUCETS

	Unit	Total
Kitchen		
Minimum	EA	180
Average	EA	350
Maximum	EA	430
Bath		
Minimum	EA	180
Average	EA	360
Maximum	EA	510
Lavatory, domestic		
Minimum	EA	180
Average	EA	400
Maximum	EA	600
Washroom		
Minimum	EA	210
Average	EA	400
Maximum	EA	650
Handicapped		
Minimum	EA	240
Average	EA	500
Maximum	EA	750
Shower		
Minimum	EA	210
Average	EA	440
Maximum	EA	650
For trim and rough-in		
Minimum	EA	200
Average	EA	260
Maximum	EA	490

15440.18 HYDRANTS

	Unit	Total
Wall hydrant		
8" thick	EA	460
12" thick	EA	550

15440.20 LAVATORIES

	Unit	Total
Lavatory, counter top, porcelain enamel on cast iron		
Minimum	EA	310
Average	EA	430
Maximum	EA	710
Wall hung, china		
Minimum	EA	380
Average	EA	450
Maximum	EA	960

	Unit	Total		Unit	Total

15440.20 LAVATORIES (Cont.)
Handicapped

	Unit	Total
Minimum	EA.	570
Average	EA.	690
Maximum	EA.	1,100

For trim and rough-in

	Unit	Total
Minimum	EA.	360
Average	EA.	560
Maximum	EA.	750

15440.30 SHOWERS
Shower, fiberglass, 36"x34"x84"

	Unit	Total
Minimum	EA.	980
Average	EA.	1,400
Maximum	EA.	1,750

Steel, 1 piece, 36"x36"

	Unit	Total
Minimum	EA.	940
Average	EA.	1,400
Maximum	EA.	1,550

Receptor, molded stone, 36"x36"

	Unit	Total
Minimum	EA.	410
Average	EA.	660
Maximum	EA.	1,050

For trim and rough-in

	Unit	Total
Minimum	EA.	480
Average	EA.	690
Maximum	EA.	1,050

15440.40 SINKS
Service sink, 24"x29"

	Unit	Total
Minimum	EA.	780
Average	EA.	980
Maximum	EA.	1,450

Kitchen sink, single, stainless steel, single bowl

	Unit	Total
Minimum	EA.	400
Average	EA.	460
Maximum	EA.	770

Double bowl

	Unit	Total
Minimum	EA.	460
Average	EA.	550
Maximum	EA.	910

Porcelain enamel, cast iron, single bowl

	Unit	Total
Minimum	EA.	320
Average	EA.	400
Maximum	EA.	600

Double bowl

	Unit	Total
Minimum	EA.	420
Average	EA.	580
Maximum	EA.	840

Mop sink, 24"x36"x10"

	Unit	Total
Minimum	EA.	600
Average	EA.	720
Maximum	EA.	970

Washing machine box

	Unit	Total
Minimum	EA.	320
Average	EA.	440
Maximum	EA.	600

For trim and rough-in

	Unit	Total
Minimum	EA.	480
Average	EA.	730
Maximum	EA.	950

15440.60 WATER CLOSETS
Water closet flush tank, floor mounted

	Unit	Total
Minimum	EA.	470
Average	EA.	840
Maximum	EA.	1,300

Handicapped

	Unit	Total
Minimum	EA.	560
Average	EA.	960
Maximum	EA.	1,900

For trim and rough-in

	Unit	Total
Minimum	EA.	350
Average	EA.	440
Maximum	EA.	620

15440.70 WATER HEATERS
Water heater, electric

	Unit	Total
6 gal	EA.	450
10 gal	EA.	460
15 gal	EA.	460
20 gal	EA.	620

15440.70 WATER HEATERS (Cont.)

	Unit	Total
30 gal	EA.	640
40 gal	EA.	680
52 gal	EA.	770

Oil fired

	Unit	Total
20 gal	EA.	1,600
50 gal	EA.	2,400

15450.40 STORAGE TANKS
Hot water storage tank, cement lined

	Unit	Total
10 gallon	EA.	660
70 gallon	EA.	1,750

15555.10 BOILERS
Cast iron, gas fired, hot water

	Unit	Total
115 mbh	EA.	4,000
175 mbh	EA.	4,650
235 mbh	EA.	5,550

Steam

	Unit	Total
115 mbh	EA.	4,250
175 mbh	EA.	4,950
235 mbh	EA.	5,650

Electric, hot water

	Unit	Total
115 mbh	EA.	5,000
175 mbh	EA.	5,400
235 mbh	EA.	6,000

Steam

	Unit	Total
115 mbh	EA.	6,000
175 mbh	EA.	7,100
235 mbh	EA.	7,650

Oil fired, hot water

	Unit	Total
115 mbh	EA.	4,350
175 mbh	EA.	5,350
235 mbh	EA.	7,050

Steam

	Unit	Total
115 mbh	EA.	4,350
175 mbh	EA.	5,350
235 mbh	EA.	6,700

15610.10 FURNACES
Electric, hot air

	Unit	Total
40 mbh	EA.	1,100
60 mbh	EA.	1,200
80 mbh	EA.	1,300
100 mbh	EA.	1,450
125 mbh	EA.	1,650

Gas fired hot air

	Unit	Total
40 mbh	EA.	1,100
60 mbh	EA.	1,150
80 mbh	EA.	1,300
100 mbh	EA.	1,400
125 mbh	EA.	1,500

Oil fired hot air

	Unit	Total
40 mbh	EA.	1,400
60 mbh	EA.	2,100
80 mbh	EA.	2,100
100 mbh	EA.	2,200
125 mbh	EA.	2,250

15670.10 CONDENSING UNITS
Air cooled condenser, single circuit

	Unit	Total
3 ton	EA.	1,550
5 ton	EA.	2,350

With low ambient dampers

	Unit	Total
3 ton	EA.	1,800
5 ton	EA.	2,750

15780.20 ROOFTOP UNITS
Packaged, single zone rooftop unit, with roof curb

	Unit	Total
2 ton	EA.	4,000
3 ton	EA.	4,200
4 ton	EA.	4,600

15830.10 RADIATION UNITS
Baseboard radiation unit

	Unit	Total
1.7 mbh/lf	L.F.	95.50
2.1 mbh/lf	L.F.	130

15830.70 UNIT HEATERS
Steam unit heater, horizontal

	Unit	Total
12,500 btuh, 200 cfm	EA.	600
17,000 btuh, 300 cfm	EA.	760

	Unit	Total		Unit	Total
15855.10 AIR HANDLING UNITS			**15940.10 DIFFUSERS**		
Air handling unit, medium pressure, single zone			Ceiling diffusers, round, baked enamel finish		
1500 cfm	EA	4,350	6" dia.	EA	65.00
3000 cfm	EA	5,900	8" dia.	EA	79.25
Rooftop air handling units			10" dia.	EA	85.00
4950 cfm	EA	12,100	12" dia.	EA	100
7370 cfm	EA	15,400	Rectangular		
15870.20 EXHAUST FANS			6x6"	EA	68.00
Belt drive roof exhaust fans			9x9"	EA	88.00
640 cfm, 2618 fpm	EA	1,100	12x12"	EA	120
940 cfm, 2604 fpm	EA	1,400	15x15"	EA	140
15890.10 METAL DUCTWORK			18x18"	EA	170
Rectangular duct			**15940.40 REGISTERS AND GRILLES**		
Galvanized steel			Lay in flush mounted, perforated face, return		
Minimum	Lb.	6.12	6x6/24x24	EA	71.50
Average	Lb.	7.51	8x8/24x24	EA	71.50
Maximum	Lb.	11.25	9x9/24x24	EA	75.75
Aluminum			10x10/24x24	EA	79.75
Minimum	Lb.	13.75	12x12/24x24	EA	79.75
Average	Lb.	17.50	Rectangular, ceiling return, single deflection		
Maximum	Lb.	23.00	10x10	EA	58.00
Fittings			12x12	EA	62.75
Minimum	EA	26.50	14x14	EA	70.25
Average	EA	39.75	16x8	EA	62.75
Maximum	EA	73.75	16x16	EA	62.75
15890.30 FLEXIBLE DUCTWORK			Wall, return air register		
Flexible duct, 1.25" fiberglass			12x12	EA	63.00
5" dia.	L.F.	6.19	16x16	EA	86.00
6" dia.	L.F.	6.88	18x18	EA	99.00
7" dia.	L.F.	7.93	Ceiling, return air grille		
8" dia.	L.F.	8.36	6x6	EA	47.25
10" dia.	L.F.	10.50	8x8	EA	58.00
12" dia.	L.F.	11.25	10x10	EA	66.25
Flexible duct connector, 3" wide fabric	L.F.	12.00	Ceiling, exhaust grille, aluminum egg crate		
15910.10 DAMPERS			6x6	EA	38.50
Horizontal parallel aluminum backdraft damper			8x8	EA	42.25
12" x 12"	EA	69.50	10x10	EA	44.25
16" x 16"	EA	73.25	12x12	EA	55.00

DIVISION 16 ELECTRICAL

	Unit	Total		Unit	Total

16050.30 BUS DUCT

Bus duct, 100a, plug-in					
10', 600v	EA.	410			
With ground	EA.	590			
Circuit breakers, with enclosure					
1 pole					
15a-60a	EA.	300			
70a-100a	EA.	340			
2 pole					
15a-60a	EA.	410			
70a-100a	EA.	500			
Circuit breaker, adapter cubicle					
225a	EA.	4,200			
400a	EA.	4,900			
Fusible switches, 240v, 3 phase					
30a	EA.	660			
60a	EA.	810			
100a	EA.	1,050			
200a	EA.	1,850			

16110.20 CONDUIT SPECIALTIES

Rod beam clamp, 1/2"	EA.	9.06			
Hanger rod					
3/8"	L.F.	3.88			
1/2"	L.F.	6.35			
All thread rod					
1/4"	L.F.	2.35			
3/8"	L.F.	3.05			
1/2"	L.F.	4.06			
5/8"	L.F.	6.60			
Hanger channel, 1-1/2"					
No holes	EA.	5.82			
Holes	EA.	6.72			
Channel strap					
1/2"	EA.	4.37			
3/4"	EA.	4.73			
Conduit penetrations, roof and wall, 8" thick					
1/2"	EA.	41.00			
3/4"	EA.	41.00			
1"	EA.	53.50			
Threaded rod couplings					
1/4"	EA.	4.60			
3/8"	EA.	4.67			
1/2"	EA.	4.85			
5/8"	EA.	5.65			
3/4"	EA.	5.87			
Hex nuts					
1/4"	EA.	3.48			
3/8"	EA.	3.56			
1/2"	EA.	3.81			
5/8"	EA.	4.35			
3/4"	EA.	4.68			
Square nuts					
1/4"	EA.	3.47			
3/8"	EA.	3.59			
3/8"	EA.	3.76			
5/8"	EA.	3.89			
3/4"	EA.	4.31			
Flat washers					
1/4"	EA.	14			
3/8"	EA.	20			
1/2"	EA.	28			
5/8"	EA.	55			
3/4"	EA.	77			
Lockwashers					
1/4"	EA.	09			
3/8"	EA.	15			
1/2"	EA.	19			
5/8"	EA.	33			
3/4"	EA.	55			

16110.21 ALUMINUM CONDUIT

Aluminum conduit					
1/2"	L.F.	3.59			
3/4"	L.F.	4.72			
1"	L.F.	6.20			
90 deg. elbow					
1/2"	EA.	25.50			
3/4"	EA.	34.25			
1"	EA.	44.75			

16110.21 ALUMINUM CONDUIT (Cont.)

Coupling					
1/2"	EA.	6.73			
3/4"	EA.	9.08			
1"	EA.	12.00			

16110.22 EMT CONDUIT

EMT conduit					
1/2"	L.F.	2.57			
3/4"	L.F.	3.69			
1"	L.F.	5.04			
90 deg. elbow					
1/2"	EA.	11.25			
3/4"	EA.	12.50			
1"	EA.	16.00			
Connector, steel compression					
1/2"	EA.	7.34			
3/4"	EA.	8.62			
1"	EA.	10.00			
Coupling, steel, compression					
1/2"	EA.	6.35			
3/4"	EA.	7.22			
1"	EA.	8.89			
1 hole strap, steel					
1/2"	EA.	2.88			
3/4"	EA.	2.93			
1"	EA.	3.08			
Connector, steel set screw					
1/2"	EA.	5.81			
3/4"	EA.	6.53			
1"	EA.	7.88			
Insulated throat					
1/2"	EA.	6.20			
3/4"	EA.	7.17			
1"	EA.	8.81			
Connector, die cast set screw					
1/2"	EA.	4.68			
3/4"	EA.	5.18			
1"	EA.	6.27			
Insulated throat					
1/2"	EA.	5.50			
3/4"	EA.	6.44			
1"	EA.	8.28			
Coupling, steel set screw					
1/2"	EA.	4.76			
3/4"	EA.	5.83			
1"	EA.	7.79			
Diecast set screw					
1/2"	EA.	3.41			
3/4"	EA.	3.86			
1"	EA.	4.63			
1 hole malleable straps					
1/2"	EA.	3.04			
3/4"	EA.	3.19			
1"	EA.	3.52			
EMT to rigid compression coupling					
1/2"	EA.	10.50			
3/4"	EA.	12.25			
1"	EA.	18.50			
Set screw couplings					
1/2"	EA.	7.70			
3/4"	EA.	8.23			
1"	EA.	12.25			
Set screw offset connectors					
1/2"	EA.	8.99			
3/4"	EA.	9.78			
1"	EA.	15.25			
Compression offset connectors					
1/2"	EA.	10.50			
3/4"	EA.	11.50			
1"	EA.	16.75			
Type "LB" set screw condulets					
1/2"	EA.	25.75			
3/4"	EA.	32.25			
1"	EA.	44.75			
Type "T" set screw condulets					
1/2"	EA.	32.75			
3/4"	EA.	43.00			
1"	EA.	53.25			

	Unit	Total
16110.22 EMT CONDUIT (Cont.)		
Type "C" set screw condulets		
1/2"	EA.	27.50
3/4"	EA.	33.25
1"	EA.	45.75
Type "LL" set screw condulets		
1/2"	EA.	27.50
3/4"	EA.	33.00
1"	EA.	45.75
Type "LR" set screw condulets		
1/2"	EA.	27.50
3/4"	EA.	33.00
1"	EA.	45.75
Type "LB" compression condulets		
1/2"	EA.	44.25
3/4"	EA.	69.50
1"	EA.	79.50
Type "T" compression condulets		
1/2"	EA.	59.50
3/4"	EA.	73.00
1"	EA.	110
Condulet covers		
1/2"	EA.	9.80
3/4"	EA.	10.25
1"	EA.	10.75
Clamp type entrance caps		
1/2"	EA.	24.25
3/4"	EA.	28.50
1"	EA.	37.00
Slip fitter type entrance caps		
1/2"	EA.	22.25
3/4"	EA.	26.25
1"	EA.	34.50
16110.23 FLEXIBLE CONDUIT		
Flexible conduit, steel		
3/8"	L.F.	2.53
1/2	L.F.	2.60
3/4"	L.F.	3.48
1"	L.F.	4.21
Flexible conduit, liquid tight		
3/8"	L.F.	3.67
1/2"	L.F.	3.89
3/4"	L.F.	5.22
1"	L.F.	6.52
Connector, straight		
3/8"	EA.	8.56
1/2"	EA.	8.79
3/4"	EA.	10.25
1"	EA.	14.50
Straight insulated throat connectors		
3/8"	EA.	12.00
1/2"	EA.	12.00
3/4"	EA.	15.50
1"	EA.	18.50
90 deg connectors		
3/8"	EA.	14.75
1/2"	EA.	14.75
3/4"	EA.	19.25
1"	EA.	27.50
90 degree insulated throat connectors		
3/8"	EA.	15.75
1/2"	EA.	15.75
3/4"	EA.	20.50
1"	EA.	29.25
Flexible aluminum conduit		
3/8"	L.F.	2.41
1/2"	L.F.	2.49
3/4"	L.F.	3.33
1"	L.F.	3.91
Connector, straight		
3/8"	EA.	7.90
1/2"	EA.	8.37
3/4"	EA.	8.99
1"	EA.	15.00
Straight insulated throat connectors		
3/8"	EA.	7.13
1/2"	EA.	8.34
3/4"	EA.	8.49

	Unit	Total
16110.23 FLEXIBLE CONDUIT (Cont.)		
1"	EA.	13.00
90 deg connectors		
3/8"	EA.	11.75
1/2"	EA.	13.00
3/4"	EA.	15.00
1"	EA.	20.25
90 deg insulated throat connectors		
3/8"	EA.	12.25
1/2"	EA.	13.50
3/4"	EA.	15.75
1"	EA.	21.25
16110.24 GALVANIZED CONDUIT		
Galvanized rigid steel conduit		
1/2"	L.F.	5.15
3/4"	L.F.	6.09
1"	L.F.	7.92
1-1/4"	L.F.	10.75
1-1/2"	L.F.	12.50
2"	L.F.	14.75
90 degree ell		
1/2"	EA.	27.00
3/4"	EA.	31.25
1"	EA.	42.00
1-1/4"	EA.	52.50
1-1/2"	EA.	61.25
2"	EA.	76.25
Couplings, with set screws		
1/2"	EA.	8.51
3/4"	EA.	10.75
1"	EA.	16.25
1-1/4"	EA.	25.25
1-1/2"	EA.	32.25
2"	EA.	63.75
Split couplings		
1/2"	EA.	17.25
3/4"	EA.	22.50
1"	EA.	26.50
1-1/4"	EA.	36.25
1-1/2"	EA.	46.00
2"	EA.	85.50
Erickson couplings		
1/2"	EA.	35.00
3/4"	EA.	39.75
1"	EA.	53.75
1-1/4"	EA.	82.50
1-1/2"	EA.	96.75
2"	EA.	150
Seal fittings		
1/2"	EA.	61.00
3/4"	EA.	71.75
1"	EA.	89.75
1-1/4"	EA.	100
1-1/2"	EA.	130
2"	EA.	160
Entrance fitting, (weather head), threaded		
1/2"	EA.	38.50
3/4"	EA.	44.00
1"	EA.	52.00
1-1/4"	EA.	66.50
1-1/2"	EA.	85.25
2"	EA.	110
Locknuts		
1/2"	EA.	3.54
3/4"	EA.	3.58
1"	EA.	3.72
1-1/4"	EA.	3.87
1-1/2"	EA.	4.82
2"	EA.	5.21
Plastic conduit bushings		
1/2"	EA.	8.59
3/4"	EA.	10.25
1"	EA.	13.50
1-1/4"	EA.	15.75
1-1/2"	EA.	18.25
2"	EA.	23.75
Conduit bushings, steel		
1/2"	EA.	8.79

	Unit	Total

16110.24 GALVANIZED CONDUIT (Cont.)

	Unit	Total
3/4"	EA.	10.50
1"	EA.	13.75
1-1/4"	EA.	16.25
1-1/2"	EA.	19.00
2"	EA.	25.00
Pipe cap		
1/2"	EA.	3.91
3/4"	EA.	3.96
1"	EA.	4.33
1-1/4"	EA.	7.03
1-1/2"	EA.	7.98
2"	EA.	8.31
Threaded couplings		
1/2"	EA.	5.25
3/4"	EA.	6.29
1"	EA.	8.82
1-1/4"	EA.	10.25
1-1/2"	EA.	12.00
2"	EA.	14.25
Threadless couplings		
1/2"	EA.	12.50
3/4"	EA.	14.25
1"	EA.	17.50
1-1/4"	EA.	22.00
1-1/2"	EA.	27.75
2"	EA.	36.50
Threadless connectors		
1/2"	EA.	9.44
3/4"	EA.	12.75
1"	EA.	16.75
1-1/4"	EA.	24.75
1-1/2"	EA.	35.00
2"	EA.	55.25
Setscrew connectors		
1/2"	EA.	7.90
3/4"	EA.	9.48
1"	EA.	12.25
1-1/4"	EA.	18.00
1-1/2"	EA.	24.00
2"	EA.	41.00
Clamp type entrance caps		
1/2"	EA.	29.00
3/4"	EA.	35.50
1"	EA.	43.75
1-1/4"	EA.	49.75
1-1/2"	EA.	70.75
2"	EA.	84.25
"LB" condulets		
1/2"	EA.	30.25
3/4"	EA.	37.25
1"	EA.	47.25
1-1/4"	EA.	63.50
1-1/2"	EA.	80.50
2"	EA.	110
"T" condulets		
1/2"	EA.	37.75
3/4"	EA.	44.25
1"	EA.	55.25
1-1/4"	EA.	70.25
1-1/2"	EA.	84.00
2"	EA.	120
"X" condulets		
1/2"	EA.	47.75
3/4"	EA.	52.50
1"	EA.	70.25
1-1/4"	EA.	82.75
1-1/2"	EA.	98.25
2"	EA.	170
Blank steel condulet covers		
1/2"	EA.	9.43
3/4"	EA.	10.00
1"	EA.	11.25
1-1/4"	EA.	14.00
1-1/2"	EA.	14.25
2"	EA.	18.25

16110.24 GALVANIZED CONDUIT (Cont.)

	Unit	Total
Solid condulet gaskets		
1/2"	EA.	5.62
3/4"	EA.	5.81
1"	EA.	6.19
1-1/4"	EA.	8.88
1-1/2"	EA.	9.07
2"	EA.	9.51
One-hole malleable straps		
1/2"	EA.	3.05
3/4"	EA.	3.20
1"	EA.	3.43
1-1/4"	EA.	4.86
1-1/2"	EA.	5.09
2"	EA.	6.76
One-hole steel straps		
1/2"	EA.	2.77
3/4"	EA.	2.81
1"	EA.	2.91
1-1/4"	EA.	3.68
1-1/2"	EA.	3.80
2"	EA.	3.94
Grounding locknuts		
1/2"	EA.	7.43
3/4"	EA.	7.98
1"	EA.	9.15
1-1/4"	EA.	10.00
1-1/2"	EA.	10.25
2"	EA.	12.25
Insulated grounding metal bushings		
1/2"	EA.	14.25
3/4"	EA.	17.00
1"	EA.	19.75
1-1/4"	EA.	25.50
1-1/2"	EA.	31.75
2"	EA.	38.75

16110.25 PLASTIC CONDUIT

	Unit	Total
PVC conduit, schedule 40		
1/2"	L.F.	2.65
3/4"	L.F.	2.80
1"	L.F.	3.81
1-1/4"	L.F.	4.25
1-1/2"	L.F.	5.22
2"	L.F.	5.75
Couplings		
1/2"	EA.	3.77
3/4"	EA.	3.86
1"	EA.	4.14
1-1/4"	EA.	5.01
1-1/2"	EA.	5.41
2"	EA.	5.87
90 degree elbows		
1/2"	EA.	8.33
3/4"	EA.	10.00
1"	EA.	11.00
1-1/4"	EA.	13.75
1-1/2"	EA.	18.25
2"	EA.	22.25
Terminal adapters		
1/2"	EA.	7.30
3/4"	EA.	7.68
1"	EA.	7.93
1-1/4"	EA.	12.25
1-1/2"	EA.	12.75
2"	EA.	13.50
LB conduit body		
1/2"	EA.	17.50
3/4"	EA.	18.75
1	EA.	19.50
1-1/4"	EA.	30.75
1-1/2"	EA.	32.75
2"	EA.	42.25
PVC cement		
1 pint	EA.	15.00
1 quart	EA.	22.00

	Unit	Total
16110.27 PLASTIC COATED CONDUIT		
Rigid steel conduit, plastic coated		
1/2"	L.F.	7.72
3/4"	L.F.	9.03
1"	L.F.	12.00
1-1/4"	L.F.	15.00
1-1/2"	L.F.	18.25
2"	L.F.	23.00
90 degree elbows		
1/2"	EA.	37.75
3/4"	EA.	43.50
1"	EA.	50.25
1-1/4"	EA.	58.50
1-1/2"	EA.	72.00
2"	EA.	96.75
Couplings		
1/2"	EA.	8.87
3/4"	EA.	10.50
1"	EA.	12.75
1-1/4"	EA.	15.00
1-1/2"	EA.	19.25
2"	EA.	23.75
1 hole conduit straps		
3/4"	EA.	11.75
1"	EA.	12.00
1-1/4"	EA.	16.50
1-1/2"	EA.	17.25
2"	EA.	23.50
16110.28 STEEL CONDUIT		
Intermediate metal conduit (IMC)		
1/2"	L.F.	3.55
3/4"	L.F.	4.56
1"	L.F.	6.20
1-1/4"	L.F.	7.61
1-1/2"	L.F.	9.92
2"	L.F.	12.00
90 degree ell		
1/2"	EA.	29.50
3/4"	EA.	34.00
1"	EA.	46.00
1-1/4"	EA.	58.25
1-1/2"	EA.	68.50
2"	EA.	89.00
Couplings		
1/2"	EA.	6.49
3/4"	EA.	7.81
1"	EA.	11.00
1-1/4"	EA.	13.00
1-1/2"	EA.	15.75
2"	EA.	19.00
16110.35		
Single Raceway		
3/4" x 17/32" Conduit	L.F.	4.34
Mounting Strap	EA.	4.01
Connector	EA.	4.16
Elbow		
45 degree	EA.	11.00
90 degree	EA.	5.77
internal	EA.	6.40
external	EA.	6.17
Switch	EA.	46.50
Utility Box	EA.	40.00
Receptacle	EA.	50.25
3/4" x 21/32" Conduit	L.F.	4.57
Mounting Strap	EA.	4.26
Connector	EA.	4.29
Elbow		
45 degree	EA.	12.75
90 degree	EA.	5.94
internal	EA.	6.86
external	EA.	6.86
Switch	EA.	46.50
Utility Box	EA.	40.00
Receptacle	EA.	50.25

	Unit	Total
16120.41 ALUMINUM CONDUCTORS		
Type XHHW, stranded aluminum, 600v		
#8	L.F.	.62
#6	L.F.	.71
#4	L.F.	.90
#2	L.F.	1.12
1/0	L.F.	1.56
2/0	L.F.	1.87
3/0	L.F.	2.26
4/0	L.F.	2.47
Type S.E.U. cable		
#8/3	L.F.	3.01
#6/3	L.F.	3.21
#4/3	L.F.	4.05
#2/3	L.F.	4.84
#1/3	L.F.	5.79
1/0-3	L.F.	6.32
2/0-3	L.F.	7.00
3/0-3	L.F.	9.05
4/0-3	L.F.	9.45
Type S.E.R. cable with ground		
#8/3	L.F.	3.50
#6/3	L.F.	4.16
16120.43 COPPER CONDUCTORS		
Copper conductors, type THW, solid		
#14	L.F.	.39
#12	L.F.	.52
#10	L.F.	.69
THHN-THWN, solid		
#14	L.F.	.39
#12	L.F.	.52
#10	L.F.	.69
Stranded		
#14	L.F.	.39
#12	L.F.	.52
#10	L.F.	.69
#8	L.F.	1.03
#6	L.F.	1.38
#4	L.F.	1.90
#2	L.F.	2.52
#1	L.F.	3.11
1/0	L.F.	3.74
2/0	L.F.	4.63
3/0	L.F.	5.82
4/0	L.F.	7.05
Bare stranded wire		
#8	L.F.	.96
#6	L.F.	1.39
#4	L.F.	1.79
#2	L.F.	2.52
#1	L.F.	3.17
Type "BX" solid armored cable		
#14/2	L.F.	2.49
#14/3	L.F.	3.17
#14/4	L.F.	3.87
#12/2	L.F.	2.72
#12/3	L.F.	3.40
#12/4	L.F.	4.19
#10/2	L.F.	3.61
#10/3	L.F.	4.55
#10/4	L.F.	6.15
Steel type, metal clad cable, solid, with ground		
#14/2	L.F.	1.95
#14/3	L.F.	2.47
#14/4	L.F.	3.07
#12/2	L.F.	2.10
#12/3	L.F.	2.93
#12/4	L.F.	3.71
#10/2	L.F.	3.11
#10/3	L.F.	4.07
#10/4	L.F.	5.64
16120.47 SHEATHED CABLE		
Non-metallic sheathed cable		
Type NM cable with ground		
#14/2	L.F.	1.35
#12/2	L.F.	1.61
#10/2	L.F.	2.05
#8/2	L.F.	2.73

	Unit	Total			Unit	Total

16120.47 SHEATHED CABLE (Cont.)

#6/2	L.F.	3.87
#14/3	L.F.	2.22
#12/3	L.F.	2.55
#10/3	L.F.	3.03
#8/3	L.F.	3.90
#6/3	L.F.	5.19
#4/3	L.F.	9.02
#2/3	L.F.	12.50
Type U.F. cable with ground		
#14/2	L.F.	1.48
#12/2	L.F.	1.89
#14/3	L.F.	1.90
#12/3	L.F.	2.33
Type S.F.U. cable, 3 conductor		
#8	L.F.	3.63
#6	L.F.	5.12
Type SER cable, 4 conductor		
#6	L.F.	6.82
#4	L.F.	8.75
Flexible cord, type STO cord		
#18/2	L.F.	1.02
#18/3	L.F.	1.20
#18/4	L.F.	1.61
#16/2	L.F.	1.13
#16/3	L.F.	1.03
#16/4	L.F.	1.34
#14/2	L.F.	1.67
#14/3	L.F.	1.63
#14/4	L.F.	1.99
#12/2	L.F.	2.10
#12/3	L.F.	1.73
#12/4	L.F.	2.39
#10/2	L.F.	2.58
#10/3	L.F.	2.55
#10/4	L.F.	3.74

16130.40 BOXES

Round cast box, type SEH		
1/2"	EA.	43.25
3/4"	EA.	48.00
SEHC		
1/2"	EA.	47.25
3/4"	EA.	52.00
SEHL		
1/2"	EA.	47.75
3/4"	EA.	53.75
SEHT		
1/2"	EA.	54.25
3/4"	EA.	59.50
SEHX		
1/2"	EA.	61.75
3/4"	EA.	69.50
Blank cover	EA.	14.50
1/2", hub cover	EA.	14.25
Cover with gasket	EA.	16.75
Rectangle, type FS boxes		
1/2"	EA.	33.50
3/4"	EA.	37.75
1"	EA.	45.00
FSA		
1/2"	EA.	41.75
3/4"	EA.	44.00
FSC		
1/2"	EA.	34.75
3/4"	EA.	40.50
1"	EA.	49.00
FSL		
1/2"	EA.	41.50
3/4"	EA.	45.00
FSR		
1/2"	EA.	42.25
3/4"	EA.	46.25
FSS		
1/2"	EA.	34.75
3/4"	EA.	39.25
FSLA		
1/2"	EA.	31.00
3/4"	EA.	35.50

16130.40 BOXES (Cont.)

FSCA		
1/2"	EA.	46.25
3/4"	EA.	49.00
FSCC		
1/2"	EA.	40.75
3/4"	EA.	54.25
FSCT		
1/2"	EA.	40.75
3/4"	EA.	50.75
1"	EA.	52.50
FST		
1/2"	EA.	53.75
3/4"	EA.	58.75
FSX		
1/2"	EA.	64.50
3/4"	EA.	70.25
FSCD boxes		
1/2"	EA.	60.50
3/4"	EA.	69.00
Rectangle, type FS, 2 gang boxes		
1/2"	EA.	45.25
3/4"	EA.	49.25
1"	EA.	57.00
FSC, 2 gang boxes		
1/2"	EA.	46.50
3/4"	EA.	52.50
1"	EA.	64.50
FSS, 2 gang boxes		
3/4"	EA.	51.00
FS, tandem boxes		
1/2"	EA.	51.00
3/4"	EA.	54.75
FSC, tandem boxes		
1/2"	EA.	59.50
3/4"	EA.	64.75
FS, three gang boxes		
3/4"	EA.	65.50
1"	EA.	72.50
FSS, three gang boxes, 3/4"	EA.	79.25
Weatherproof cast aluminum boxes, 1 gang, 3 outlets		
1/2"	EA.	33.25
3/4"	EA.	40.25
2 gang, 3 outlets		
1/2"	EA.	45.75
3/4"	EA.	48.75
1 gang, 4 outlets		
1/2"	EA.	52.50
3/4"	EA.	61.00
2 gang, 4 outlets		
1/2"	EA.	53.00
3/4"	EA.	61.75
1 gang, 5 outlets		
1/2"	EA.	58.00
3/4"	EA.	64.75
2 gang, 5 outlets		
1/2"	EA.	65.50
3/4"	EA.	74.25
2 gang, 6 outlets		
1/2"	EA.	76.00
3/4"	EA.	80.75
2 gang, 7 outlets		
1/2"	EA.	87.25
3/4"	EA.	98.75
Weatherproof and type FS box covers,		
blank, 1 gang	EA.	12.75
Tumbler switch, 1 gang	EA.	15.75
1 gang, single recept	EA.	13.50
Duplex recept	EA.	14.75
Despard	EA.	14.75
Red pilot light	EA.	33.00
SW and		
Single recept	EA.	23.50
Duplex recept	EA.	21.75
2 gang		
Blank	EA.	15.25
Tumbler switch	EA.	16.25
Single recept	EA.	16.25

	Unit	Total		Unit	Total

16130.40 BOXES (Cont.)

Duplex recept	EA	16.25
3 gang		
Blank	EA	20.25
Tumbler switch	EA	22.00
4 gang		
Tumbler switch	EA	28.00
Box covers		
Surface	EA	28.75
Sealing	EA	30.25
Dome	EA	36.75
1/2" nipple	EA	43.25
3/4" nipple	EA	44.25

16130.60 PULL AND JUNCTION BOXES

4"		
Octagon box	EA	11.00
Box extension	EA	9.56
Plaster ring	EA	7.03
Cover blank	EA	5.31
Square box	EA	12.50
Box extension	EA	8.65
Plaster ring	EA	6.52
Cover blank	EA	5.27
Switch and device boxes		
2 gang	EA	22.25
3 gang	EA	33.25
4 gang	EA	45.00
Device covers		
2 gang	EA	15.50
3 gang	EA	16.00
4 gang	EA	20.25
Handy box	EA	11.25
Extension	EA	7.32
Switch cover	EA	5.73
Switch box with knockout	EA	15.00
Weatherproof cover, spring type	EA	15.25
Cover plate, dryer receptacle 1 gang plastic	EA	8.20
For 4" receptacle, 2 gang	EA	9.39
Duplex receptacle cover plate, plastic	EA	4.62
4", vertical bracket box, 1-1/2" with		
RMX clamps	EA	16.50
BX clamps	EA	17.00
4", octagon device cover		
1 switch	EA	8.01
1 duplex recept	EA	8.01
4", octagon swivel hanger box, 1/2" hub	EA	14.75
3/4" hub	EA	16.50
4" octagon adjustable bar hangers		
18-1/2"	EA	8.38
26-1/2"	EA	8.84
With clip		
18-1/2"	EA	7.07
26-1/2"	EA	7.52
4", square face bracket boxes, 1-1/2"		
RMX	EA	18.00
BX	EA	18.75
4" square to round plaster rings	EA	6.70
2 gang device plaster rings	EA	6.79
Surface covers		
1 gang switch	EA	6.43
2 gang switch	EA	6.48
1 single recept	EA	7.68
1 20a twist lock recept	EA	8.61
1 30a twist lock recept	EA	9.92
1 duplex recept	EA	6.26
2 duplex recept	EA	6.26
Switch and duplex recept	EA	7.80
4" plastic round boxes, ground straps		
Box only	EA	11.25
Box w/clamps	EA	15.00
Box w/16" bar	EA	19.25
Box w/24" bar	EA	20.75
4" plastic round box covers		
Blank cover	EA	4.99
Plaster ring	EA	5.65
4" plastic square boxes		
Box only	EA	11.00
Box w/clamps	EA	14.75

16130.60 PULL AND JUNCTION BOXES (Cont.)

Box w/hanger	EA	18.75
Box w/nails and clamp	EA	19.50
4" plastic square box covers		
Blank cover	EA	4.96
1 gang ring	EA	5.18
2 gang ring	EA	5.68
Round ring	EA	5.33

16130.80 RECEPTACLES

Contractor grade duplex receptacles, 15a 120v		
Duplex	EA	14.75
125 volt, 20a, duplex, standard grade	EA	24.00
Ground fault interrupter type	EA	55.00
250 volt, 20a, 2 pole, single, ground type	EA	31.50
120/208v, 4 pole, single receptacle, twist lock		
20a	EA	44.75
50a	EA	64.25
125/250v, 3 pole, flush receptacle		
30a	EA	41.50
50a	EA	46.75
60a	EA	93.00
Dryer receptacle, 250v, 30a/50a, 3 wire	EA	36.00
Clock receptacle, 2 pole, grounding type	EA	24.00
125v, 20a single recept. grounding type		
Standard grade	EA	25.00
125/250v, 3 pole, 3 wire surface recepts		
30a	EA	38.25
50a	EA	40.25
60a	EA	68.25
Cord set, 3 wire, 6' cord		
30a	EA	36.25
50a	EA	43.00
125/250v, 3 pole, 3 wire cap		
30a	EA	43.00
50a	EA	56.50
60a	EA	68.00

16199.10 UTILITY POLES & FITTINGS

Wood pole, creosoted		
25'	EA	610
30'	EA	740
Treated, wood preservative, 6"x6"		
8'	EA	130
10'	EA	190
12'	EA	210
14'	EA	280
16'	EA	330
18'	EA	380
20'	EA	450
Aluminum, brushed, no base		
8'	EA	720
10'	EA	860
15'	EA	940
20'	EA	1,150
25'	EA	1,500
Steel, no base		
10'	EA	860
15'	EA	960
20'	EA	1,250
25'	EA	1,450
Concrete, no base		
13'	EA	1,250
16'	EA	1,750
18'	EA	2,100
25'	EA	2,500

16350.10 CIRCUIT BREAKERS

Load center circuit breakers, 240v		
1 pole, 10-60a	EA	33.25
2 pole		
10-60a	EA	65.25
70-100a	EA	160
110-150a	EA	300
Load center, G.F.I. breakers, 240v		
1 pole, 15-30a	EA	160
Tandem breakers, 240v		
1 pole, 15-30a	EA	58.00
2 pole, 15-30a	EA	92.75

	Unit	Total		Unit	Total

16365.10 FUSES
Fuse, one-time, 250v

30a	EA. 5.74
60a	EA. 7.39
100a	EA. 20.25

16395.10 GROUNDING
Ground rods, copper clad, 1/2" x

6'	EA. 57.75
8'	EA. 67.00

5/8" x

6'	EA. 66.25
8'	EA. 89.75

Ground rod clamp

5/8"	EA. 13.75

Ground rod couplings

1/2"	EA. 16.75
5/8"	EA. 20.75

Ground rod, driving stud

1/2"	EA. 14.75
5/8"	EA. 16.25

Ground rod clamps, #8-2 to

1" pipe	EA. 22.00
2" pipe	EA. 27.75

16430.20 METERING
Outdoor wp meter sockets, 1 gang, 240v, 1 phase
Includes sealing ring, 100a

	EA. 150
150a	EA. 180
200a	EA. 210

Die cast hubs, 1-1/4"

	EA. 28.50
1-1/2"	EA. 29.75
2"	EA. 31.50

16470.10 PANELBOARDS
Indoor load center, 1 phase 240v main lug only

30a - 2 spaces	EA. 150
100a - 8 spaces	EA. 240
150a - 16 spaces	EA. 400
200a - 24 spaces	EA. 650
200a - 42 spaces	EA. 700

Main circuit breaker

100a - 8 spaces	EA. 410
100a - 16 spaces	EA. 440
150a - 16 spaces	EA. 630
150a - 24 spaces	EA. 730
200a - 24 spaces	EA. 710
200a - 42 spaces	EA. 930

120/208v, flush, 3 ph., 4 wire, main only
100a

12 circuits	EA. 1,100
20 circuits	EA. 1,450
30 circuits	EA. 2,000

225a

30 circuits	EA. 2,100
42 circuits	EA. 2,650

16490.10 SWITCHES
Photo electric switches
1000 watt

105-135v	EA. 82.00

Dimmer switch and switch plate

600w	EA. 51.25

Time clocks with skip, 40a, 120v

SPST	EA. 140

Contractor grade wall switch 15a, 120v

Single pole	EA. 12.50
Three way	EA. 16.25
Four way	EA. 27.75

Specification grade toggle switches, 20a, 120-277v

Single pole	EA. 16.75
Double pole	EA. 28.25
3 way	EA. 26.00
4 way	EA. 48.00

Combination switch and pilot light, single pole	EA. 32.25
3 way	EA. 38.50
Combination switch and receptacle, single pole	EA. 37.50
3 way	EA. 41.50

Switch plates, plastic ivory

1 gang	EA. 5.71
2 gang	EA. 7.55
3 gang	EA. 9.35

16490.10 SWITCHES (Cont.)

4 gang	EA. 13.25
5 gang	EA. 14.50
6 gang	EA. 16.50

Stainless steel

1 gang	EA. 8.51
2 gang	EA. 11.00
3 gang	EA. 15.00
4 gang	EA. 21.25
5 gang	EA. 24.25
6 gang	EA. 29.25

Brass

1 gang	EA. 11.25
2 gang	EA. 19.50
3 gang	EA. 27.75
4 gang	EA. 32.25
5 gang	EA. 38.75
6 gang	EA. 46.00

16510.05 INTERIOR LIGHTING
Recessed fluorescent fixtures, 2'x2'

2 lamp	EA. 110
4 lamp	EA. 130

Surface mounted incandescent fixtures

40w	EA. 140
75w	EA. 140
100w	EA. 150
150w	EA. 180

Pendant

40w	EA. 130
75w	EA. 140
100w	EA. 150
150w	EA. 160

Contractor grade recessed down lights

100 watt housing only	EA. 130
150 watt housing only	EA. 160
100 watt trim	EA. 86.25
150 watt trim	EA. 120

Recessed incandescent fixtures

40w	EA. 230
75w	EA. 240
100w	EA. 250
150w	EA. 270

Light track single circuit

2'	EA. 71.75
4'	EA. 78.75
8'	EA. 130
12'	EA. 190

Fittings and accessories

Dead end	EA. 25.00
Starter kit	EA. 37.25
Conduit feed	EA. 29.50
Straight connector	EA. 27.25
Center feed	EA. 37.75
L-connector	EA. 29.50
T-connector	EA. 36.25
X-connector	EA. 45.25
Cord and plug	EA. 38.75
Rigid corner	EA. 52.00
Flex connector	EA. 42.75
2 way connector	EA. 110
Spacer clip	EA. 4.77
Grid box	EA. 17.75
T-bar clip	EA. 5.47
Utility hook	EA. 16.00

Fixtures, square

R-20	EA. 49.00
R-30	EA. 70.50
40w flood	EA. 110
40w spot	EA. 110
100w flood	EA. 120
100w spot	EA. 98.25
Mini spot	EA. 47.00
Mini flood	EA. 95.50
Quartz, 500w	EA. 230
R-20 sphere	EA. 75.25
R-30 sphere	EA. 44.25
R-20 cylinder	EA. 56.00
R-30 cylinder	EA. 63.25

	Unit	Total		Unit	Total
16510.05 INTERIOR LIGHTING (Cont.)			**16850.10 ELECTRIC HEATING**		
R-40 cylinder	EA	64.25	Baseboard heater		
R-30 wall wash	EA	95.50	2', 375w	EA	110
R-40 wall wash	EA	120	3', 500w	EA	120
16510.10 LIGHTING INDUSTRIAL			4', 750w	EA	130
Surface mounted fluorescent, wrap around lens			5', 935w	EA	170
1 lamp	EA	140	6', 1125w	EA	200
2 lamps	EA	200	7', 1310w	EA	220
Wall mounted fluorescent			8', 1500w	EA	250
2-20w lamps	EA	120	9', 1680w	EA	280
2-30w lamps	EA	130	10', 1875w	EA	330
2-40w lamps	EA	140	Unit heater, wall mounted		
Strip fluorescent			750w	EA	280
4'			1500w	EA	330
1 lamp	EA	87.75	Thermostat		
2 lamps	EA	97.25	Integral	EA	70.75
8'			Line voltage	EA	71.75
1 lamp	EA	110	Electric heater connection	EA	18.50
2 lamps	EA	150	Fittings		
Compact fluorescent			Inside corner	EA	51.00
2-7w	EA	220	Outside corner	EA	53.25
2-13w	EA	270	Receptacle section	EA	54.25
16670.10 LIGHTNING PROTECTION			Blank section	EA	60.75
Lightning protection			Radiant ceiling heater panels		
Copper point, nickel plated, 12'			500w	EA	360
1/2" dia.	EA	110	750w	EA	390
5/8" dia.	EA	120	Unit heater thermostat	EA	85.25
16750.20 SIGNALING SYSTEMS			Mounting bracket	EA	99.50
Contractor grade doorbell chime kit			Relay	EA	110
Chime	EA	100	**16910.40 CONTROL CABLE**		
Doorbutton	EA	26.25	Control cable, 600v, #14 THWN, PVC jacket		
			2 wire	L.F	.85
			4 wire	L.F	1.21

Part Four

Metro Area Multipliers

The costs presented in this Costbook attempt to represent national averages. Costs, however, vary among regions, states and even between adjacent localities. In order to more closely approximate the probable costs for specific locations throughout the U. S., this table of Metro Area Multipliers is provided. These adjustment factors are used to modify costs obtained from this book to help account for regional variations of construction costs and to provide a more accurate estimate for specific areas. The factors are formulated by comparing costs in a specific area to the costs presented in this Costbook. An example of how to use these factors is shown below. Whenever local current costs are known, whether material prices or labor rates, they should be used when more accuracy is required.

Cost Obtained from Costbook Pages   **X** $\dfrac{\textbf{\textit{Metro Area Multiplier}}}{\textbf{\textit{Divided by 100}}}$ **=**   *Adjusted Cost*

For example, a project estimated to cost $1,000,000 using the Costbook can be adjusted to more closely approximate the cost in Los Angeles where the Multiplier is 115:

$$1,000,000 \ X \ \frac{115}{100} = 1,150,000$$

Metro Area Multipliers

State	Metropolitan Area	Multiplier	State	Metropolitan Area	Multiplier
AK	ANCHORAGE	137	FL	MIAMI	96
AL	ANNISTON	90		NAPLES	98
	AUBURN-OPELIKA	90		OCALA	95
	BIRMINGHAM	88		ORLANDO	95
	DOTHAN	86		PANAMA CITY	83
	GADSDEN	86		PENSACOLA	88
	HUNTSVILLE	88		SARASOTA-BRADENTON	89
	MOBILE	90		TALLAHASSEE	85
	MONTGOMERY	86		TAMPA-ST. PETERSBURG-CLEARWATER	93
	TUSCALOOSA	88		WEST PALM BEACH-BOCA RATON	96
AR	FAYETTEVILLE-SPRINGDALE-ROGERS	80	GA	ALBANY	85
	FORT SMITH	86		ATHENS	87
	JONESBORO	85		ATLANTA	95
	LITTLE ROCK-NORTH LITTLE ROCK	87		AUGUSTA	83
	PINE BLUFF	87		COLUMBUS	84
	TEXARKANA	86		MACON	87
AZ	FLAGSTAFF	98		SAVANNAH	89
	PHOENIX-MESA	98	HI	HONOLULU	135
	TUCSON	97	IA	CEDAR RAPIDS	97
	YUMA	99		DAVENPORT	100
CA	BAKERSFIELD	112		DES MOINES	101
	FRESNO	114		DUBUQUE	95
	LOS ANGELES-LONG BEACH	115		IOWA CITY	100
	MODESTO	111		SIOUX CITY	95
	OAKLAND	120		WATERLOO-CEDAR FALLS	94
	ORANGE COUNTY	114	ID	BOISE CITY	99
	REDDING	111		POCATELLO	97
	RIVERSIDE-SAN BERNARDINO	112	IL	BLOOMINGTON-NORMAL	106
	SACRAMENTO	114		CHAMPAIGN-URBANA	105
	SALINAS	116		CHICAGO	115
	SAN DIEGO	113		DECATUR	104
	SAN FRANCISCO	125		KANKAKEE	107
	SAN JOSE	122		PEORIA-PEKIN	106
	SAN LUIS OBISPO	110		ROCKFORD	106
	SANTA CRUZ-WATSONVILLE	116		SPRINGFIELD	104
	SANTA ROSA	117	IN	BLOOMINGTON	101
	STOCKTON-LODI	113		EVANSVILLE	100
	VALLEJO-FAIRFIELD-NAPA	116		FORT WAYNE	100
	VENTURA	112		GARY	108
	SANTA BARBARA	115		INDIANAPOLIS	104
CO	BOULDER-LONGMONT	96		KOKOMO	100
	COLORADO SPRINGS	101		LAFAYETTE	100
	DENVER	101		MUNCIE	100
	FORT COLLINS-LOVELAND	94		SOUTH BEND	101
	GRAND JUNCTION	96		TERRE HAUTE	100
	GREELEY	94	KS	KANSAS CITY	99
	PUEBLO	98		LAWRENCE	94
CT	BRIDGEPORT	112		TOPEKA	93
	DANBURY	112		WICHITA	93
	HARTFORD	111	KY	LEXINGTON	93
	NEW HAVEN-MERIDEN	112		LOUISVILLE	96
	NEW LONDON-NORWICH	109		OWENSBORO	94
	STAMFORD-NORWALK	115	LA	ALEXANDRIA	88
	WATERBURY	111		BATON ROUGE	92
DC	WASHINGTON	104		HOUMA	92
DE	DOVER	104		LAFAYETTE	90
	WILMINGTON-NEWARK	105		LAKE CHARLES	92
FL	DAYTONA BEACH	92		MONROE	88
	FORT LAUDERDALE	96		NEW ORLEANS	95
	FORT MYERS-CAPE CORAL	88		SHREVEPORT-BOSSIER CITY	89
	FORT PIERCE-PORT ST. LUCIE	95	MA	BARNSTABLE-YARMOUTH	115
	FORT WALTON BEACH	97		BOSTON	118
	GAINESVILLE	90		BROCKTON	114
	JACKSONVILLE	94		FITCHBURG-LEOMINSTER	111
	LAKELAND-WINTER HAVEN	90		LAWRENCE	114
	MELBOURNE-TITUSVILLE-PALM BAY	98		LOWELL	109

Metro Area Multipliers

State	Metropolitan Area	Multiplier
MA	NEW BEDFORD	114
	PITTSFIELD	110
	SPRINGFIELD	110
	WORCESTER	109
MD	BALTIMORE	98
	CUMBERLAND	96
	HAGERSTOWN	91
ME	BANGOR	95
	LEWISTON-AUBURN	96
	PORTLAND	97
MI	ANN ARBOR	106
	DETROIT	111
	FLINT	105
	GRAND RAPIDS-MUSKEGON-HOLLAND	100
	JACKSON	104
	KALAMAZOO-BATTLE CREEK	96
	LANSING-EAST LANSING	104
	SAGINAW-BAY CITY-MIDLAND	102
MN	DULUTH	106
	MINNEAPOLIS-ST. PAUL	111
	ROCHESTER	106
	ST. CLOUD	108
MO	COLUMBIA	99
	JOPLIN	95
	KANSAS CITY	102
	SPRINGFIELD	97
	ST. JOSEPH	99
	ST. LOUIS	99
MS	BILOXI-GULFPORT-PASCAGOULA	87
	JACKSON	86
MT	BILLINGS	98
	GREAT FALLS	99
	MISSOULA	97
NC	ASHEVILLE	82
	CHARLOTTE	83
	FAYETTEVILLE	84
	GREENSBORO-WINSTON-SALEM-HIGH POINT	83
	GREENVILLE	83
	HICKORY-MORGANTON-LENOIR	79
	RALEIGH-DURHAM-CHAPEL HILL	83
	ROCKY MOUNT	79
	WILMINGTON	84
ND	BISMARCK	95
	FARGO	96
	GRAND FORKS	94
NE	LINCOLN	92
	OMAHA	96
NH	MANCHESTER	100
	NASHUA	98
	PORTSMOUTH	92
NJ	ATLANTIC-CAPE MAY	110
	BERGEN-PASSAIC	113
	JERSEY CITY	112
	MIDDLESEX-SOMERSET-HUNTERDON	108
	MONMOUTH-OCEAN	111
	NEWARK	114
	TRENTON	111
	VINELAND-MILLVILLE-BRIDGETON	109
NM	ALBUQUERQUE	96
	LAS CRUCES	92
	SANTA FE	98
NV	LAS VEGAS	108
	RENO	107

State	Metropolitan Area	Multiplier
NY	ALBANY-SCHENECTADY-TROY	102
	BINGHAMTON	100
	BUFFALO-NIAGARA FALLS	108
	ELMIRA	95
	GLENS FALLS	94
	JAMESTOWN	102
	NASSAU-SUFFOLK	115
	NEW YORK	132
	ROCHESTER	106
	SYRACUSE	103
	UTICA-ROME	103
OH	AKRON	103
	CANTON-MASSILLON	101
	CINCINNATI	98
	CLEVELAND-LORAIN-ELYRIA	106
	COLUMBUS	101
	DAYTON-SPRINGFIELD	99
	LIMA	101
	MANSFIELD	97
	STEUBENVILLE	104
	TOLEDO	103
	YOUNGSTOWN-WARREN	100
OK	ENID	88
	LAWTON	89
	OKLAHOMA CITY	91
	TULSA	90
OR	EUGENE-SPRINGFIELD	106
	MEDFORD-ASHLAND	104
	PORTLAND	108
	SALEM	107
PA	ALLENTOWN-BETHLEHEM-EASTON	105
	ALTOONA	103
	ERIE	103
	HARRISBURG-LEBANON-CARLISLE	101
	JOHNSTOWN	104
	LANCASTER	101
	PHILADELPHIA	114
	PITTSBURGH	104
	READING	103
	SCRANTON-WILKES-BARRE-HAZLETON	102
	STATE COLLEGE	98
	WILLIAMSPORT	100
	YORK	102
RI	PROVIDENCE	109
SC	AIKEN	91
	CHARLESTON-NORTH CHARLESTON	85
	COLUMBIA	85
	FLORENCE	82
	GREENVILLE-SPARTANBURG-ANDERSON	84
	MYRTLE BEACH	90
SD	RAPID CITY	89
	SIOUX FALLS	90
TN	CHATTANOOGA	88
	JACKSON	87
	JOHNSON CITY	84
	KNOXVILLE	88
	MEMPHIS	93
	NASHVILLE	92

State	Metropolitan Area	Multiplier
TX	ABILENE	86
	AMARILLO	87
	AUSTIN-SAN MARCOS	88
	BEAUMONT-PORT ARTHUR	88
	BROWNSVILLE-HARLINGEN-SAN BENITO	89
	BRYAN-COLLEGE STATION	87
	CORPUS CHRISTI	85
	DALLAS	91
	EL PASO	85
	FORT WORTH-ARLINGTON	91
	GALVESTON-TEXAS CITY	90
	HOUSTON	92
	LAREDO	78
	LONGVIEW-MARSHALL	83
	LUBBOCK	88
	MCALLEN-EDINBURG-MISSION	84
	ODESSA-MIDLAND	85
	SAN ANGELO	83
	SAN ANTONIO	89
	TEXARKANA	86
	TYLER	86
	VICTORIA	86
	WACO	86
	WICHITA FALLS	86
UT	PROVO-OREM	94
	SALT LAKE CITY-OGDEN	93
VA	CHARLOTTESVILLE	88
	LYNCHBURG	87
	NORFOLK-VA BEACH-NEWPORT NEWS	91
	RICHMOND-PETERSBURG	91
	ROANOKE	85

State	Metropolitan Area	Multiplier
VT	BURLINGTON	98
WA	BELLINGHAM	112
	BREMERTON	110
	OLYMPIA	108
	RICHLAND-KENNEWICK-PASCO	106
	SEATTLE-BELLEVUE-EVERETT	111
	SPOKANE	108
	TACOMA	111
	YAKIMA	106
WI	APPLETON-OSHKOSH-NEENAH	101
	EAU CLAIRE	100
	GREEN BAY	101
	JANESVILLE-BELOIT	101
	KENOSHA	103
	LA CROSSE	99
	MADISON	100
	MILWAUKEE-WAUKESHA	107
	RACINE	103
	WAUSAU	100
WV	CHARLESTON	99
	HUNTINGTON	100
	PARKERSBURG	98
	WHEELING	100
WY	CASPER	92
	CHEYENNE	93

A

ABS DWV Pipe 177
Accessories 149
Addition 28, 42, 66, 68, 76, 86
    Renovation 28
Administration Building 25, 45
    Hub 57
Air Handling Unit 181
Allowances 134
Aluminum Conductors 185
    Conduit 182
    Windows 158
Americans with Disabilities Act (ADA) 13
Aquatic Center 119
Area Multipliers 3, 191
ASHRAE 90.1 69
    (2004 Standards). 25
    (2007 Standards) 17
Asphalt Shingles 154
    Surfaces 139
Assembly Space 21
Athletic Business Facility 119
Auditorium 69
Audubon Center 115

B

Backflow Preventers 169
Bank 19, 35
Base Course 136
Baths 179
    Accessories 167
Batt Insulation 154
Beam Concrete 143
    Formwork 142
    Reinforcing 142
Bituminous Dampproofing 154
Blinds 168
Blocking 149
Bard Insulation 154
Boilers 180
Bond Beam & Lintels 146
Boxes 186
Brass Fittings 175
    I.P.S. 173
Brick Masonry 145
Building Demolition 136
    Selective 136
    Excavation 137
Built-Up Asphalt Roofing 155
Bulk Excavation 137
Bus Duct 182

C

C.I. Pipe, Above Ground 169
    Below Ground 170
California Energy Efficiency Standards 13
Campground Property 41
Campus Building Addittion 87
    Master Plan 87
Cancer Care Research Facility 81
Carbon Dioxide Monitors 69
Carpet Padding 163
Cart Barn 39
Casework 168

Caulking 156
Ceiling Framing 149
    and Walls 162
Central College 69
Ceramic Tile 161
Chain Link Fence 139
Children's Center 19, 37
Christian Life Center 123
Chrome Plated Fitting 176
Circuit Breakers 187
Class A Office Building 25
Cleanouts 178
Clearing and Grubbing 136
Clinical Spaces 81
    Waste Water Reduction 89
Closers 159
Collaborative of High Performance School 45
College 43
    of Design 67
Column 153
    Concrete 143
    Formwork 142
    Reinforcing 142
Commercial Building 17
    Businesses 33
    Construction 5
Commons 57, 59, 87, 121
    Building 57
Community Center 19, 57
    Health Center 85
Composite Structural Steel 72
Concrete 133
    Accessories 143
    Admixtures 143
    Finishes 143
    Frame Type II 80
    Masonry Unit 146
    Paving 139
Condensing Unit 180
Conduit Specialties 182
Construction 5, 191
    Aids 134
    Headquarters 107
Control Cable 189
Copper Conductor 185
    Fittings 170
    Pipe 170
Core Drilling 136
Counter Top 168
County Courthouse 21
criminal courts facility 21
Criminal Justice Center 21
CSI Divisions 19
Curb & Gutter 136
    Formwork 142
Curing Concrete 143
Customer Center 109

D

Dampers 181
Dampproofing.. 154
Daylighting........................45, 47, 49, 53, 61, 67, 85, 105
Design-Build 22
Diffusers 181
Displacement Ventilation 49, 61
Disposals & Accessories 179
Door 133
    Trim 159
Doors, Windows & Glass...................................... 133
Drainage Fields 139
    Piping... 136
Drains, Roof & Floor 179
Ductile Iron Pipe 138
Ductwork Insulation 169

E

Electric Heating 189
Elementary School 19, 45, 49, 51, 53, 61
Elevated Slab Concrete.. 143
    Reinforcing.. 142
EMT Conduit 182
Energy Corporation 101
Energy Efficiency 31
    Manufacturing Facility 75
    Performance 17
Energy Recovery Coils... 69
Energy Star 10, 15, 37, 97, 124, 127, 131
Environmental Education Center 27
    Labs 71
Environmental Protection Agency (EPA) 15
Equipment... 133, 135
    Pad Reinforcing 142
        Concrete 143
        Formwork 142
Executive Briefing Centers 109
Exhaust Fans 181
Expanded Facility 87
EXT Painting, Buildings 163
    Misc. 164
    Sitework 163
Exterior Insulation 49, 61

F

Facility Management 87
Faculty Offices 69
Faucets 179
Fence.. 136
Fertilizing 140
Field Staff.. 134
Finish Carpentry 152
Finishes... 133
Fire and Rescue Station 14
Fitness Area 119
Flagpole 167
Flashing & Trim 156
Flats 129
Flexible Conduit 183
    Ductwork 181
Flex-Theatre.. 71
Floor Framing 149
    Sheating 151
Footing Concrete... 143
    Formwork 142

Footing Reinforcing.. 142
Foundation Reinforcing... 142
Fuel Cell Facility 75
Furnaces.. 180
Furnishings 133
Furring 150
Fuses.. 188
Future of LEED® 18

G

Galvanized Conduit.. 183
    Steel Pipe 178
Gas Piping 136
General Requirements... 133
Geothermal Well 31
Glass Block 146
Glazed Curtain Walls 159
Glazing 159
Golf Course Community 39
Government Services Administration (GSA), ... 17
Grade Beam Concrete... 144
    Formwork 142
    Reinforcing 142
Gravel & Stone 138
Green Articles 3
    Building Costs 14
    Economy 11
    Globes 10
    Home...125, 127
    Initiatives...13, 14
    Revolution 15
    Roof 16, 29, 63, 76, 100, 103
Green Roofing 16
Grounding 188
Grouting.. 144
Guardrail 136
Gulf Freeway Building... 103
Gymnasium..................37, 45, 51, 53, 58, 71, 111, 117
Gypsum Board 161
    Lath 160

H

Hand Excavation... 137
Hauling Material 138
Hazard Analyses.. 101
Health Center... 87
Healthcare Facility 89
Heavy Timber 151
    Framing... 26
High Performance Building Regulations 10
High School 71
    Campus Operating Costs 71
    Sustainable Practices 71
Hinges 159
Historic Building 91, 109
Hose Bibbs 178
Hospice Care..19, 83
Hydrant 136, 179

Index

I

Impervious Surfaces 69
Indoor Environments 12
Industrial Wastewater Treatment 75
Infusion Center 81
Innovation and Design 17
Interior Painting, Building 165
    Lighting 188

J

Job Corps Center 47
    Requirements 134
Judicial Complex 21

L

Laker Turf Building 113
Land Remediation Project 75
Landscape Accessories 141
Lavatories 179
Lawn Irrigation 139
Leadership in Environmental and Energy Design 12
LEED® Checklist 9
    Consultant 9
    Core and Shell 18
    Energy and Atmosphere 10
    Estimator 9
    for Existing Buildings 18
    for Schools 18
    New Construction 18
    Registration 13
    Energy Recovery Credit 43
    Silver Certification 10
    Cost 11
    Criteria 12
    EBOM 18
Library 23, 31, 45
Lighting Industrial 189
    Protection 189
Locksets 159
Loose Fill Insulation 154
Low VOC Paints 27, 29
    Fume Solutions 16
Lutheran Church 120, 121

M

Maintenance Equipment 168
Manholes 139
Masonry 133
    Accessories 145
    Control Joints 145
    Flashing 145
    Grout 145
Material and Resource Credit 10
Measuring Sustainability 16
Mechanical 133
Medical Plaza 79
Mennonite School 63
Metal Anchors 148
    Door Frames 157
    Ductwork 181
    Framing 148
    Lath 160
    Lintels 148
    Roofing 156
    Shingles 155
    Siding Panels 155

Metal Studs 160
Metering 188
Metro Area Multipliers 191
Mezzanine 109
Middle School 73
Millwork 153
Miscellaneous Formwork 142
Mobilization 134
Municipal Facilities 13
    Green Building Initiatives 13

N

National Registry of Historic Places 37
    Resource-Protection Project 105
Nature Center 29

O

Office & Shop 97
Office Tower 93
Old Mill Building (Adaptive Re-use) 107
Ornamental Metal 148

P

Paint 166
    Preparation 163
Panel Work 153
    Boards 188
Parging Masonry 147
Pavement Demolition 136
Pavers Masonry 146
Pavilion 29
Peak Demand Reduction 49
Pediatric Clinic 89
Pest Control 167
Photovoltaic Array 69
    Solutions 16
Pile Cap Concrete 144
    Formwork 142
    Reinforcing 142
Pipe Hnagers, Light 169
Plaster 160
    Accessories 160
Plastic Coated Conduit 184, 185
    Pipe 138
    Siding 155
    Skylights 156
Plywood Siding 155
Pneumatic Concrete 143
Portland Cement 161
Precast Specialties 144
    Cored Slab 122
    Panels 74
Pre-Engineered 112
    Buildings 168
        Component 12
Prestressed Piling 138
Pretreatment Department 25
Project's Carbon Footprint 25
Public Structure 13
Pull & Junction Boxes 187

Q

Quarry Tile 161

R

Radiation Units 180
Railings 148
Rainwater Haervesting 69
Receptacles 187
Recreational Courts 140
Recycled Content Materials 23
    Solutions 16
Reduced Cooling Load 49, 61
Reduced HVAC Equipment 49, 61
Reflective Roof 15
Refractories 147
Regional Headquarters 91
Registers & Grilles 181
Renewable Energy 17
    Business 17
    Water Resources (ReWa) 25
Renewables & LEED 18
Renovation 19, 28, 32, 36, 54, 68, 82, 86, 94, 108
Research Floors 81
Residential Equipment 168
Resilient Base & Accessories 163
    Sheet Flooring 162
    Tile 162
Restoration & Cleaning 147
Retail Building 19, 41
Retrofitting Renewables 18
RIPRAP 138
Roadway Excavation 137
Roof Framing 150
    Sheating 151
Rooftop Units 180

S

Sanctuary 121
Sanitary Piping 136
    Sewers 139
Saw Cutting Pavement 136
School Community Building 37
Science Center 69
    Complex 19, 43
Seeding 140
Sensible Solar Load 69
Septic Tanks 136, 139
Sheated Cable 185
Shell-Only Building 102
Shower 180
    Stall 167
Shrub & Tree Maintenance 141
Sidewalks 144
Signaling System 189
Signs 134
Single-Ply Roofing 156
Sink 180
Site Development 18
Sitework & Demolition 133
Slab/Mat Concrete 144
    Formwork 142
    Reinforcing 142
Slate Shingles 155
Sleepers 151
Soffits 151
Soil Boring 136
    Stabilization 138

Soil Treatment 138
Solar House 131
Special Construction 133
    Doors 158
Specialties 133, 169
Sprayed Insulation 154
Square Foot Costs 3
Stairwork 153
State Office Building 105
State-of-theArt science classrooms 87
    Facility 23
    Middle School 73
Steel Conduit 185
    Frame 24, 30, 42, 62, 74, 79, 121, 156, 167
    Piles 138
    Pipe 178
    Siding 155
Stick Frame 126
Stone 147
Storage Tanks 168, 180
Strainers 179
Structural Steel 148
    Braced Frame 44
    Welding 148
Sun Shades 49
Surveying 134
Sustainability 5, 11
    Buildings 5
    Facility 71
    High School Building 71
    Roofing 15
    Site Development 11
Switches 188

T

Temporary Facilities 134
Terrazzo 161
Testing 134
Thermal & Moisture Protection 133
Tilt-Up 19, 38, 78, 106
Title-24 13
Topsoil 140
Toxicity Reduction 31
Traps 179
Treatment & Collection 13
Tree Cutting & Clearing 136
Trenching 137
Type 2 Non-Combustible 50

U

U.S. Green Building Council 5
Underdrain 139
Unit Heaters 180
Unit Masonry Flooring 162
Unit-In-Place Costs 3, 7, 133
University 55, 81
    Research Building 81
Unprotected Non-Combustible 46
Urban Wildlife Refuge 27
USGBC 9, 15-18, 89, 95
Utility Excavation 137

Index

V

Vacuum Breakers 179
Valves..178
Vapor Barriers 154
Vaults 168
Vents & Wall Louvers..167
Vibration Control 169
Vitrified Clay Pipe 139

W

Wall Concrete 144
    Covering...166
    Formwork..142
    Framing...151
    Reinforcing..142
    Sheating..151
Warehouse Maintenance Building........................ 97
Waste Reduction 31
Water Closet...180
    Heaters...180
    Piping 136
    Treatment Process.................................... 25
Waterproofing 154

Water-Source Heat Pump 87
Weatherstripping 159
Weed Control 141
Welfare Fund Building 99
Wellness Center..117
Wells 138
Windows 133
    Treatment 168
Wood & Plastics 133
    Timber Piles 138
    Decking...151
    Doors..157
    Flooring...162
    Frame...........................34, 98, 124, 128, 130, 157
    Shakes..155
    Shingles..155
Wood Siding...155
    Treatment 153
    Trusses...152
    Windows..158

Y

YMCA.......................................19, 110, 111, 118, 119